Advances in Intelligent Vehicles

Advances in Intelligent Vehicles

Edited by
Yaobin Chen
Indiana University—Purdue University Indianapolis

Lingxi Li
Indiana University—Purdue University Indianapolis

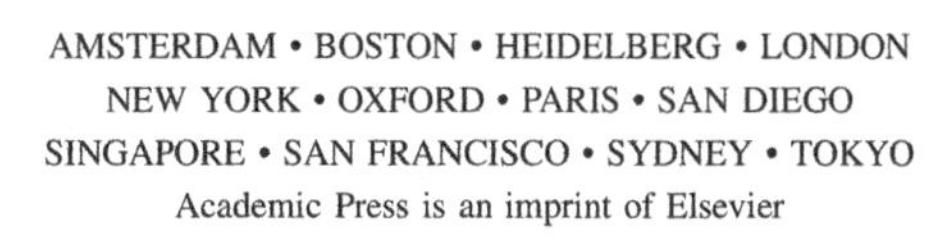
AMSTERDAM • BOSTON • HEIDELBERG • LONDON
NEW YORK • OXFORD • PARIS • SAN DIEGO
SINGAPORE • SAN FRANCISCO • SYDNEY • TOKYO
Academic Press is an imprint of Elsevier

Academic Press is an imprint of Elsevier
The Boulevard, Langford Lane, Kidlington, Oxford OX5 1GB, UK
225 Wyman Street, Waltham, MA 02451, USA

First edition 2014

British Library Cataloguing in Publication Data
A catalogue record for this book is available from the British Library

Library of Congress Cataloging-in-Publication Data
A catalog record for this book is available from the Library of Congress

ISBN–13: 978-0-12-397199-9

For information on all Academic Press publications
visit our web site at store.elsevier.com

Printed and bound in the US

14 15 16 17 10 9 8 7 6 5 4 3 2 1

Contents

Preface

Advances in Intelligent Vehicles embraces recent advancement of research and development in intelligent vehicle technologies that enhance the safety, reliability, and performance of vehicles and vehicular networks and systems. Intelligent vehicles consist of semi-autonomous or autonomous driving systems that are controlled based on multiple sensors and sensor networks in a complex driving environment with high-level interactions between vehicles and infrastructure. The overall system of intelligent vehicles can be viewed as a cyber-physical system comprising vehicles and sensing/actuating devices (physical systems) and control, communication, and computation components (cyber systems).

The purpose of this book is to provide researchers and engineers with up-to-date research results and state-of-the-art technologies in the area of intelligent vehicles and intelligent transportation systems. The book consists of 10 chapters contributed by leading experts in the field. These chapters cover a variety of important topics within the scope of intelligent vehicles, such as novel design and testing of intelligent vehicles (Chapters 1–3), collision avoidance (Chapter 4), human factors and the study of driver behavior (Chapters 5–7), driver assistance systems (Chapters 8 and 9), and active safety systems (Chapter 10). These representative works discuss the latest technologies that are employed in intelligent vehicles and promote the better understanding of the current development and the future research trend in this area.

Chapter 1 studies the design and development of narrow vehicles, which can be seen as a potential solution to traffic congestion and pollution problems. Chapter 2 focuses on procedures and tools used in the testing stages of the intelligent vehicle design process and develops a procedure that uses both simulation environments and small-scale indoor testbeds before the outdoor tests and makes use of different levels of virtualization for sensors, agents, scenarios, and environments. Chapter 3 presents a mathematical framework of modeling connected vehicles, where vehicles are able to communicate with each other and the roadside. With Field Theory as a basis, this chapter discusses strategies of integrating the effects introduced by connected vehicles. Chapter 4 studies the problem of collision avoidance for autonomous vehicles and develops strategies that guarantee the avoidance of a dynamic obstacle. Two examples are also presented to illustrate the performance of the control policies. Chapter 5 carefully examines the effect of human factors on driver behavior, which leads to the development of one simulator with Hardware-in-the-Loop (HIL) and Driver-in-the-Loop (DIL), and an instrumented vehicle to measure the information on driver actions, vehicle states, and relative relationship with other vehicles. A comparative analysis of the differences is carried out in the driving simulator and the real-world environment. Chapter 6 studies the effect of factors (e.g. vehicle, road, environment, traffic situation, etc.) other than human factors on driver behavior characteristics. Driver experiments are conducted in diverse category roads and some representative parameters, and test methods are adopted to analyze and compare the effects of a variety of factors. Chapter 7 studies influences of human factors for the safe human system and discusses the potential application of suitable heuristics (where there is a role of human intellect in progress and/or interconnected knowledge, experiences, and intuition) for these situations. Chapter 8 presents a robust vision-based road environment perception system used for navigating an intelligent vehicle through challenging traffic scenarios and discusses ways of utilizing and integration of the intensity evidence and geometry cues, as well as temporal supports of the road scenes for the detection and tracking of the targets in probabilistic models. Chapter 9 focuses on the study of driver assistance systems and presents a configuration of the Intersection Driver Assistance System (IDAS) and an algorithm of multi-objective IDAS for intersection support, which includes two functions as driving support and traffic signal violation warning. The effects of IDAS on intersection driving are analyzed by random traffic simulation and field tests. Finally, Chapter 10 discusses technologies and issues related to the design of a Human–Machine Interface (HMI) for Advanced Driver Assistance Systems (ADAS). In particular, it presents challenges for HMI design to meet the dynamic requirements of drivers in different traffic situations.

The editors believe that this book presents the most recent advances and research development on a variety of topics related to intelligent vehicles. The materials covered in this book will greatly help disseminate research results and findings and promote future research in this area.

Finally, we would like to take this opportunity to thank all of our colleagues and students who have contributed to this book. We would also like to thank the editorial staff members from the Zhejiang University Press and Elsevier. Without their help, this book would not have been possible. We are also grateful to anonymous reviewers for their constructive comments and suggestions.

Yaobin Chen, Lingxi Li

Department of Electrical and Computer Engineering, Purdue School of Engineering and Technology Indiana University–Purdue University Indianapolis (IUPUI), Indianapolis, IN 46202, USA

CHAPTER 1

Modeling and Control of a New Narrow Vehicle

Toshio Fukuda*, Jian Huang[†], Takayuki Matsuno[‡], Kosuke Sekiyama*

**Department of Micro-Nano Systems Engineering, Nagoya University, Japan*

[†]*Department of Control Science and Engineering, Huazhong University of Science and Technology, China*

[‡]*Graduate School of Natural Science and Technology, Okayama University, Japan*

Chapter Outline

Advances in Intelligent Vehicles. http://dx.doi.org/10.1016/B978-0-12-397199-9.00001-X

1.1 Introduction

Parking, pollution, and congestion problems caused by cars in urban areas have made life uncomfortable and inconvenient. To improve living conditions, development of an intelligent, self-balanced, and less polluting narrow vehicle might offer a good opportunity. In terms of this idea, many autonomous robots and intelligent vehicles have been designed based on Mobile Wheeled Inverted Pendulum (MWIP) models [1–7], such as PMP [1], iBot [2], Segway [3], and so on.

The MWIP models have attracted much attention in the field of control theory because of the nonlinear and underactuated with inherent unstable dynamics. Many previous attempts used linear [8,9] or feedback linearization methods [10–13] for modeling and control. These rely on a rather precise description of nonlinear functions and show a lack of robustness to model errors and external disturbances.

There are also some other control methods implemented with MWIP models. Lin et al. [6] adopted adaptive control for self-balancing and yaw motion control of two-wheeled mobile vehicles. The nested saturation control design technique is applied to derive a control law for two-wheeled vehicles [7]. Adaptive robust dynamic balance and motion control are utilized to handle the parametric and functional uncertainties [14]. Jung and Kim [15] presented a method for online learning and control of an MWIP by using neural networks.

Sliding Mode Control (SMC) might be a comparatively appropriate approach to deal with uncertain MWIP systems because SMC is less sensitive to the parameter variations and noise disturbances. It has been proved that SMC algorithms can robustly stabilize a class of underactuated mechanical systems such as a mobile robot [16]. Park et al. [17] proposed an adaptive neural SMC method for trajectory tracking of non-holonomic wheeled mobile robots with model uncertainties and external disturbances. We proposed a velocity control method for the MWIP based on sliding mode and a novel sliding surface [18]. Ashrafiuon and Erwin [19] proposed an SMC approach for underactuated multibody systems. Tsai et al. [20] proposed an adaptive sliding mode controller to a hierarchical tracking control in triwheeled mobile robots. Terminal Sliding Mode Control (TSMC) of finite time mechanisms is an example of a variable structure control idea whose formation and development is based on introducing a nonlinear function into sliding hyperplanes. Compared to a linear sliding mode surface, terminal sliding mode has no switching control term and the chattering can be effectively alleviated [21]. On the other hand, TSMC can improve the transient performance substantially. TSMC has already been used successfully in control applications [22–26]. Compared to conventional SMC, TSMC provides faster, finite time convergence, and higher control precision. Nonlinear terminal sliding mode surface functions such as cubic polynomials [26] can also be applied.

While the MWIP system has been successfully applied in many fields, there is still much room for improvement. For instance, drivers can only stand on the Segway vehicles during driving, which is not comfortable for a long-term operation. Another deficiency of Segway is that the body will not always be upright during its operation. To overcome these shortcomings, a new narrow vehicle called the UW-Car is introduced in this study. The novel structure includes an MWIP base and a movable seat driven by a linear motor along the straight direction of motion. The adjustable seat can guarantee the vehicle body will always be upright during driving, which is discussed further in the subsequent sections. The mechanism of a UW-Car is shown in Figure 1.1.

It is well known that the brake system is one of the most important parts relating to the safety of vehicles, the main purpose of which is to reduce the speed or stop driving, or keep the stopped vehicle at rest. In the driving procedure, in order to keep a safe distance between vehicles, precise control of the braking procedure is particularly important. Therefore, studying the braking of mobile robots based on the MWIP structure is of great significance from both the practical and theoretical points of view. Kidane et al. [27] proposed a tilt brake algorithm of a narrow commuter vehicle that was verified for different low-speed maneuvers. We propose an optimal braking strategy for the UW-Car, which can guarantee that an optimal braking period is obtained and the vehicle's body is always upright during the braking process.

For the UW-Car, achieving a high acceleration performance is difficult due to its non-holonomic characteristic. Conventional methods may realize quick acceleration while bringing unsuitable vibrations at the start and end of acceleration [28]. To solve this problem, we proposed a control method using the desired trajectory of acceleration and dynamics canceling inputs so as to provide the high acceleration performance highlighted in this chapter.

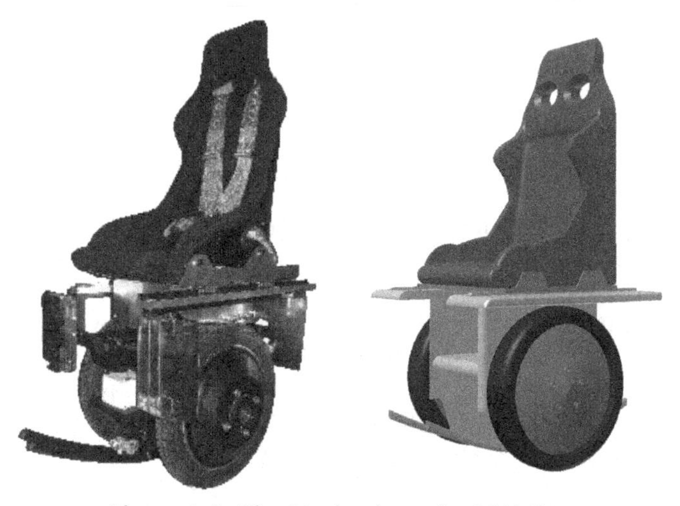

Figure 1.1: The Mechanism of a UW-Car.

1.2 Modeling of the UW-Car

1.2.1 Dynamic Model on Flat Ground

In this study, a novel transportation system called the UW-Car is investigated, which is different from normal MWIP systems. The UW-Car system is modeled as a one-dimensional inverse pendulum rotating along the wheels' axis with a movable seat above. The seat moves forward and back on top of the MWIP along the direction of motion. The structure of the UW-Car system is described in Figure 1.2, where θ_w and θ_1 are the wheel's rotational angle and the body inclination angle respectively. λ denotes the displacement of the seat. We assume that the system moves on flat ground. To simplify the model derivation, we divide the UW-Car system into three parts: the body, the seat, and the wheels. Some notation should first be clarified.

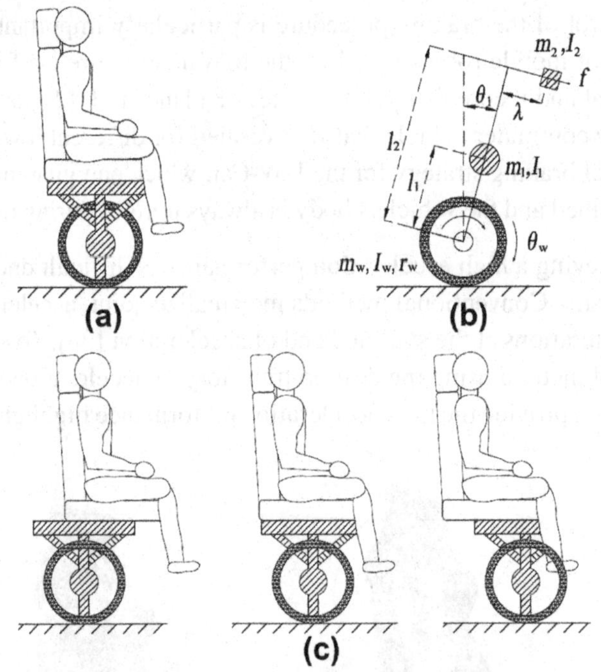

Figure 1.2: Prototype and Model of a UW-Car System

m_1, m_2, m_w — masses of the body, the seat and the wheel.

I_b, I_s, I_w — moments of inertia of the body, the seat and the wheel.

ℓ_1 — length between the wheel axle and the center of gravity of the body.

ℓ_2 — length between the wheel axle and the plane of the movable seat.

r_w — radius of the wheel.

D_1 — viscous resistance in the driving system.

D_2 — viscous resistance of the seat moving.

D_w — viscous resistance of the ground.

τ_w — torque of the motor driving wheels.

f — force for the linear motor driving the seat.

It should be pointed out that only straight movement is considered here. Hence a three-dimensional model [11] is not required. The following three-dimensional vector is used to describe the dynamic model of a UW-Car system on flat ground: $\mathbf{q}^{(1)} = [\theta_{\mathrm{w}} \quad \theta_1 \quad \lambda]^{\mathrm{T}}$.

To simplify the representation, we assume $S_1 = \sin\theta_1$, $C_1 = \cos\theta_1$. The coordinates of the wheels, the body, and the seat are denoted by $\mathbf{x_w} = [x_{\mathrm{w}} \quad y_{\mathrm{w}}]^{\mathrm{T}}$, $\mathbf{x}_1 = [x_1 \quad y_1]^{\mathrm{T}}$, $\mathbf{x}_2 = [x_2 \quad y_2]^{\mathrm{T}}$ respectively.

The coordinate system of a UW-Car is depicted in Figure 1.2(b). The positions and velocities of three parts of a UW-Car system are given by

$$\begin{cases} x_{\mathrm{w}} = r_{\mathrm{w}}\theta_{\mathrm{w}} \\ y_{\mathrm{w}} = 0 \end{cases} \qquad \begin{cases} \dot{x}_{\mathrm{w}} = r_{\mathrm{w}}\dot{\theta}_{\mathrm{w}} \\ \dot{y}_{\mathrm{w}} = 0 \end{cases} \tag{1.1}$$

$$\begin{cases} x_1 = r_{\mathrm{w}}\theta_{\mathrm{w}} + l_1 S_1 \\ y_1 = l_1 C_1 \end{cases} \qquad \begin{cases} \dot{x}_1 = r_{\mathrm{w}}\dot{\theta}_{\mathrm{w}} + l_1 C_1 \dot{\theta}_1 \\ \dot{y}_1 = -l_1 S_1 \dot{\theta}_1 \end{cases} \tag{1.2}$$

$$\begin{cases} x_2 = r_{\mathrm{w}}\theta_{\mathrm{w}} + l_2 S_1 + \lambda C_1 \\ y_2 = l_2 C_1 - \lambda S_1 \end{cases} \qquad \begin{cases} \dot{x}_2 = r_{\mathrm{w}}\dot{\theta}_{\mathrm{w}} + (l_2 C_1 - \lambda S_1)\dot{\theta}_1 + \dot{\lambda} C_1 \\ \dot{y}_2 = -(l_2 S_1 + \lambda C_1)\dot{\theta}_1 - \dot{\lambda} S_1 \end{cases}. \tag{1.3}$$

Lagrange's motion equation is used to analyze the dynamics of the system. The kinetic, potential, and dissipated energy are computed as follows.

Kinetic energy is represented by $T = T_{\mathrm{w}} + T_1 + T_2$, where T_{w}, T_1, and T_2 are the kinetic energies of the wheels, body and seat respectively:

$$T_{\mathrm{w}} = \frac{1}{2} m_{\mathrm{w}} \dot{\mathbf{x}}_{\mathbf{w}}^{\mathrm{T}} \dot{\mathbf{x}}_{\mathbf{w}} + \frac{1}{2} I_{\mathrm{w}} \dot{\theta}_{\mathrm{w}}^2 = \frac{1}{2} m_{\mathrm{w}} r_{\mathrm{w}}^2 \dot{\theta}_{\mathrm{w}}^2 + \frac{1}{2} I_{\mathrm{w}} \dot{\theta}_{\mathrm{w}}^2 \tag{1.4}$$

$$T_1 = \frac{1}{2} \dot{\mathbf{x}}_1^{\mathrm{T}} \dot{\mathbf{x}}_1 + \frac{1}{2} I_{\mathrm{b}} \dot{\theta}_1^2 = \frac{1}{2} m_1 \left(r_{\mathrm{w}} \dot{\theta}_{\mathrm{w}} + l_1 C_1 \dot{\theta}_1 \right)^2 + \frac{1}{2} m_1 l_1^2 S_1^2 \dot{\theta}_1^2 + \frac{1}{2} I_{\mathrm{b}} \dot{\theta}_1^2 \tag{1.5}$$

$$\begin{aligned} T_2 = \frac{1}{2} m_2 \dot{\mathbf{x}}_2^{\mathrm{T}} \dot{\mathbf{x}}_2 + \frac{1}{2} I_{\mathrm{s}} \dot{\theta}_1^2 &= \frac{1}{2} m_2 \Big(\dot{\lambda}^2 + l_2^2 \dot{\theta}_1^2 + \lambda^2 \dot{\theta}_1^2 + r_{\mathrm{w}}^2 \dot{\theta}_{\mathrm{w}}^2 \\ &+ 2 r_{\mathrm{w}} C_1 \dot{\theta}_{\mathrm{w}} \dot{\lambda} + 2 r_{\mathrm{w}} (l_2 C_1 - \lambda S_1) \dot{\theta}_{\mathrm{w}} \dot{\theta}_1 + 2 l_2 \dot{\theta}_1 \dot{\lambda} \Big) + \frac{1}{2} I_{\mathrm{s}} \dot{\theta}_1^2. \end{aligned} \tag{1.6}$$

The potential energy of a UW-Car system is written as follows:

$$U = m_1 g l_1 C_1 + m_2 g (l_2 C_1 - \lambda S_1). \tag{1.7}$$

The energy is dissipated due to the effect of the friction between the wheels and the ground, in the driving system and with the seat moving. The dissipated energy is

$$D = \frac{1}{2} D_{\mathrm{w}} \dot{\theta}_{\mathrm{w}}^2 + \frac{1}{2} D_1 \dot{\theta}_1^2 + \frac{1}{2} D_2 \dot{\lambda}^2. \tag{1.8}$$

The equations of motion are derived by the application of Lagrange's equation:

$$\frac{\mathrm{d}}{\mathrm{d}t}\left(\frac{\partial T}{\partial \dot{\mathbf{q}}^{(1)}}\right)-\frac{\partial T}{\partial \mathbf{q}^{(1)}}+\frac{\partial U}{\partial \mathbf{q}^{(1)}}+\frac{\partial D}{\partial \dot{\mathbf{q}}^{(1)}}=\mathbf{E}\left(\mathbf{q}^{(1)}\right)\boldsymbol{\tau},\ \text{where } \boldsymbol{\tau}=[\tau_{\mathrm{w}}\ \ f]^{\mathrm{T}}.$$

Lagrange's motion equation leads to a second-order underactuated model with six state variables and two inputs given by

$$\begin{cases} M_{11}\ddot{\theta}_{\mathrm{w}}+(M_{12}C_1-m_2r_{\mathrm{w}}\lambda S_1)\ddot{\theta}_1+m_2r_{\mathrm{w}}C_1\ddot{\lambda}=\tau_{\mathrm{w}}-D_{\mathrm{w}}\dot{\theta}_{\mathrm{w}} \\ \qquad +(M_{12}S_1+m_2r_{\mathrm{w}}\lambda C_1)\dot{\theta}_1^2+2m_2r_{\mathrm{w}}S_1\dot{\theta}_1\dot{\lambda} \\ \left(M_{12}C_1-m_2r_{\mathrm{w}}\lambda S_1\right)\ddot{\theta}_{\mathrm{w}}+\left(M_{22}+m_2\lambda^2\right)\ddot{\theta}_1+m_2l_2\ddot{\lambda}=-\tau_{\mathrm{w}}, \\ \qquad -D_1\dot{\theta}_1+G_1S_1+m_2g\lambda C_1-2m_2\lambda\dot{\theta}_1\dot{\lambda} \\ m_2r_{\mathrm{w}}C_1\ddot{\theta}_{\mathrm{w}}+m_2l_2\ddot{\theta}_1+m_2\ddot{\lambda}=f-D_2\dot{\lambda}+m_2\lambda\dot{\theta}_1^2+m_2gS_1 \end{cases} \tag{1.9}$$

where M_{11}, M_{12}, M_{22}, and G_1 are given in the Appendix at the end of the chapter.

The vector form of the Lagrange equations of a UW-Car system is given by

$$\mathbf{M}^{(1)}\left(\mathbf{q}^{(1)}\right)\ddot{\mathbf{q}}^{(1)}+\mathbf{N}^{(1)}\left(\mathbf{q}^{(1)},\dot{\mathbf{q}}^{(1)}\right)=\mathbf{B}^{(1)}\boldsymbol{\tau}, \tag{1.10}$$

where matrices $\mathbf{M}^{(1)}$, $\mathbf{N}^{(1)}$, and $\mathbf{B}^{(1)}$ are also given in the Appendix.

1.2.2 Analysis of Equilibria in Set-Point Velocity Control

Note that in the velocity control, we usually do not care about the exact position of the UW-Car. Therefore, let us choose the state variables as $\mathbf{x}=[x_1\ \ x_2\ \ x_3\ \ x_4\ \ x_5]^{\mathrm{T}}=[\theta_1\ \ \lambda\ \ \dot{\theta}_w\ \ \dot{\theta}_1\ \ \dot{\lambda}]^{\mathrm{T}}$, the state model of a UW-Car system can be represented by

$$\begin{cases} \dot{x}_1=x_4 \\ \dot{x}_2=x_5 \\ M_{11}\dot{x}_3+(M_{12}C_1-m_2r_{\mathrm{w}}x_2S_1)\dot{x}_4+m_2r_{\mathrm{w}}C_1\dot{x}_5=\tau_{\mathrm{w}}-D_{\mathrm{w}}x_3 \\ \qquad +\left(M_{12}S_1+m_2r_{\mathrm{w}}x_2C_1\right)x_4^2+2m_2r_{\mathrm{w}}S_1x_4x_5. \\ \left(M_{12}C_1-m_2r_wx_2S_1\right)\dot{x}_3+\left(M_{22}+m_2x_2^2\right)\dot{x}_4+m_2l_2\dot{x}_5=-\tau_{\mathrm{w}} \\ \qquad -D_1x_4+G_1S_1+m_2gC_1x_2-2m_2x_2x_4x_5 \\ m_2r_{\mathrm{w}}C_1\dot{x}_3+m_2l_2\dot{x}_4+m_2\dot{x}_5=f-D_2x_5+m_2x_2x_4^2+m_2gS_1 \end{cases} \tag{1.11}$$

The vector form of the state model is given by

$$\mathbf{R}(\mathbf{x})\cdot\dot{\mathbf{x}} = \mathbf{H}(\mathbf{x}) + \mathbf{K}\boldsymbol{\tau}, \tag{1.12}$$

where

$$\mathbf{K}(\mathbf{q}) = \begin{bmatrix} 0 & 0 & 1 & -1 & 0 \\ 0 & 0 & 0 & 0 & 1 \end{bmatrix}^{\mathrm{T}}.$$

The matrices $\mathbf{R}(\mathbf{x})$ and $\mathbf{H}(\mathbf{x})$ are given in the Appendix.

Supposing $\mathbf{x}^* = \begin{bmatrix} x_1^* & x_2^* & x_3^* & x_4^* & x_5^* \end{bmatrix}^{\mathrm{T}}$ is the desired equilibrium of system (1.9), the following equation can be obtained:

$$x_2^* = \frac{D_\omega x_3^* - (m_1 l_1 + m_2 l_2) g \sin\left(x_4^*\right)}{m_2 g \cos\left(x_4^*\right)}. \tag{1.13}$$

In the case of a velocity control problem, the desired velocity $\dot{\theta}_{\mathrm{w}}^* = x_3^*$ is always given in advance. We expect that the equilibrium satisfies $x_1^* = x_4^* = x_5^* = 0$. This means that the UW-Car moves at a constant velocity x_3^* without any body inclination while keeping the seat in a fixed position. Accordingly, x_2^* can be rewritten as

$$x_2^* = \lambda^* = \frac{D_\omega x_3^*}{m_2 g} = \frac{D_\omega \dot{\theta}_{\mathrm{w}}^*}{m_2 g}. \tag{1.14}$$

Normally, x_2^* is small because the viscous parameter D_{w} is usually small.

1.2.3 Dynamic Model in Rough Terrain

In this subsection, the modeling problem of a UW-Car system moving in a rough terrain is investigated (Figure 1.3). The rough terrain is described by a differentiable function $f(x,y) = 0$.

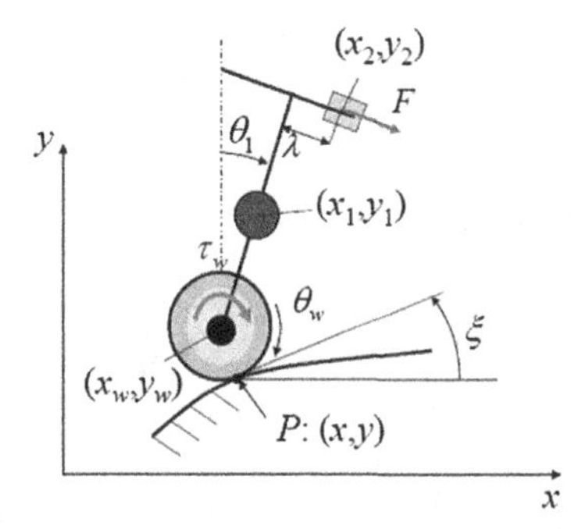

Figure 1.3: A UW-Car System in Rough Terrain.

The wheel−ground contact point is $P = f(x,y)^{\mathrm{T}}$. The gradient angle of the ground at point P is denoted by ξ.

The following six-dimensional vector will be used for the description of a UW-Car system's dynamic model:

$$\mathbf{q}^{(2)} = [x \quad y \quad \xi \quad \theta_{\mathrm{w}} \quad \theta_1 \quad \lambda]^{\mathrm{T}}. \tag{1.15}$$

To simplify the representation, we assume:

$$S_\xi = \sin\theta_\xi,\ C_\xi = \cos\theta_\xi,\ S_1 = \sin\theta_1,\ C_1 = \cos\theta_1 \tag{1.16}$$

$$\begin{gathered} f_x = \frac{\partial f}{\partial x},\ f_y = \frac{\partial f}{\partial y},\ f_{xx} = \frac{\partial f}{\partial x}\left(\frac{\partial f}{\partial x}\right),\ f_{yy} = \frac{\partial f}{\partial y}\left(\frac{\partial f}{\partial y}\right), \\ f_{xy} = \frac{\partial f}{\partial y}\left(\frac{\partial f}{\partial x}\right),\ f_{yx} = \frac{\partial f}{\partial x}\left(\frac{\partial f}{\partial y}\right). \end{gathered} \tag{1.17}$$

The positions and velocities of three parts of a UW-Car system are given by

$$\begin{cases} x_{\mathrm{w}} = x - r_{\mathrm{w}} S_\xi \\ y_{\mathrm{w}} = y + r_{\mathrm{w}} C_\xi \end{cases},\quad \begin{cases} \dot{x}_{\mathrm{w}} = \dot{x} - r_{\mathrm{w}} C_\xi \dot{\xi} \\ \dot{y}_{\mathrm{w}} = \dot{y} - r_{\mathrm{w}} S_\xi \dot{\xi} \end{cases} \tag{1.18}$$

$$\begin{cases} x_1 = x_{\mathrm{w}} + l_1 S_1 \\ y_1 = y_{\mathrm{w}} + l_1 C_1 \end{cases},\quad \begin{cases} \dot{x}_1 = \dot{x} - r_{\mathrm{w}} C_\xi \dot{\xi} + l_1 C_1 \dot{\theta}_1 \\ \dot{y}_1 = \dot{y} - r_{\mathrm{w}} S_\xi \dot{\xi} - l_1 S_1 \dot{\theta}_1 \end{cases} \tag{1.19}$$

$$\begin{cases} x_2 = x_{\mathrm{w}} + l_2 S_1 + \lambda C_1 \\ y_2 = y_{\mathrm{w}} + l_2 C_1 - \lambda S_1 \end{cases},\quad \begin{cases} \dot{x}_2 = \dot{x} - r_{\mathrm{w}} C_\xi \dot{\xi} + (l_2 C_1 - \lambda S_1)\dot{\theta}_1 + C_1 \dot{\lambda} \\ \dot{y}_2 = \dot{y} - r_{\mathrm{w}} S_\xi \dot{\xi} - (l_2 S_1 + \lambda C_1)\dot{\theta}_1 - S_1 \dot{\lambda} \end{cases}. \tag{1.20}$$

The Larangian equation of motion is used for the derivation of the dynamic equation. The kinetic, potential, and dissipated energy and their contributions to the dynamic equation are computed as follows.

The kinetic energy of the wheel, the body, and the seat can then be computed as

$$T_{\mathrm{w}} = \frac{1}{2} m_{\mathrm{w}} \dot{\mathbf{x}}_{\mathrm{w}}^{\mathrm{T}} \dot{\mathbf{x}}_{\mathrm{w}} + \frac{1}{2} I_{\mathrm{w}} \dot{\theta}_{\mathrm{w}}^2 = \frac{1}{2} \dot{\mathbf{q}}^{(2)\mathrm{T}} \mathbf{M}_{\mathrm{w}} \dot{\mathbf{q}}^{(2)} \tag{1.21}$$

$$T_1 = \frac{1}{2} m_1 \dot{\mathbf{x}}_1^{\mathrm{T}} \dot{\mathbf{x}}_1 + \frac{1}{2} I_1 \dot{\theta}_1^2 = \frac{1}{2} \dot{\mathbf{q}}^{(2)\mathrm{T}} \mathbf{M}_1 \dot{\mathbf{q}}^{(2)} \tag{1.22}$$

$$T_2 = \frac{1}{2} m_2 \dot{\mathbf{x}}_2^{\mathrm{T}} \dot{\mathbf{x}}_2 + \frac{1}{2} I_2 \dot{\theta}_1^2 = \frac{1}{2} \dot{\mathbf{q}}^{(2)\mathrm{T}} \mathbf{M}_2 \dot{\mathbf{q}}^{(2)}. \tag{1.23}$$

The symmetric matrices $\mathbf{M}_\mathrm{w}$, $\mathbf{M}_1$, and $\mathbf{M}_2$ are given in the Appendix at the end of the chapter. The contributions of kinetic energy to the Lagrangian equation are given by

$$\frac{\mathrm{d}}{\mathrm{d}t}\left(\frac{\partial}{\partial \dot{\mathbf{q}}^{(2)}}T_\mathrm{w}\right)-\frac{\partial T_\mathrm{w}}{\partial \mathbf{q}^{(2)}}=\mathbf{M}_\mathrm{w}\ddot{\mathbf{q}}^{(2)}+\mathbf{dM}_\mathrm{w}\dot{\mathbf{q}}^{(2)}+\mathbf{N}_\mathrm{w}\left(\mathbf{q}^{(2)},\dot{\mathbf{q}}^{(2)}\right) \tag{1.24}$$

$$\frac{\mathrm{d}}{\mathrm{d}t}\left(\frac{\partial}{\partial \dot{\mathbf{q}}^{(2)}}T_1\right)-\frac{\partial T_1}{\partial \mathbf{q}^{(2)}}=\mathbf{M}_1\ddot{\mathbf{q}}^{(2)}+\mathbf{dM}_1\dot{\mathbf{q}}^{(2)}+\mathbf{N}_1\left(\mathbf{q}^{(2)},\dot{\mathbf{q}}^{(2)}\right) \tag{1.25}$$

$$\frac{\mathrm{d}}{\mathrm{d}t}\left(\frac{\partial}{\partial \dot{\mathbf{q}}^{(2)}}T_2\right)-\frac{\partial T_2}{\partial \mathbf{q}^{(2)}}=\mathbf{M}_2\ddot{\mathbf{q}}^{(2)}+\mathbf{dM}_2\dot{\mathbf{q}}^{(2)}+\mathbf{N}_2\left(\mathbf{q}^{(2)},\dot{\mathbf{q}}^{(2)}\right), \tag{1.26}$$

where $\mathbf{dM}_w$, $\mathbf{dM}_1$, $\mathbf{dM}_2$, $\mathbf{N}_w$, $\mathbf{N}_1$, and $\mathbf{N}_2$ are also given in the Appendix.

The sum of gravitational potential energy of a UW-Car system is

$$U=m_\mathrm{w}g\cdot\left(y+r_\mathrm{w}C_\xi\right)+m_1g\cdot\left(y+r_\mathrm{w}C_\xi+l_1C_1\right)+m_2g\cdot\left(y+r_\mathrm{w}C_\xi+l_2C_1\quad\lambda S_1\right). \tag{1.27}$$

The contribution of this potential energy to the Lagrange motion equations is

$$\mathbf{dU}=\frac{\partial U}{\partial \mathbf{q}^{(2)}}[0\quad mg\quad -mgr_wS_\xi\quad 0\quad -(m_1l_1+m_2l_2)gS_1-m_2g\lambda C_1\quad -m_2gS_1]^\mathrm{T}, \tag{1.28}$$

where m is given by $m=m_\mathrm{w}+m_1+m_2$.

The energy will be dissipated due to the effect of friction between the wheels and the ground. The dissipated energy is

$$D=\frac{1}{2}D_\mathrm{w}\dot{\theta}_\mathrm{w}^2. \tag{1.29}$$

Combining (1.21)−(1.29), the Lagrangian equations of a UW-Car system are represented as follows:

$$\mathbf{M}^{(2)}\left(\mathbf{q}^{(2)}\right)\cdot\ddot{\mathbf{q}}^{(2)}+\mathbf{N}^{(2)}\left(\mathbf{q}^{(2)},\dot{\mathbf{q}}^{(2)}\right)=\mathbf{B}^{(2)}\boldsymbol{\tau}+\mathbf{A}^\mathrm{T}\boldsymbol{\mu}, \tag{1.30}$$

where

$$\mathbf{M}^{(2)}\left(\mathbf{q}^{(2)}\right)=\mathbf{M}_w\left(\mathbf{q}^{(2)}\right)+\mathbf{M}_1\left(\mathbf{q}^{(2)}\right)+\mathbf{M}_2\left(\mathbf{q}^{(2)}\right)$$

$$\mathbf{N}^{(2)}\left(\mathbf{q}^{(2)},\dot{\mathbf{q}}^{(2)}\right) = (\mathbf{dM}_w + \mathbf{dM}_1 + \mathbf{dM}_2)\dot{\mathbf{q}}^{(2)} + \mathbf{N}_w + \mathbf{N}_1 + \mathbf{N}_2 + \mathbf{dU} + \frac{\partial D}{\partial \dot{\mathbf{q}}^{(2)}}\mathbf{B}^{(2)}$$

$$= \begin{bmatrix} 0 & 0 & 0 & 1 & -1 & 0 \\ 0 & 0 & 0 & 0 & 0 & 1 \end{bmatrix}^{\mathrm{T}}, \boldsymbol{\tau} = [\tau_{\mathrm{w}} \quad F]^{\mathrm{T}}.$$

$\boldsymbol{\mu} = [\mu_1 \quad \mu_2 \quad \mu_3]^{\mathrm{T}}$ is the Lagrangian multipliers vector. Matrix $\mathbf{A}$ comes from some constraint equations that are illustrated as follows.

The first constrained equation is derived from the nonslip condition of the rolling wheels, which is given by

$$\begin{cases} \dot{x} - r_{\mathrm{w}} C_\xi \dot{\theta}_{\mathrm{w}} = 0 \\ \dot{y} - r_{\mathrm{w}} S_\xi \dot{\theta}_{\mathrm{w}} = 0. \end{cases} \tag{1.31}$$

The other constraint equation comes from the terrain function $f(x,y) = 0$. For any differentiable function f, obviously we have

$$\nabla f\big|_P \cdot [C_\xi \quad S_\xi] = 0 \Rightarrow f_x \cdot C_\xi + f_y \cdot S_\xi = 0, \tag{1.32}$$

where $\nabla f|_P$ is the gradient of the function f at point P. Differentiating both sides of (1.32), it follows that

$$F_x \cdot \dot{x} + F_y \cdot \dot{y} + F_\xi \cdot \dot{\xi} = 0, \tag{1.33}$$

where

$$F_x = f_{xx} C_\xi + f_{yx} S_\xi \tag{1.34a}$$

$$F_y = f_{xy} C_\xi + f_{yy} S_\xi \tag{1.34b}$$

$$F_\xi = f_y C_\xi - f_x S_\xi. \tag{1.34c}$$

Combining (1.31) and (1.33), matrix $\mathbf{A}$ can be computed as

$$\mathbf{A} = [\mathbf{A}_1 \quad \mathbf{A}_2] = \left[\begin{array}{ccc|ccc} 1 & 0 & 0 & -r_{\mathrm{w}} C_\xi & 0 & 0 \\ 0 & 1 & 0 & -r_{\mathrm{w}} S_\xi & 0 & 0 \\ F_x & F_y & F_\xi & 0 & 0 & 0 \end{array}\right]. \tag{1.35}$$

Dividing $\mathbf{q}^{(2)}$ into two parts:

$$\mathbf{q}^{(2)} = \left[\mathbf{q}^{(3)} \quad \mathbf{q}^{(1)}\right]^{\mathrm{T}}, \ \mathbf{q}^{(3)} = [x \quad y \quad \xi]^{\mathrm{T}}, \ \mathbf{q}^{(1)} = [\theta_{\mathrm{w}} \quad \theta_1 \quad \lambda]^{\mathrm{T}}. \tag{1.36}$$

By using the technique proposed in Ref. [29], the reduced-order dynamic equation of a UW-Car system in a rough terrain is finally obtained as follows:

$$\overline{\mathbf{M}}^{(2)} \cdot \ddot{\mathbf{q}}^{(1)} + \overline{\mathbf{N}}^{(2)}\left(\mathbf{q}^{(2)}, \dot{\mathbf{q}}^{(2)}\right) = \overline{\mathbf{B}}^{(2)}\boldsymbol{\tau}, \tag{1.37}$$

where

$$\overline{\mathbf{M}}^{(2)} = \mathbf{C}^{\mathrm{T}}\mathbf{M}^{(2)}\mathbf{C}, \ \ \overline{\mathbf{N}} = \mathbf{C}^{\mathrm{T}}\left(\mathbf{M}^{(2)}\dot{\mathbf{C}}\dot{\mathbf{q}}^{(1)} + \mathbf{N}^{(2)}\right), \ \ \overline{\mathbf{B}}^{(2)} = \mathbf{C}^{\mathrm{T}}\mathbf{B}^{(2)}. \tag{1.38a}$$

$$\mathbf{C} = \begin{bmatrix} -\mathbf{A}_1^{-1}\mathbf{A}_2 \\ \mathbf{I}_{3\times 3} \end{bmatrix}, \quad \mathbf{C}^{\mathrm{T}}\mathbf{A}^{\mathrm{T}} = 0. \tag{1.38b}$$

1.3 Velocity Control Using a Sliding Mode Approach

1.3.1 Velocity Control of a UW-Car System on Flat Ground

To ensure that the UW-Car system is driven steadily, a special sliding surface and sliding mode controller design scheme are proposed in this subsection.

There are two basic requirements of our TSMC controllers:

1. Only the situation of the UW-Car running on flat ground is considered.
2. The body should be kept upright and the seat should vibrate as little as possible while the UW-Car system is running.

In the rest of this chapter, "^" denotes that the terms are evaluated based on parameters of the nominal system moving on flat ground without any uncertainties and disturbances.

Assuming matrix $\mathbf{M}^{(1)}(\mathbf{q}^{(1)})$ is invertible, we start by rewriting the general model (1.10) as

$$\ddot{\mathbf{q}}^{(1)} = \mathbf{F}\left(\dot{\mathbf{q}}^{(1)}, \mathbf{q}^{(1)}\right) + \mathbf{G}\left(\dot{\mathbf{q}}^{(1)}, \mathbf{q}^{(1)}\right)\boldsymbol{\tau}, \tag{1.39}$$

where

$$\mathbf{F} = \begin{bmatrix} F_1 & F_2 & F_3 \end{bmatrix}^{\mathrm{T}}, \ \ \mathbf{G} = \begin{bmatrix} G_{11} & G_{21} & G_{31} \\ G_{12} & G_{22} & G_{32} \end{bmatrix}^{\mathrm{T}},$$

and the nominal system is given by

$$\ddot{\mathbf{q}}^{(1)} = \hat{\mathbf{F}}\left(\dot{\mathbf{q}}^{(1)}, \mathbf{q}^{(1)}\right) + \hat{\mathbf{G}}\left(\dot{\mathbf{q}}^{(1)}, \mathbf{q}^{(1)}\right)\boldsymbol{\tau}. \tag{1.40}$$

Considering the underactuated feature of a UW-Car system, we have to reduce the order to obtain the controller.

Choosing a new state variable $\mathbf{q}_1 = [\theta_1 \quad \lambda]^{\mathrm{T}}$, the following subsystem is then investigated in the TSMC controller design:

$$\ddot{\mathbf{q}}_1 = \mathbf{F}_1(\mathbf{q}_1, \dot{\mathbf{q}}_1) + \mathbf{G}_1(\mathbf{q}_1, \dot{\mathbf{q}}_1)\boldsymbol{\tau}, \tag{1.41}$$

where

$$\mathbf{F}_1 = \begin{bmatrix} F_2 \\ F_3 \end{bmatrix}, \quad \mathbf{G}_1 = \begin{bmatrix} G_{21} & G_{22} \\ G_{31} & G_{32} \end{bmatrix}.$$

Similarly, the nominal subsystem can be represented as:

$$\ddot{\mathbf{q}}_1 = \hat{\mathbf{F}}_1(\mathbf{q}_1, \dot{\mathbf{q}}_1) + \hat{\mathbf{G}}_1(\mathbf{q}_1, \dot{\mathbf{q}}_1)\boldsymbol{\tau}. \tag{1.42}$$

It is assumed that the following equations are satisfied:

$$\left|\hat{F}_i - F_i\right| \le \tilde{F}_i, \quad \tilde{\mathbf{F}} = \begin{bmatrix} \tilde{F}_2 & \tilde{F}_3 \end{bmatrix}, \quad \mathbf{G}_1 = (\mathbf{I} + \boldsymbol{\Delta})\hat{\mathbf{G}}_1, \tag{1.43}$$

where F_i and $\hat{F}_i$ represent the ith element of matrices $\mathbf{F}$ and $\hat{\mathbf{F}}$ with i =2, 3, G_{ij} and $\hat{G}_{ij}$ represent the (i, j)th element of matrices $\mathbf{G}$ and $\hat{\mathbf{G}}$ with $i = 2, 3, j = 1, 2$. $\mathbf{I}$ is a 2×2 identity matrix and $\boldsymbol{\Delta}$ is composed of Δ_{ij} satisfying the following inequality:

$$\left|\Delta_{ij}\right| \le D_{ij}, \quad D_{ij} > 0, \quad \|D\| < 1.$$

According to the terminal sliding mode control method proposed in Ref. [26], the sliding surface was defined as follows:

$$\mathbf{s}(t) = \dot{\mathbf{e}}(t) + \mathbf{C}\mathbf{e}(t) - \mathbf{w}(t), \tag{1.44}$$

where $\mathbf{e}(t) = \mathbf{x}(t) - \mathbf{x}_{\mathrm{d}}(t)$, with $\mathbf{x}_{\mathrm{d}}(t)$ is the reference value. $\mathbf{C} = \mathrm{diag}(c_1, c_2, \ldots, c_m)$, $c_i > 0$, $\mathbf{w}(t) = \dot{\mathbf{v}}(t) + \mathbf{C}\mathbf{v}(t)$.

The Terminal Sliding Mode Control method is applied to the subsystem (1.41). The inclination angle is expected to be zero and the desired position of seat λ^* can be obtained from (1.14). We define the following sliding surfaces:

$$\begin{aligned} s_1(t) &= \dot{\theta}_1 + c_1\left(\theta_1(t) - \theta_1^*\right) - \dot{v}_1(t) - c_1 v_1(t) \\ s_2(t) &= \dot{\lambda} + c_2\left(\lambda(t) - \lambda^*\right) - \dot{v}_2(t) - c_2 v_2(t), \end{aligned} \tag{1.45}$$

where c_1 and c_2 are positive constants. The augmenting functions v_1 and v_2 are designed as cubic polynomials that guarantee assumption 1 in Ref. [26] holds.

Proposition 1. Suppose the model uncertainties of a UW-Car system (1.39) satisfy (1.43), the sliding surfaces (1.45) will be achieved while the inclination angle θ_1 will converge to zero and the position of the seat will reach λ^* in finite time, if the following control law is applied to the system:

$$\boldsymbol{\tau} = \hat{\mathbf{G}}_1^{-1}[-\hat{\mathbf{F}}_1 - \mathbf{C}_1\dot{\mathbf{e}}_1 + \ddot{\mathbf{v}} + \mathbf{C}_1\dot{\mathbf{v}} - \mathbf{k}\,\mathrm{sgn}(\mathbf{s})], \tag{1.46}$$

where

$$\mathbf{C}_1 = \mathrm{diag}(\,c_1 \quad c_2\,),\ \ \mathbf{e}_1 = [\,\theta_1 - \theta_1^* \quad \lambda - \lambda^*\,]^{\mathrm{T}}$$
$$\mathbf{v} = [\,v_1 \quad v_2\,]^{\mathrm{T}},\ \mathrm{sgn}(\mathbf{s}) = \mathrm{diag}(\,\mathrm{sgn}(s_1) \quad \mathrm{sgn}(s_2)\,)$$
$$\mathbf{k} = (\mathbf{I} - \mathbf{D})^{-1}(\tilde{\mathbf{F}} + \mathbf{D}\left| -\hat{\mathbf{F}} - \mathbf{C}_1\dot{\mathbf{e}}_1 + \ddot{\mathbf{v}} + \mathbf{C}_1\dot{\mathbf{v}}\right| + \boldsymbol{\Gamma}),\ \ \boldsymbol{\Gamma} = [\,\gamma_1 \quad \gamma_2\,],\ \ \gamma_i > 0.$$

Proof. The proof of convergence is proposed in Ref. [26] according to Theorems 2 and 3.

Note that although the proposed TSMC controllers do not aim at controlling the velocity of the UW-Car, the UW-Car will finally converge to a constant speed when it is stabilized at an equilibrium $\mathbf{q}_1^* = [\,0 \quad \lambda^*\,]^{\mathrm{T}}$. This can be easily understood from (1.14).

Proposition 2. Suppose the following inequality is satisfied:

$$M_{11} + M_{12}C_1 - m_2 r_w \lambda S_1 > 0. \tag{1.47}$$

The proposed TSMC controller guarantees that the angle velocity $\dot{\theta}_{\mathrm{w}}$ can converge to $\dot{\theta}_{\mathrm{w}}^*$, when the UW-Car system is running on flat ground.

Proof. Obviously the sliding surface is achieved in this special case. Adding the first two equations of (1.9), it follows that:

$$\begin{aligned}&\left(M_{11} + M_{12}C_1 - m_2 r_{\mathrm{w}}\lambda S_1\right)\ddot{\theta}_{\mathrm{w}} + \left(m_2 r_{\mathrm{w}} C_1 + m_2 l_2\right)\ddot{\lambda} + \left(M_{12}C_1\right.\\ &\quad \left. + M_{22} - m_2 r_{\mathrm{w}}\lambda S_1 + m_2\lambda^2\right)\ddot{\theta}_1 = -D_{\mathrm{w}}\dot{\theta}_{\mathrm{w}} - D_1\dot{\theta}_1 + 2m_2 r_{\mathrm{w}} S_1\dot{\theta}_1\dot{\lambda}\\ &\quad + \left(M_{12}S_1 + m_2 r_{\mathrm{w}}\lambda C_1\right)\dot{\theta}_1^2 + G_1 S_1 + m_2 g\lambda C_1 - 2m_2\lambda\dot{\theta}_1\dot{\lambda}.\end{aligned} \tag{1.48}$$

Note that (1.48) represents the internal dynamics model of a UW-Car, which has nothing to do with the control input τ_{w} and f. State variables of the dynamic model are always on the sliding surface and satisfy the following equations:

$$\begin{aligned}&\dot{\theta}_1(t) + c_1\theta_1(t) - \dot{v}_1(t) - c_1 v_1(t) = 0\\ &\dot{\lambda} + c_2\left(\lambda(t) - \lambda^*\right) - \dot{v}_2(t) - c_2 v_2(t) = 0.\end{aligned} \tag{1.49}$$

Given known initial states $\theta_1(0)$, $\dot{\theta}_1(0)$, $\lambda(0)$, $\dot{\lambda}(0)$ and the sliding mode surface functions v_1 and v_2, θ_1, $\dot{\theta}_1$, $\ddot{\theta}_1$, λ, $\dot{\lambda}$ and $\ddot{\lambda}$ are obtained directly from (1.49). Substituting these solutions into (1.48), it follows that

$$Q(t)\ddot{\theta}_w + D_w\dot{\theta}_w = P(t), \tag{1.50}$$

where $P(t)$ and $Q(t)$ are known time-dependent functions and $Q(t)$ satisfies

$$Q(t) = M_{11} + M_{12}C_1 - m_2 r_w \lambda S_1.$$

Equation (1.50) is a first-order linear differential function with respect to the angle velocity $\dot{\theta}_w$. Since $D_w > 0$, the stability of (1.50) is only related to $Q(t)$. Because the inequality (1.47) is satisfied, the solution of (1.50) is asymptotically stable. From (1.49), it follows that θ_1, $\dot{\theta}_1$, λ, and $\dot{\lambda}$ converge to the equilibrium that results in the convergence of $\dot{\theta}_w$ to the desired angle velocity $\dot{\theta}_w^*$.

1.3.2 Optimal Braking Controller Designed Using Sliding Mode Control

A UW-Car system moving smoothly with a constant velocity has been investigated in the above section. A braking strategy aimed at obtaining the shortest braking distance is discussed in this subsection.

We assume that the UW-Car system has been stabilized in the equilibrium $\mathbf{q}_1^0 = [0 \quad \lambda]^T$ with a corresponding steady velocity V_0 before braking. A switching TSMC control strategy is proposed to allow the UW-Car system to brake within the shortest distance. Because of the inertia of the seat, the system cannot brake smoothly in just one time. We have to adjust the position of the seat to guarantee that the velocity will ultimately decrease to zero. In short, the whole procedure is divided into the following two phases.

- **Phase 1**: Decrease the speed of a UW-Car to zero as soon as possible.
- **Phase 2:** Adjust the position of the seat to stop the system.

The two phases are depicted in Figures 1.4(a) and 1.5 respectively.

In the procedure of braking, we designed three TSMC controllers TSMC1, TSMC2, and TSMC3 according to the analysis of the braking procedure and Eq. (1.46).

In phase 1, we assume that the UW-Car system has been stabilized at equilibrium $\mathbf{q}_1^0 = [0 \quad \lambda]^T$ with constant velocity V_0 before braking. TSMC1 is applied to decrease the velocity to V_1 ($V_1 < V_0$) at equilibrium $\mathbf{q}_1^{(1)} = [0 \quad -\lambda_1]^T$ at time $t = T_{f0}$, and keep the position of the seat until the velocity is reduced to zero in the end of phase 1 at $t = T_{f1}$. The distance that the UW-Car moves forward is denoted by FS, which can be computed as the area depicted in Figure 1.4(b):

$$FS = r_w \cdot \text{area}(A_1 + A_2). \tag{1.51}$$

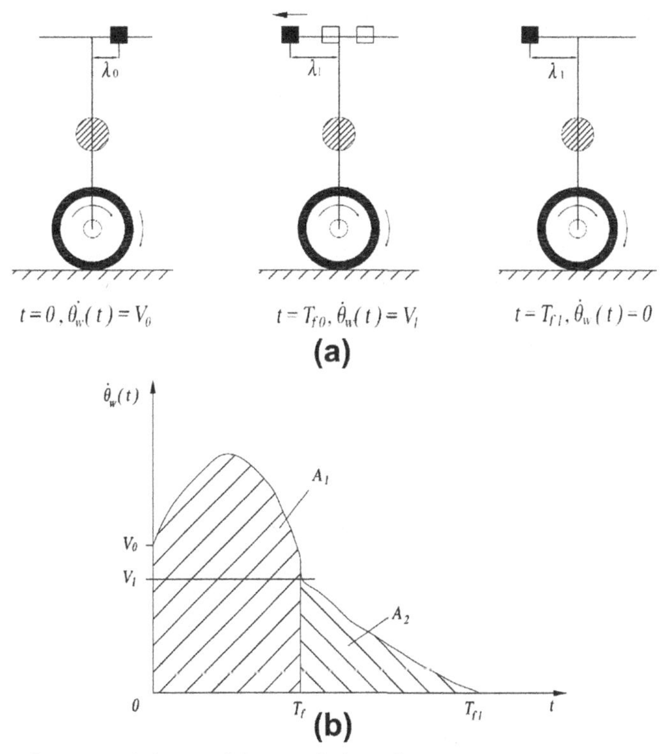

Figure 1.4: Phase 1 of the Braking and the Distance a UW-Car Moves Forward.

In phase 2, at the beginning, the UW-Car stays at equilibrium $\mathbf{q}_1^{(1)} = [0 \quad -\lambda_1]^{\mathrm{T}}$ with velocity $\dot{\theta}_{\mathrm{w}} = 0$, which is the last state of phase 1. From time T_{f1}, the system will be controlled by TSMC2 until time T_{f2} and reaches the equilibrium $\mathbf{q}_1^{(2)} = [0 \quad \lambda_2]^{\mathrm{T}}$ with an appropriate instant velocity V_2, which guarantees that the system is finally stabilized at stable equilibrium $\mathbf{q}_1^{(3)} = [0 \quad 0]^{\mathrm{T}}$. From time T_{f2} to time T_{f3}, the UW-Car system is controlled by TSMC3 and the system finally stops at time $t = T_{f3}$ at the equilibrium $\mathbf{q}_1^{(3)} = [0 \quad 0]^{\mathrm{T}}$.

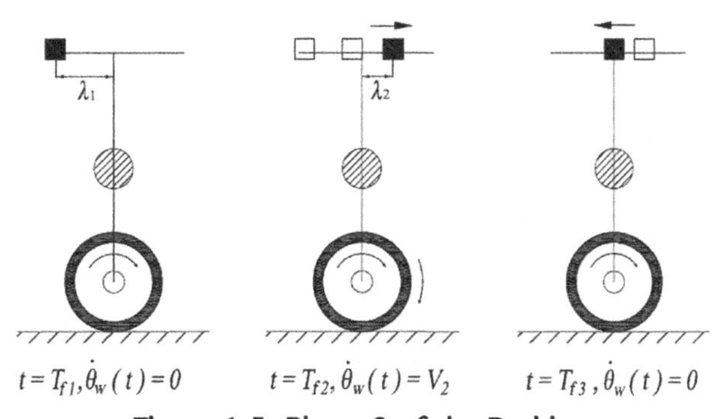

Figure 1.5: Phase 2 of the Braking.

The switching control diagram is shown in Figure 1.6.

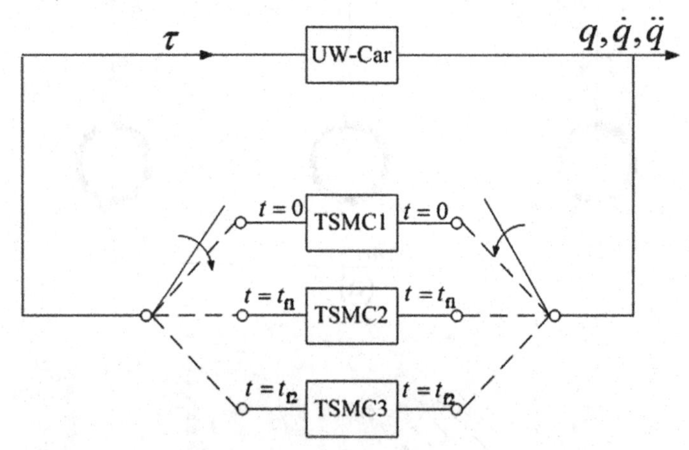

Figure 1.6: The Structure of Switching TSMCs Used for Braking a UW-Car.

The controllers are based on sliding surfaces given by

$$\begin{aligned} s_1(t) &= \dot{\theta}_1 + c_1\left(\theta_1(t) - \theta_1^*\right) - \dot{v}_1^{(i)}(t) - c_1 v_1^{(i)}(t) \\ s_2(t) &= \dot{\lambda} + c_2(\lambda(t) - \lambda_i) - \dot{v}_2^{(i)}(t) - c_2 v_2^{(i)}(t), \end{aligned} \tag{1.52}$$

where

$$v_2^{(1)}(t) = \begin{cases} a_0^{(1)} + a_1^{(1)}t + a_2^{(1)}t^2 + a_3^{(1)}t^3, & \text{if } 0 \le t \le T_{f0} \\ 0, & \text{if } t > T_{f0} \end{cases} \tag{1.53}$$

$$v_2^{(2)}(t) = \begin{cases} a_0^{(2)} + a_1^{(2)}(t - T_{f1}) + a_2^{(2)}\left(t - T_{f1}\right)^2 & \\ +a_3^{(2)}\left(t - T_{f1}\right)^3, & T_{f1} \le t \le T_{f2} \\ 0, & t > T_{f2} \end{cases} \tag{1.54}$$

$$v_2^{(3)}(t) = \begin{cases} a_0^{(3)} + a_1^{(3)}(t - T_{f2}) + a_2^{(3)}\left(t - T_{f2}\right)^2 & \\ +a_3^{(3)}\left(t - T_{f2}\right)^3, & T_{f2} \le t \le T_{f3} \\ 0, & t > T_{f3} \end{cases} \tag{1.55}$$

$$\begin{aligned} & a_0^{(i)} = \Delta_i,\ a_1^{(i)} = 0,\ a_2^{(i)} = -3\Delta_i/T_i^2,\ a_3^{(i)} = 2\Delta_i/T_i^3 \\ & \Delta_i = \lambda_{i-1} - \lambda_i, T_1 = T_{f0},\ T_{j+1} = T_{f(j+1)} - T_{fj} \quad i = 1,\ 2,\ 3,\ j = 1,\ 2. \end{aligned} \tag{1.56}$$

Note that the initial position of the seat is denoted as λ_0 and the final position as λ_3.

Proposition 3 (calculation of braking distance). The distance that the UW-Car moves forward in phase 1 is computed as:

$$FS = r_{\mathrm{w}} \cdot (\mathrm{area}(A_1) + \mathrm{area}(A_2)), \tag{1.57}$$

where

$$\begin{aligned}
\mathrm{Area}(A_1) = {} & \frac{z_1^{(1)}}{24p} T_{f0}^4 + \frac{1}{6}\left(\frac{z_2^{(1)}}{p} - \frac{z_1^{(1)}}{p^2}\right) T_{f0}^3 \\
& + \frac{1}{2}\left(\frac{z_3^{(1)}}{p} - \frac{z_2^{(1)}}{p^2} + \frac{z_1^{(1)}}{p^3}\right) T_{f0}^2 + \left(\frac{z_4^{(1)}}{p} - \frac{z_3^{(1)}}{p^2} + \frac{z_2^{(1)}}{p^3} - \frac{z_1^{(1)}}{p^4}\right) T_{f0} \\
& + \frac{1}{p}\left(\frac{z_4^{(1)}}{p} - \frac{z_3^{(1)}}{p^2} + \frac{z_2^{(1)}}{p^3} - \frac{z_1^{(1)}}{p^4} - V_0\right)\left(\mathrm{e}^{-pT_{f0}} - 1\right)
\end{aligned}$$

$$\mathrm{Area}(A_2) = \frac{q_2\lambda_1}{p^2} \ln\left(1 - \frac{pV_1}{q_2\lambda_1}\right) + \frac{V_1}{p}$$

$$\begin{aligned}
V_1 = \dot{\theta}(T_{f0}) = {} & \frac{z_1^{(1)}}{6p} T_{f0}^3 + \frac{1}{2}\left(\frac{z_2^{(1)}}{p} - \frac{z_1^{(1)}}{p^2}\right) T_{f0}^2 + \left(\frac{z_3^{(1)}}{p} - \frac{z_2^{(1)}}{p^2} + \frac{z_1^{(1)}}{p^3}\right) T_{f0} \\
& + \left(\frac{z_4^{(1)}}{p} - \frac{z_3^{(1)}}{p^2} + \frac{z_2^{(1)}}{p^3} - \frac{z_1^{(1)}}{p^4}\right) + \left(\frac{z_4^{(1)}}{p} - \frac{z_3^{(1)}}{p^2} + \frac{z_2^{(1)}}{p^3} - \frac{z_1^{(1)}}{p^4} - V_0\right) \mathrm{e}^{-pT_{f0}}
\end{aligned}$$

$$p = \frac{D_{\mathrm{w}}}{(m_{\mathrm{w}} + m_1 + m_2) r_{\mathrm{w}}^2 + I_{\mathrm{w}} + m_1 l_1 r_{\mathrm{w}} + m_2 l_2 r_{\mathrm{w}}}$$

$$q_1 = \frac{m_2 (r_{\mathrm{w}} + l_2)}{(m_{\mathrm{w}} + m_1 + m_2) r_{\mathrm{w}}^2 + I_{\mathrm{w}} + m_1 l_1 r_{\mathrm{w}} + m_2 l_2 r_{\mathrm{w}}}$$

$$q_2 = \frac{m_2 g}{(m_{\mathrm{w}} + m_1 + m_2) r_{\mathrm{w}}^2 + I_{\mathrm{w}} + m_1 l_1 r_{\mathrm{w}} + m_2 l_2 r_{\mathrm{w}}}$$

$$z_1^{(1)} = 6c_2^2 a_3^{(1)} q_1 \quad z_2^{(1)} = 2c_2^2 a_2^{(1)} q_1 \quad z_3^{(1)} = -6a_3^{(1)} q_1$$

$$z_4^{(1)} = \left(c_2^2 \lambda_0 - 2a_2^{(1)}\right) q_1 \qquad c_2 = \sqrt{q_2/q_1}\,.$$

Proof. From time 0 to time T_{f0}, the system state $\mathbf{q}_1$ is always on the sliding surface of TSMC1. Thus, the dynamics of the whole system can be represented by the following equations:

$$\begin{aligned}
&\dot{\lambda} + c_2(\lambda - \lambda_1) - \dot{v}_2^{(1)}(t) - c_2 v_2^{(1)}(t) = 0 \\
&(M_{11} + M_{12})\ddot{\theta}_{\mathrm{w}} + m_2(r_{\mathrm{w}} + l_2)\ddot{\lambda} = -D_{\mathrm{w}}\dot{\theta}_{\mathrm{w}} + m_2 g\lambda.
\end{aligned} \tag{1.58}$$

The second equation of (1.58) is derived by substituting $\theta_1 = 0$, $\dot{\theta}_1 = 0$ and $\ddot{\theta}_1 = 0$ into the internal dynamics (1.48). Since the functions $v_2^{(1)}(t)$ and $\dot{v}_2^{(1)}(t)$ are known, it is apparent that $\lambda(t)$ can be easily solved from the first equation of (1.58). Substituting the solution of $\lambda(t)$ and its derivatives into the second equation, it is rewritten as a linear first-order differential equation with respect to the angle velocity $\dot{\theta}_{\mathrm{w}}(t)$. With initial conditions $\lambda(0) = 0$, $\dot{\lambda}(0) = 0$, $\dot{\theta}_{\mathrm{w}}(0) = V_0$, $\dot{\theta}_{\mathrm{w}}$ can also be solved as follows:

$$\begin{aligned}
\dot{\theta}_\omega =& \frac{z_1^{(1)}}{6p}t^3 + \frac{1}{2}\left(\frac{z_2^{(1)}}{p} - \frac{z_1^{(1)}}{p^2}\right)t^2 + \left(\frac{z_3^{(1)}}{p} - \frac{z_2^{(1)}}{p^2} + \frac{z_1^{(1)}}{p^3}\right)t \\
&+ \left(\frac{z_4^{(1)}}{p} - \frac{z_3^{(1)}}{p^2} + \frac{z_2^{(1)}}{p^3} - \frac{z_1^{(1)}}{p^4}\right) + \left(\frac{z_4^{(1)}}{p} - \frac{z_3^{(1)}}{p^2} + \frac{z_2^{(1)}}{p^3} - \frac{z_1^{(1)}}{p^4} - V_0\right)\mathrm{e}^{-pt}.
\end{aligned} \tag{1.59}$$

Therefore, area(A_1) is computed as

$$\begin{aligned}
\mathrm{Area}(A_1) =& \int_0^{T_{f0}} \dot{\theta}_{\mathrm{w}}(t)\mathrm{d}t = \frac{z_1^{(1)}}{24p}T_{f0}^4 + \frac{1}{6}\left(\frac{z_2^{(1)}}{p} - \frac{z_1^{(1)}}{p^2}\right)T_{f0}^3 \\
&+ \frac{1}{2}\left(\frac{z_3^{(1)}}{p} - \frac{z_2^{(1)}}{p^2} + \frac{z_1^{(1)}}{p^3}\right)T_{f0}^2 + \left(\frac{z_4^{(1)}}{p} - \frac{z_3^{(1)}}{p^2} + \frac{z_2^{(1)}}{p^3} - \frac{z_1^{(1)}}{p^4}\right)T_{f0} \\
&+ \frac{1}{p}\left(\frac{z_4^{(1)}}{p} - \frac{z_3^{(1)}}{p^2} + \frac{z_2^{(1)}}{p^3} - \frac{z_1^{(1)}}{p^4} - V_0\right)\left(\mathrm{e}^{-pT_{f0}} - 1\right).
\end{aligned} \tag{1.60}$$

From time T_{f1} to time T_{f2}, the dynamic of the UW-Car system can be represented by the following equations with initial velocity $\dot{\theta}(T_{f0}) = V_1$:

$$\begin{aligned}
&\dot{\lambda} + c_2(\lambda - \lambda_1) = 0 \\
&(M_{11} + M_{12})\ddot{\theta}_{\mathrm{w}} + m_2(r_{\mathrm{w}} + l_2)\ddot{\lambda} = -D_{\mathrm{w}}\dot{\theta}_{\mathrm{w}} + m_2 g\lambda.
\end{aligned} \tag{1.61}$$

Similarly, area(A_2) is computed as follows:

$$\text{area}(A_2) = \int_{T_{f0}}^{T_{f1}} \dot{\theta}_w(t)\mathrm{d}t = \frac{q_2\lambda_1}{p^2}\ln\left(1 - \frac{pV_1}{q_2\lambda_1}\right) + \frac{V_1}{p}. \tag{1.62}$$

It is apparent that the area presented in Figure 1.3(b) can be computed as area(A_1) + area(A_2). Therefore, Eq. (1.51) can be represented by Eq. (1.57).

Proposition 4. The velocity $\dot{\theta}_w(t)$ is reduced to V_2 at time $t = T_{f2}$, which guarantees that the velocity of the UW-Car system will become zero; meanwhile system state $\mathbf{q}_1$ converges to the equilibrium $\mathbf{q}_1^{(3)} = [0 \quad 0]^T$ at the end of phase 2.

Proof. The constant velocity V_2 is chosen as

$$\begin{aligned} V_2 = \Bigg[&-\frac{z_1^{(2)}}{6p}T_3^3 - \frac{1}{2}\left(\frac{z_2^{(2)}}{p} - \frac{z_1^{(2)}}{p^2}\right)T_3^2 - \left(\frac{z_3^{(2)}}{p} - \frac{z_2^{(2)}}{p^2} + \frac{z_1^{(2)}}{p^3}\right)T_3 \\ &-\left(\frac{z_4^{(2)}}{p} - \frac{z_3^{(2)}}{p^2} + \frac{z_2^{(2)}}{p^3} - \frac{z_1^{(2)}}{p^4}\right)\Bigg]\mathrm{e}^{pT_3} - \left(\frac{z_4^{(2)}}{p} - \frac{z_3^{(2)}}{p^2} + \frac{z_2^{(2)}}{p^3} - \frac{z_1^{(2)}}{p^4}\right), \end{aligned} \tag{1.63}$$

where

$$z_1^{(2)} = 6c_2^2a_3^{(3)}q_1 \quad z_2^{(2)} = 2c_2^2a_2^{(3)}q_1 \quad z_3^{(2)} = -6a_3^{(3)}q_1 \quad z_4^{(2)} = \left(c_2^2\lambda_2 - 2a_2^{(3)}q_1\right).$$

Obviously only the nominal system (1.16) should be studied. TSMC2 ensures the velocity of the UW-Car system can reach V_2 at time $t = T_{f2}$ while keeping the body inclination angle and the seat position at the equilibrium $\mathbf{q}_1^{(2)} = [0 \quad \lambda_2]^T$. From time T_{f2}, the system is controlled by TSMC3 with the sliding surface (1.55). The system state $\mathbf{q}_1$ is always on the sliding surface, and the dynamics of the system can be represented by the following equations when $T_{f2} \le t \le T_{f3}$:

$$\begin{aligned} &\dot{\lambda} + c_2(\lambda - \lambda_2) - \dot{v}_2^{(3)}(t) - c_2v_2^{(3)}(t) = 0 \\ &(M_{11} + M_{12})\ddot{\theta}_\omega + m_2(r_w + l_2)\ddot{\lambda} = -D_\omega\dot{\theta}_\omega + m_2g\lambda. \end{aligned} \tag{1.64}$$

The final velocity can be computed by Eq. (1.64) with initial conditions $\dot{\lambda}(T_{f2}) = 0$, $\lambda(T_{f2}) = \lambda_2$, $\dot{\theta}_\omega(T_{f2}) = V_2$. The solution of $\dot{\theta}_\omega(t)$ is obtained as follows:

$$\begin{aligned} \dot{\theta}_\omega(t) = &\frac{z_1^{(2)}}{6p}(t - T_{f2})^3 + \frac{1}{2}\left(\frac{z_2^{(2)}}{p} - \frac{z_1^{(2)}}{p^2}\right)(t - T_{f2})^2 \\ &+\left(\frac{z_3^{(2)}}{p} - \frac{z_2^{(2)}}{p^2} + \frac{z_1^{(2)}}{p^3}\right)(t - T_{f2}) + \left(\frac{z_4^{(2)}}{p} - \frac{z_3^{(2)}}{p^2} + \frac{z_2^{(2)}}{p^3} - \frac{z_1^{(2)}}{p^4}\right) \\ &+\left(\frac{z_4^{(2)}}{p} - \frac{z_3^{(2)}}{p^2} + \frac{z_2^{(2)}}{p^3} - \frac{z_1^{(2)}}{p^4} - V_2\right)\mathrm{e}^{-p(t-T_{f2})}. \end{aligned} \tag{1.65}$$

Substituting Eq. (1.63) into Eq. (1.65), obviously $\dot{\theta}(T_{f3})$ equals zero. Meanwhile, the system state $\lambda(t)$ converges to zero at time $t = T_{f3}$.

In order to obtain an optimal braking scheme, the braking distance FS has to be minimized. Thus, the following optimization problem should be solved:

$$\min_{\lambda_1, T_{f0}} (FS) \quad \text{s.t. } |\lambda_1| \leq \lambda_{\max}, \quad T_{f0} > 0. \tag{1.66}$$

Genetic algorithms (GAs), which are among the most effective optimal research methods, is adopted to optimize parameters λ_1 and T_{f0}. The optimal λ_1^* and T_{f0}^* are regarded as parameters of TSMC1.

As the main result of this subsection, the optimal braking scheme using the three TSMC controllers is illustrated as follows:

- **Step 1** (Optimization). An offline GA method is used to search the optimal parameters (λ_1^* and T_{f0}^*) according to (1.66).
- **Step 2** (Braking). In phase 1, the optimal parameters are computed by a GA in step 1 as the parameters of controller TSMC1. Apply the controller TSMC1 to the UW-Car system at $t = 0$ to ensure the velocity is zero as soon as possible and the distance the UW-Car moves forward is minimized.
- **Step 3** (Adjustment 2-1). Apply the controller TSMC2 to the UW-Car system at the beginning of phase 2 to reach an appropriate initial velocity V_2.
- **Step 4** (Adjustment 2-2). At time $t = T_{f2}$, switch controller TSMC2 to controller TSMC3, and the UW-Car system finally brakes steady at time $t = T_{f3}$.

1.3.3 Velocity Control of a UW-Car System in Rough Terrain

Due to the complexity of the rough terrain function $f(x,y)$, it is necessary to obtain the nominal model of (1.37) before controller design. Assume that the nominal case of (1.37) is the model of a UW-Car system moving on flat ground, which can be obtained by selecting function f as

$$f(x,y) = \hat{f}(x,y) = y = 0. \tag{1.67}$$

The corresponding nominal cases of matrices **A** and **C** can be easily computed as follows:

$$\hat{\mathbf{A}} = \begin{bmatrix} \hat{\mathbf{A}}_1 & \hat{\mathbf{A}}_2 \end{bmatrix} = \left[\begin{array}{ccc|ccc} 1 & 0 & 0 & -r_w & 0 & 0 \\ 0 & 1 & 0 & 0 & 0 & 0 \\ 0 & 0 & 1 & 0 & 0 & 0 \end{array}\right] \tag{1.68}$$

$$\hat{\mathbf{C}} = \begin{bmatrix} -\hat{\mathbf{A}}_1^{-T} \hat{\mathbf{A}}_2 \\ \mathbf{I}_{3\times 3} \end{bmatrix} = \left[\begin{array}{ccc|ccc} r_w & 0 & 0 & 1 & 0 & 0 \\ 0 & 0 & 0 & 0 & 1 & 0 \\ 0 & 0 & 0 & 0 & 0 & 1 \end{array}\right]^{\mathrm{T}}. \tag{1.69}$$

Consequently, the nominal case of the reduced-order dynamic model (1.37) is described by (1.10).

Assuming matrices $\overline{\mathbf{M}}^{(2)}(\mathbf{q}^{(1)})$ and $\hat{\mathbf{M}}(\mathbf{q}^{(1)})$ are invertible, we start by rewriting the general model (1.37) as

$$\ddot{\mathbf{q}}^{(1)} = \overline{\mathbf{F}}\left(\dot{\mathbf{q}}^{(2)}, \mathbf{q}^{(2)}\right) + \overline{\mathbf{G}}\left(\dot{\mathbf{q}}^{(2)}, \mathbf{q}^{(2)}\right)\boldsymbol{\tau}, \tag{1.70}$$

where matrices $\overline{\mathbf{F}}$ and $\overline{\mathbf{G}}$ satisfy

$$\overline{\mathbf{F}}\left(\mathbf{q}^{(2)}, \dot{\mathbf{q}}^{(2)}\right) = -\overline{\mathbf{M}}^{(2)}\left(\mathbf{q}^{(2)}\right)^{-1}\overline{\mathbf{N}}^{(2)}\left(\mathbf{q}^{(2)}, \dot{\mathbf{q}}^{(2)}\right) = \begin{bmatrix} \overline{F}_1 & \overline{F}_2 & \overline{F}_3 \end{bmatrix}^{\mathrm{T}} \tag{1.71a}$$

$$\overline{\mathbf{G}}\left(\mathbf{q}^{(2)}, \dot{\mathbf{q}}^{(2)}\right) = \overline{\mathbf{M}}^{(2)}\left(\mathbf{q}^{(2)}\right)^{-1}\overline{\mathbf{B}}^{(2)} = \begin{bmatrix} \overline{G}_{11} & \overline{G}_{21} & \overline{G}_{31} \\ \overline{G}_{12} & \overline{G}_{22} & \overline{G}_{32} \end{bmatrix}^{\mathrm{T}}. \tag{1.71b}$$

Considering the underactuated feature of system (1.70), the following subsystem is investigated in SMC controller design:

$$\ddot{\mathbf{q}}_1 = \overline{\mathbf{M}}_3\left(\dot{\mathbf{q}}^{(2)}, \mathbf{q}^{(2)}\right) + \overline{\mathbf{G}}_3\left(\dot{\mathbf{q}}^{(2)}, \mathbf{q}^{(2)}\right)\boldsymbol{\tau}, \tag{1.72}$$

where vector $\mathbf{q}_1$ is chosen as $\mathbf{q}_1 = [\theta_1 \quad \lambda]^{\mathrm{T}}$. Matrices $\overline{\mathbf{M}}_3$ and $\overline{\mathbf{G}}_3$ satisfy

$$\overline{\mathbf{M}}_3 = \begin{bmatrix} \overline{F}_2 \\ \overline{F}_3 \end{bmatrix} \quad \overline{\mathbf{G}}_3 = \begin{bmatrix} \overline{G}_{21} & \overline{G}_{22} \\ \overline{G}_{31} & \overline{G}_{32} \end{bmatrix}. \tag{1.73}$$

$\overline{F}_i$ and $\hat{F}_i$ represent the ith element of matrices $\overline{\mathbf{F}}$ and $\hat{\mathbf{F}}$ with $i = 2, 3$. $\overline{G}_{ij}$ and $\hat{G}_{ij}$ represent the (i,j)th element of matrices $\overline{\mathbf{G}}$ and $\hat{\mathbf{G}}$ with $i = 2, 3$ and $j = 1, 2$.

It is assumed that the following equations are satisfied:

$$\left|\hat{F}_i - \overline{F}_i\right| \le \tilde{F}_i, \quad \tilde{\mathbf{F}} = \begin{bmatrix} \tilde{F}_2 & \tilde{F}_3 \end{bmatrix}, \quad \overline{\mathbf{G}}_3 = (\mathbf{I} + \boldsymbol{\Delta})\hat{\mathbf{G}}_3, \tag{1.74}$$

where $i = 2, 3$, $\mathbf{I}$ is a 2×2 identity matrix, and $\boldsymbol{\Delta}$ is composed of Δ_{ij} satisfying the following inequality:

$$\left|\Delta_{ij}\right| \le D_{ij}, \quad D_{ij} > 0, \text{ and } |\mathbf{D}| < 1. \tag{1.75}$$

The Terminal Sliding Mode Control method proposed in Ref. [26] was applied to the subsystem (1.72). Given the desired velocity $\dot{\theta}_{\mathrm{w}}^{*}$ of the UW-Car system on flat ground, the desired position of the seat λ^* can then be obtained from (1.14). We define the following sliding surfaces:

$$\begin{aligned} s_1(t) &= \dot{\theta}_1(t) + c_1\theta_1(t) - \dot{v}_1(t) - c_1 v_1(t) \\ s_2(t) &= \dot{\lambda} + c_2\left(\lambda - \lambda^*\right) - \dot{v}_2(t) - c_2 v_2(t), \end{aligned} \tag{1.76}$$

where c_1 and c_2 are positive constants. The augmenting function $\mathbf{v}$ is designed as a cubic polynomial that guarantees assumption 1 in Ref. [26] holds.

Proposition 5. Suppose the model uncertainties of system (1.72) satisfy (1.74) and (1.75). Then the sliding surfaces (1.76) will be achieved and the inclination angle θ_1 and the position of the seat λ will converge to zero and λ^* in finite time, if the following controller is applied to the system:

$$\tau = \hat{\mathbf{G}}_3^{-1}\left[-\hat{\mathbf{M}}_3 - \mathbf{C}_1\dot{\mathbf{e}}_1 + \ddot{\mathbf{v}} + \mathbf{C}_1\dot{\mathbf{v}} - \mathbf{k}\mathrm{sgn}(\mathbf{s})\right], \tag{1.77}$$

where

$$\mathbf{C}_1 = \mathrm{diag}(c_1 \quad c_2),\ \mathbf{e}_1 = [\theta_1 \quad \lambda - \lambda^*]^{\mathrm{T}},\ \mathbf{v} = [v_1 \quad v_2]^{\mathrm{T}}$$

$$\mathbf{k} = (\mathbf{I} - \mathbf{D})^{-1}\left(\mathbf{F} + \mathbf{D}\left|\hat{\mathbf{F}} + \mathbf{C}_1\dot{\mathbf{e}}_1 + \ddot{\mathbf{v}} + \mathbf{C}_1\dot{\mathbf{v}}\right| + \mathbf{\Gamma}\right)$$

$$\mathbf{\Gamma} = [\gamma_1 \quad \gamma_2],\ \gamma_i > 0,\ \mathrm{sgn}(\mathbf{s}) = \mathrm{diag}(\mathrm{sgn}(s_1) \quad \mathrm{sgn}(s_2)).$$

Proof. This proposition can be easily proven according to Theorems 2 and 3 in Ref. [26].

1.4 Stabilization of an Inverted Pendulum Cart by Consistent Trajectories in Acceleration Behavior

1.4.1 Motivation

In the conventional approach, state feedback control is applied to control an inverted pendulum to a converged state. However, state feedback control is not efficient in obtaining maximum acceleration performance while maintaining the pitch angle constant during acceleration. Shimada and Hatakeyama [28] realized quick acceleration of an inverted pendulum by inclining the angle of the pendulum for the purpose of transportation. However, unsuitable vibration of the pendulum occurred at both the start and end of acceleration. Such vibration phenomena were not considered because the pendulum was not being ridden by a person. However, in this context, the vibration during acceleration should be suppressed in consideration of the driver. As mentioned above, it is difficult to realize quick acceleration while maintaining the angle of the inverted pendulum constant. The reason for this difficulty is that the control system of the inverted pendulum is non-holonomic. Variables cannot be decided independently under a non-holonomic control system because there is a non-integrable constraint. Moreover, for the same reason, desired acceleration trajectories cannot be determined independently. Inconsistent desired trajectories would lead to the situation in which one variable affects the other variables in the constraint equation.

1.4.2 Feedback Control System

A block diagram of the proposed control method is presented in Figure 1.7. The symbols are defined as follows: $\mathbf{x} = \left[\,\theta_1\ \lambda\ \dot{\theta}_{\mathrm{w}}\ \dot{\theta}_1\ \dot{\lambda}\,\right]^{\mathrm{T}}$, $\mathbf{u} = [\tau_{\mathrm{w}}\ \ f]^{\mathrm{T}}$, $\mathbf{x}^* = \left[\,\theta_1^*\ \lambda^*\ \dot{\theta}_{\mathrm{w}}^*\ \dot{\theta}_1^*\ \dot{\lambda}^*\,\right]^{\mathrm{T}}$, and $\mathbf{a}^* = \left[\,\ddot{\theta}_{\mathrm{w}}^*\ \ddot{\theta}_1^*\ \ddot{\lambda}^*\,\right]^{\mathrm{T}}$. In addition, u_h is an input value ordered by the driver through lever inclination for motion control. Under the proposed control method, there are three components of the control input. The first is normal state feedback control, and the gain of feedback **K** will be calculated by an optimum regulator. The second is feedforward control based on the desired trajectories of acceleration. The last is feedforward inputs for dynamics canceling.

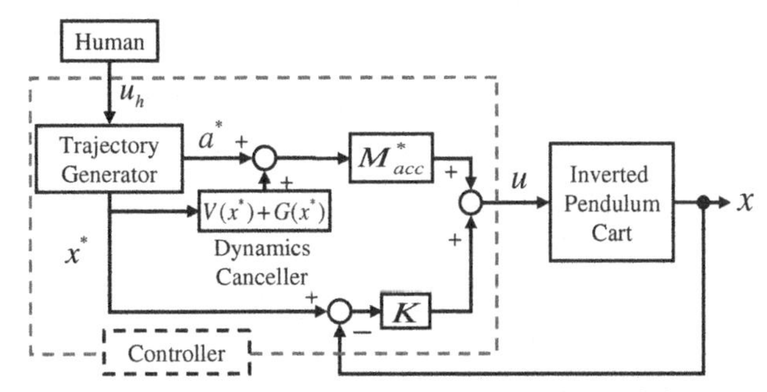

Figure 1.7: Block Diagram of the Inverted Pendulum Cart.

1.4.3 Desired Trajectory Generation

The balancing mechanism provides the possibility of quick acceleration while maintaining the pendulum angle perpendicular. However, the angle of the pendulum may be disturbed if the balancing seat motion and the acceleration of the cart are not synchronized because the inverted pendulum is an underactuated mechanical system. Such a disturbance imparts an uneasy feeling to the driver, which may cause improper operation and make the cart system unstable. State feedback control, which is adopted to control the inverted pendulum in ordinary circumstances, fails to address such a phenomenon. As such, this phenomenon is considered to be a modern control problem. First, even if some state variables are not zero, there are stabilization points with nonzero control input. Second, variables can be considered as linear projections of canonical variables. For example, $\tilde{x} = Tx$, where $\tilde{x}$ are canonical variables and T is a canonical transformation matrix. Finally, when a system designer uses methods such as pole placement or an optimum regulator, it is not possible to set the elements of the transformation matrix T as zero. Therefore, when variables move to a nonzero stabilization point, the path of the variables cannot be restricted to a certain plane using the state feedback control method. This is the reason why it is not possible to set the angle of the pendulum to zero at any time.

Here, we present the constraint condition between the angular velocity of the cart wheel and the displacement of the balancing seat in order to set the angle of the pendulum to always be zero during acceleration of the cart. First, substitution of the condition whereby $\theta_1 = 0$, $\dot{\theta}_1 = 0$, $\sin\theta_1 = 0$, $\cos\theta_1 = 1$ into Eq. (1.9) generates

$$M_{11}\ddot{\theta}_{\mathrm{w}} + M_{12}\ddot{\theta}_1 + m_2 r_{\mathrm{w}}\ddot{\lambda} = \tau_{\mathrm{w}} - (D_{\mathrm{w}} + D_1)\dot{\theta}_{\mathrm{w}} \tag{1.78}$$

and

$$M_{12}\ddot{\theta}_{\mathrm{w}} + M_{22}\ddot{\theta}_1 + m_2 l_2 \ddot{\lambda}_2 = -\tau_{\mathrm{w}} + D_1\dot{\theta}_{\mathrm{w}} + m_2 g\lambda_2. \tag{1.79}$$

By combining Eqs (1.78) and (1.79), we obtain

$$(M_{11} + M_{12})\ddot{\theta}_{\mathrm{w}} + (M_{12} + M_{22})\ddot{\theta}_1 + \{m_2 r_{\mathrm{w}} + m_2 l_2\}\ddot{\lambda} = -D_{\mathrm{w}}\dot{\theta}_{\mathrm{w}} + m_2 g\lambda. \tag{1.80}$$

Finally, by substituting $\ddot{\theta}_1 = 0$ into the previous equation, we obtain

$$(M_{11} + M_{12})\ddot{\theta}^*_{\mathrm{w}} + \{m_2 r_{\mathrm{w}} + m_2 l_2\}\ddot{\lambda}^* = -D_{\mathrm{w}}\dot{\theta}^*_{\mathrm{w}} + m_2 g\lambda^*. \tag{1.81}$$

Here, changing the expression of parameters as $\rho_1 = I^*_{\mathrm{w}} + r_{\mathrm{w}}(m_1 l_1 + m_2 l_2)$, $\rho_2 = m_2 r_{\mathrm{w}} + m_2 l_2$, $\chi_1 = D_{\mathrm{w}}$, $\chi_2 = m_2 g$ generates

$$\rho_1\ddot{\theta}^*_{\mathrm{w}} + \rho_2\ddot{\lambda}^* = -\chi_1\dot{\theta}^*_{\mathrm{w}} + \chi_2\lambda^*. \tag{1.82}$$

Adopting the Laplace transform with the condition initial value as zero generates the following equations:

$$\rho_1 s^2\Theta^*_{\mathrm{w}} + \rho_2 s^2\Lambda^* = -\chi_1 s\Theta^*_{\mathrm{w}} + \chi_2\Lambda^* \tag{1.83}$$

$$\therefore \frac{\rho_1 s + \chi_1}{-\rho_2 s^2 + \chi_2} s\Theta^*_{\mathrm{w}} = \Lambda^*. \tag{1.84}$$

Since the denominator of the fraction in the left part of the equation (1.84) includes a positive pole, it is not possible to calculate the desired trajectory of seat displacement λ^* from an arbitrary desired trajectory of the angular velocity of the motor wheel $\dot{\theta}^*_{\mathrm{w}}$ $(= L[s\Theta^*_{\mathrm{w}}])$. In order to calculate a suitable desired trajectory of seat displacement, $\dot{\theta}^*_{\mathrm{w}}$ and λ^*_2 should satisfy a certain constraint condition. First,

$$\frac{\rho_1 s + \chi_1}{-\rho_2 s^2 + \chi_2} = \frac{\rho_1 s + \chi_1}{-\rho_2\left(s + \sqrt{\frac{\chi_2}{\rho_2}}\right)\left(s - \sqrt{\frac{\chi_2}{\rho_2}}\right)}. \tag{1.85}$$

Next, assuming

$$\Theta^*_{\mathrm{w}}(s) = \frac{\mathrm{num}_{\Theta\mathrm{w}}(s)}{\mathrm{den}_{\Theta\mathrm{w}}(s)}, \tag{1.86}$$

$\Theta_{\mathrm{w}}^{*}(s)$ must satisfy the following conditions:

$$\mathrm{num}_{\Theta\mathrm{w}}(s)\big|_{s=\sqrt{\frac{\chi_2}{\rho_2}}}=0 \tag{1.87}$$

and

$$\mathrm{den}_{\Theta\mathrm{w}}(s)\big|_{s=\sqrt{\frac{\chi_2}{\rho_2}}}\neq 0. \tag{1.88}$$

Other constraints for variables are as follows:

$$\dot{\theta}_{\mathrm{w}}^{*}(0)=0 \tag{1.89}$$

$$\lim_{t\to\infty}\ddot{\theta}_{\mathrm{w}}^{*}(t)=0 \tag{1.90}$$

$$\lim_{t\to\infty}\dddot{\theta}_{\mathrm{w}}^{*}(t)=0 \tag{1.91}$$

$$\lim_{t\to\infty}\dot{\theta}_{\mathrm{w}}^{*}(t)=\dot{\theta}_{\mathrm{w}}^{f^*}. \tag{1.92}$$

Here, $\dot{\theta}_{\mathrm{w}}^{f^*}$ is the target value of the angular velocity of the tire wheel as requested by the driver. In total, there are five conditions for variables. The desired trajectory of $\dot{\theta}_{\mathrm{w}}^{*}$ is expressed as

$$\dot{\theta}_{\mathrm{w}}^{*}=\dot{\theta}_{\mathrm{w}}^{f}+\sum_{i=1}^{n}\beta_i \mathrm{e}^{-\alpha_i t} \qquad (\alpha_i>0,\ \beta_i>0). \tag{1.93}$$

Therefore, we have

$$\ddot{\theta}_{\mathrm{w}}^{*}=\sum_{i=1}^{n}-\alpha_i\beta_i \mathrm{e}^{-\alpha_i t}. \tag{1.94}$$

Since Eqs (1.93) and (1.94) satisfy the conditions of Eqs (1.91) and (1.92), the number of condition equations that must be satisfied is three. Since the three independent elements of the trajectory of $\dot{\theta}_{\mathrm{w}}^{*}$ satisfy all of the constraint conditions, n is equal to 3. Finally, we obtain

$$\dot{\theta}_{\mathrm{w}}^{*}=\dot{\theta}_{\mathrm{w}}^{f*}+\sum_{i=1}^{3}\beta_i \mathrm{e}^{-\alpha_i t}. \tag{1.95}$$

$$s\Theta_{\mathrm{w}}^{*}=\frac{\dot{\theta}_{\mathrm{w}}^{f*}}{s}+\frac{\beta_1}{s+\alpha_1}+\frac{\beta_1}{s+\alpha_2}+\frac{\beta_1}{s+\alpha_3}. \tag{1.96}$$

The constraint conditions of α_i and β_i from Eqs (1.89) and (1.90) are expressed as

$$\beta_1+\beta_1+\beta_1=-\dot{\theta}_{\mathrm{w}}^{f*} \tag{1.97}$$

and

$$\alpha_1\beta_1 + \alpha_2\beta_2 + \alpha_3\beta_3 = 0. \tag{1.98}$$

Here, Eq. (1.96) is transformed as follows:

$$\begin{aligned} s\Theta_{\mathrm{w}}^{*} &= \frac{\dot{\theta}_{\mathrm{w}}^{f*}}{s} + \frac{\beta_1}{s+\alpha_1} + \frac{\beta_1}{s+\alpha_2} + \frac{\beta_1}{s+\alpha_3} \\ &= \frac{\gamma_3 s^3 + \gamma_2 s^2 + \gamma_1 s + \gamma_0}{s(s+\alpha_1)(s+\alpha_2)(s+\alpha_3)} \end{aligned} \tag{1.99}$$

$$\gamma_3 = \dot{\theta}_{\mathrm{w}}^{f*} + \beta_1 + \beta_2 + \beta_3 \tag{1.100}$$

$$\begin{aligned} \gamma_2 &= \dot{\theta}_{\mathrm{w}}^{f*}(\alpha_1 + \alpha_2 + \alpha_3) + \beta_1(\alpha_2 + \alpha_3) + \beta_2(\alpha_1 + \alpha_3) + \beta_3(\alpha_1 + \alpha_2) \\ &= \alpha_1\left(\dot{\theta}_{\mathrm{w}}^{f*} + \beta_2 + \beta_3\right) + \alpha_2\left(\dot{\theta}_{\mathrm{w}}^{f*} + \beta_1 + \beta_3\right) + \alpha_3\left(\dot{\theta}_{\mathrm{w}}^{f*} + \beta_1 + \beta_2\right) \\ &= -\alpha_1\beta_1 - \alpha_2\beta_2 - \alpha_3\beta_3 \\ &= 0 \end{aligned} \tag{1.101}$$

$$\gamma_1 = \dot{\theta}_{\mathrm{w}}^{f*}(\alpha_1\alpha_2 + \alpha_1\alpha_3 + \alpha_2\alpha_3) + \alpha_1\alpha_2\beta_3 + \alpha_1\alpha_3\beta_2 + \alpha_2\alpha_3\beta_1 \tag{1.102}$$

$$\gamma_0 = \dot{\theta}_{\mathrm{w}}^{f*}\alpha_1\alpha_2\alpha_3. \tag{1.103}$$

Finally, we obtain

$$s\Theta_{\mathrm{w}}^{*} = \frac{\gamma_1\left(s + \frac{\gamma_0}{\gamma_1}\right)}{s(s+\alpha_1)(s+\alpha_2)(s+\alpha_3)}. \tag{1.104}$$

In order to satisfy Eq. (1.87), the necessary and sufficient condition is that Eq. (1.104) has a pole in $s = \sqrt{\chi_2/\rho_2}$:

$$\sqrt{\frac{\chi_2}{\rho_2}} = -\frac{\gamma_0}{\gamma_1} \tag{1.105}$$

$$\therefore \alpha_1\alpha_2\beta_3 + \alpha_1\alpha_3\beta_2 + \alpha_2\alpha_3\beta_1 = \eta, \tag{1.106}$$

where

$$\eta = -\dot{\theta}_{\mathrm{w}}^{f*}\left(\alpha_1\alpha_2 + \alpha_2\alpha_3 + \alpha_3\alpha_1 + \sqrt{\frac{\rho_2}{\chi_2}}\alpha_1\alpha_2\alpha_3\right). \tag{1.107}$$

From Eqs (1.97), (1.98), and (1.106), we have

$$\begin{bmatrix} 1 & 1 & 1 \\ \alpha_1 & \alpha_2 & \alpha_3 \\ \alpha_2\alpha_3 & \alpha_1\alpha_3 & \alpha_1\alpha_2 \end{bmatrix} \begin{bmatrix} \beta_1 \\ \beta_2 \\ \beta_3 \end{bmatrix} = \begin{bmatrix} -\dot{\theta}_{\mathrm{w}}^{f*} \\ 0 \\ \eta \end{bmatrix}. \tag{1.108}$$

Using the expression of the inverse matrix, we obtain:

$$\begin{bmatrix} \beta_1 \\ \beta_2 \\ \beta_3 \end{bmatrix} = \begin{bmatrix} 1 & 1 & 1 \\ \alpha_1 & \alpha_2 & \alpha_3 \\ \alpha_2\alpha_3 & \alpha_1\alpha_3 & \alpha_1\alpha_2 \end{bmatrix}^{-1} \begin{bmatrix} -\dot{\theta}_{\mathrm{w}}^{f*} \\ 0 \\ \eta \end{bmatrix}. \tag{1.109}$$

This equation indicates that β_i $(i = 1, 2, 3)$ can be calculated and that the desired trajectory of variables can be obtained if α_i $(i = 1, 2, 3)$ have been determined. The next problem is how to determine α_i. Assuming $\alpha_3 > \alpha_2 > \alpha_1$, α_3 is determined by the maximum motor torque, and α_1 is determined by the requirement for setting the time of the desired trajectory. Although determining α_2 is a controversial problem, it is appropriate to adopt the average of α_1 and α_3. Based on the above discussion, we have $\ddot{\theta}_{\mathrm{w}}^*(t)$, $\dot{\theta}_{\mathrm{w}}^*(t)$, $\ddot{\lambda}^*(t)$, $\dot{\lambda}^*(t)$, and $\lambda^*(t)$. From

$$\begin{aligned} L\left[\dot{\theta}_{\mathrm{w}}^*(t)\right] &= s\Theta_{\mathrm{w}}^* \\ &= \frac{\gamma_1 s + \gamma_0}{s(s+\alpha_1)(s+\alpha_2)(s+\alpha_3)}, \end{aligned} \tag{1.110}$$

and assuming that the initial values are zero, we obtain

$$\begin{aligned} L\left[\ddot{\theta}_{\mathrm{w}}^*(t)\right] &= s^2\Theta_{\mathrm{w}}^* \\ &= \frac{\gamma_1 s + \gamma_0}{(s+\alpha_1)(s+\alpha_2)(s+\alpha_3)}. \end{aligned} \tag{1.111}$$

Based on this result and on Eq. (1.84), we have

$$\begin{aligned} L[\lambda^*(t)] &= \Lambda^*(s) \\ &= \frac{\rho_1 s + \chi_1}{-\rho_2 s^2 + \chi_2} s\Theta_{\mathrm{w}}^*(s) \\ &= \frac{\rho_1 s + \chi_1}{-\rho_2\left(s + \sqrt{\dfrac{\chi_2}{\rho_2}}\right)} \frac{\gamma_1}{s(s+\alpha_1)(s+\alpha_2)(s+\alpha_3)} \\ &= \frac{\gamma_1\rho_1\left(s + \dfrac{\chi_1}{\rho_1}\right)}{-\rho_2 s\left(s + \sqrt{\dfrac{\chi_2}{\rho_2}}\right)(s+\alpha_1)(s+\alpha_2)(s+\alpha_3)}. \end{aligned} \tag{1.112}$$

As in the case of $\dot{\lambda}^*(t)$ and $\ddot{\lambda}^*(t)$, we obtain

$$L[\dot{\lambda}^*(t)] = s\Lambda^*(s)$$

$$= \frac{\gamma_1\rho_1\left(s+\frac{\chi_1}{\rho_1}\right)}{-\rho_2\left(s+\sqrt{\frac{\chi_2}{\rho_2}}\right)(s+\alpha_1)(s+\alpha_2)(s+\alpha_3)} \tag{1.113}$$

$$L[\ddot{\lambda}^*(t)] = s^2\Lambda^*(s)$$

$$= \frac{\gamma_1\rho_1 s\left(s+\frac{\chi_1}{\rho_1}\right)}{-\rho_2\left(s+\sqrt{\frac{\chi_2}{\rho_2}}\right)(s+\alpha_1)(s+\alpha_2)(s+\alpha_3)}. \tag{1.114}$$

1.4.4 Control Method Based on Desired Trajectory of Acceleration

In Figure 1.4, M_{acc} is calculated as follows. From Eq. (1.9), we have

$$\xi = [\theta_{\mathrm{w}} \quad \theta_1 \quad \lambda]^{\mathrm{T}} \tag{1.115}$$

$$\xi^* = [\theta_{\mathrm{w}}^* \quad \theta_1^* \quad \lambda^*]^{\mathrm{T}} \tag{1.116}$$

$$F = M(\xi)\ddot{\xi} + V(\dot{\xi},\xi) + G(\xi), \tag{1.117}$$

and

$$F = \begin{bmatrix} 1 & 0 \\ -1 & 0 \\ 0 & -1 \end{bmatrix} \begin{bmatrix} \tau_{\mathrm{w}} \\ f \end{bmatrix} = M_f \begin{bmatrix} \tau_{\mathrm{w}} \\ f \end{bmatrix}. \tag{1.118}$$

Thus, we obtain

$$M_{\mathrm{acc}} = M_f^+ M(\xi)|_{\xi=0}. \tag{1.119}$$

Here, M_f^+ is a pseudo-inverse matrix of M_f, and the values in this simulation are

$$M_f^+ = \begin{bmatrix} 0.5 & -0.5 & 0 \\ 0 & 0 & 1 \end{bmatrix}. \tag{1.120}$$

1.5 Simulation Study

In all simulations, we use the physical parameters given in Table 1.1. All algorithms represented in Section 1.3 are realized in MATLAB.

1.5.1 Set-Point Velocity Control Simulation on Flat Ground

A common-used stabilization technique is the Linear Quadratic Regulator (LQR). The plant state is chosen as a five-dimensional vector given by $x = \begin{bmatrix} \theta_1 & \lambda & \dot{\theta}_w & \dot{\theta}_1 & \dot{\lambda} \end{bmatrix}^T$. The desired equilibrium of the control system is $x^* = \begin{bmatrix} 0 & \lambda^* & \dot{\theta}_w^* & 0 & 0 \end{bmatrix}^T$. The LQR optimal control technique is used to design a linear state feedback control law $u = -\mathbf{K}\mathbf{x}$, where the gain matrix $\mathbf{K}$ minimizes the performance index:

$$J = \int_0^\infty \left(x^T Q x + u^T R u\right) dt.$$

Weighted matrices $\mathbf{Q}$ and $\mathbf{R}$ were chosen as: $\mathbf{Q} = \mathrm{diag}(10000,100,100,100,10000)$, $\mathbf{R} = \mathrm{diag}(0.01,0.01)$. Assume that the desired velocity satisfies $\dot{\theta}_w^* = 7$ rad/s and the desired distance of the seat can be computed by (1.14), $\lambda^* = 0.1888$. The initial conditions are chosen as: $\lambda = \dot{\lambda} = \theta_w = \dot{\theta}_w = \dot{\theta}_1 = 0$, $\theta_1 = 0.1$. Solving the above linear quadratic problem, the gain matrix $\mathbf{K}$ was obtained as follows:

$$\mathbf{K} = \begin{bmatrix} -1.8665 & -0.5324 & -0.0839 & -0.7319 & -0.1586 \\ 0.0171 & 0.3167 & 0.0201 & 0.0248 & 0.9951 \end{bmatrix} \times 10^3.$$

The simulation results of the velocity control of a UW-Car running on flat ground verify the ability of the designed LQR controller, and are shown in Figure 1.8.

The dashed lines represent the desired value and the solid lines are simulation trajectories. It appears that the LQR controller has the ability to guarantee the inclination angle of the body to be approximately zero in finite time, but the large overshoot and rapid change of states may make passengers feel uncomfortable.

Table 1.1: Physical Parameters of the UW-Car System

r_w	0.245	m	I_w	0.972	$kg \cdot m^2$
m_w	32.4	kg	I_1	3.79	$kg \cdot m^2$
m_1	137.6	kg	I_2	0.96	$kg \cdot m^2$
m_2	8.7	kg	D_w	2.3	$N \cdot s/m$
l_1	0.166	m	D_1	0.1	$N \cdot s/m$
l_2	0.323	m	D_2	5.0	$N \cdot s/m$

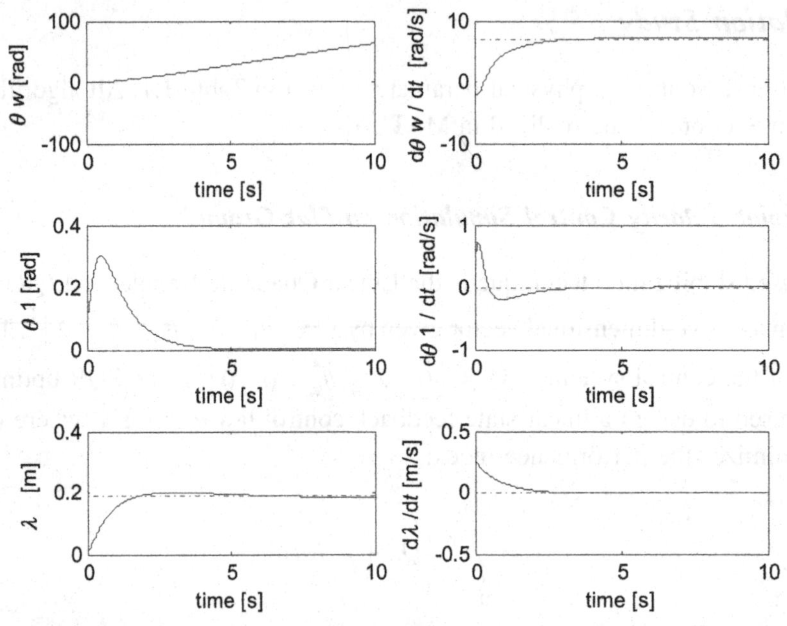

Figure 1.8: Simulation Results of a UW-Car System Using LQR Running at a Constant Velocity 7 rad/s.

Robust control methods might be a good choice for the nonlinear underactuated model. The TSMC method is investigated in the following due to its robustness to model uncertainties and external disturbances.

A TSMC controller is designed according to the analysis in Section 1.3. The augmenting function **v** is designed as a cubic polynomial:

$$v_i(t) = \begin{cases} a_{i0} + a_{i1}t + a_{i2}t^2 + a_{i3}t^3, & \text{if } 0 \le t \le T_{fi} \\ 0, & \text{if } t > T_{fi} \end{cases}, \tag{1.121}$$

where

$$a_{i0} = e_i(0),\ a_{i1} = \dot{e}_i(0),\ a_{i2} = -3\left(e_i(0)/T_{fi}^2\right) - 2\left(\dot{e}_i(0)/T_{fi}^3\right),$$
$$a_{i3} = 2\left(e_i(0)/T_{fi}^3\right) + \left(\dot{e}_i(0)/T_{fi}^2\right),\ i = 1,\ 2.$$

The TSMC controller parameters are chosen as
$c_1 = 1$, $c_2 = 3.2$, $\gamma_1 = \gamma_2 = 10$, $\tilde{F} = 0.8F$, $D_1 = D_2 = 0.75$, and $T_{f1} = 6.4$ s, $T_{f2} = 1$ s with the same initial conditions in the LQR controller. The desired angle velocity is $\dot{\theta}_w^* = 7$ rad/s and the desired distance of the seat can be computed by (1.14), $\lambda^* = 0.1888$. In order to avoid chattering associated with the sliding mode control law, we have approximated

the discontinuous sign function sgn(s) with a continuous saturation function of small boundary layers. The simulation results are shown in Figure 1.9.

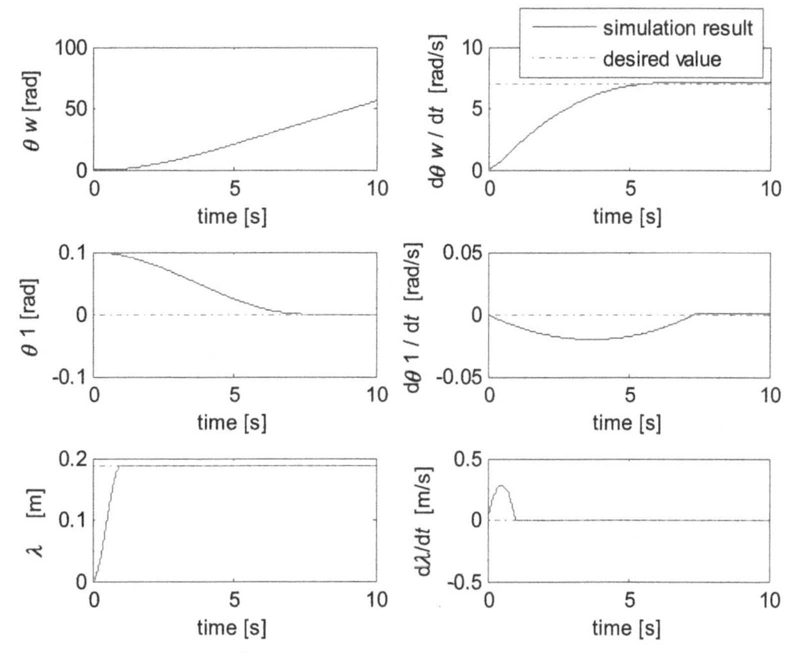

Figure 1.9: Simulation Results of a UW-Car System Using TSMC Running at a Constant Velocity 7 rad/s.

The dashed lines represent the desired value and the solid lines are simulation trajectories. It turns out that the TSMC controller guarantees the inclination angle of the body to be approximately zero and the seat to vibrate very slightly near a certain position. With the TSMC controller it takes a long time for the UW-Car system to accelerate to the desired velocity, but for other parameters such as overshoot, the speed changes are excellent.

1.5.2 Optimal Braking Control Simulation on Flat Ground

We verify the effectiveness of optimal braking control by the simulation in MATLAB.

The limitation of seat displacement is $|\lambda_{\max}| = 0.2$. Constant λ_2 is chosen as $\lambda_2 = 0.05$. For the sliding surface given by (1.53)–(1.55), the controller parameters $c_1 = 1$, $c_2 = 3.57$, $T_2 = T_3 = 1$ s, $D_1 = D_2 = 0.8$ were used. Optimal parameters are computed by GA as $\lambda^* = -0.2$, $T_{f0}^* = 0.01$. The initial conditions are: $\dot{\theta}_{\mathrm{w}} = 7$ rad/s, $\theta_1 = \dot{\lambda} = \theta_{\mathrm{w}} = \dot{\theta}_1 = 0$, $\lambda = 0.1888$, and the simulation results are shown in Figure 1.10. The distance moved forward by the UW-Car is about 4.41 m (18×0.245) and takes about 5.8 s.

Figure 1.11 shows the simulation results with parameters $\lambda_1 = -0.1$, $T_{f0} = 0.4$ and the same initial conditions in optimal braking control. The distance the UW-Car moves forward is

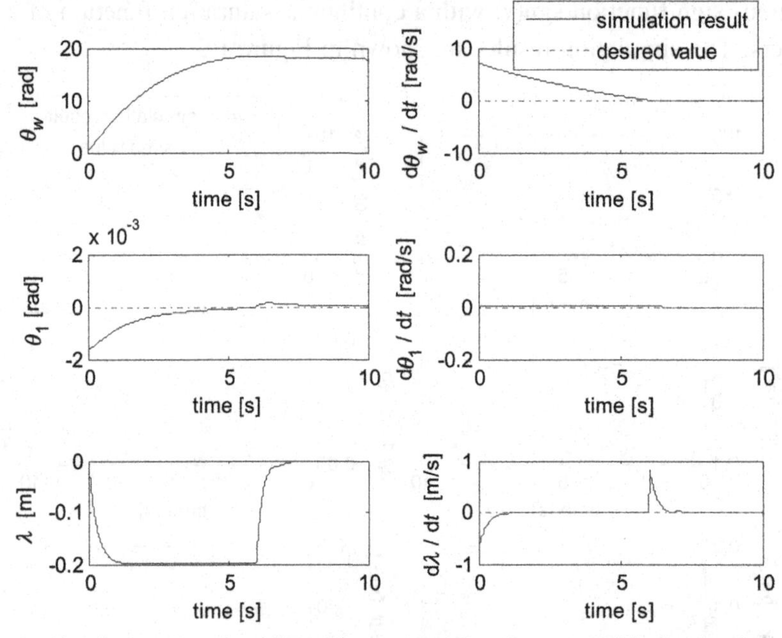

Figure 1.10: Simulation Results of Braking a UW-Car System With Optimal Parameters.

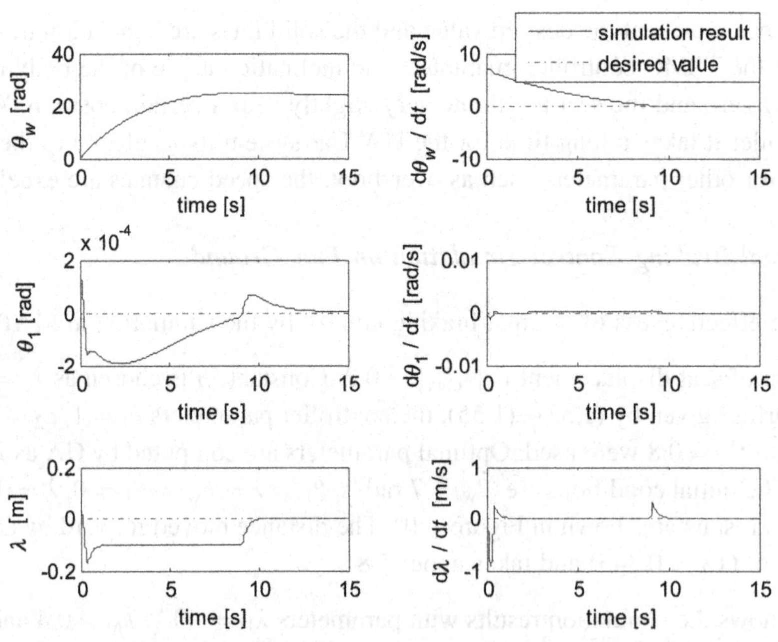

Figure 1.11: Simulation Results of Braking a UW-Car System Without Optimal Parameters.

about 6.40 m (26.11 × 0.245), in approximately 9 seconds. Compared to the two sets of figures and data, we can conclude that the optimal braking control is effective.

1.5.3 Set-Point Velocity Control Simulation in Rough Terrain

The initial states are chosen as: $\mathbf{q}^{(2)}(0) = [0.01 \quad 0 \quad 0 \quad 0 \quad 0]^{\mathrm{T}}$. To verify the ability of the designed LQR controller, a rough ground function $y = 0.04\sin(5x)$ is investigated. The simulation results of the velocity control are shown in Figure 1.12. It turns out that an LQR controller cannot guarantee the inclination angle of the body to be zero while the UW-Car system is running. The vibrations of the body and the seat become stronger when the UW-Car system passes over rough terrain, even though there are no disturbances from the outside. To make passengers feel comfortable, the robust control method might be a good choice. Simulation results using the SMC method were investigated due to its robustness to both model uncertainties and external disturbances.

According to the sliding mode control analysis in subsection 1.3.3, using the parameters in Table 1.1, the function $\tilde{\mathbf{F}}$ in (1.77) was selected as $0.2 \times \hat{\mathbf{F}}$. Matrices $\mathbf{C}_1$, $\mathbf{D}$, and $\mathbf{\Gamma}$ were selected as

$$\mathbf{C}_1 = 0.04 \times \mathbf{I},\ \mathbf{\Gamma} = 10 \times \mathbf{I},\ \mathbf{\Delta} - 0.2 \times \mathbf{I}.$$

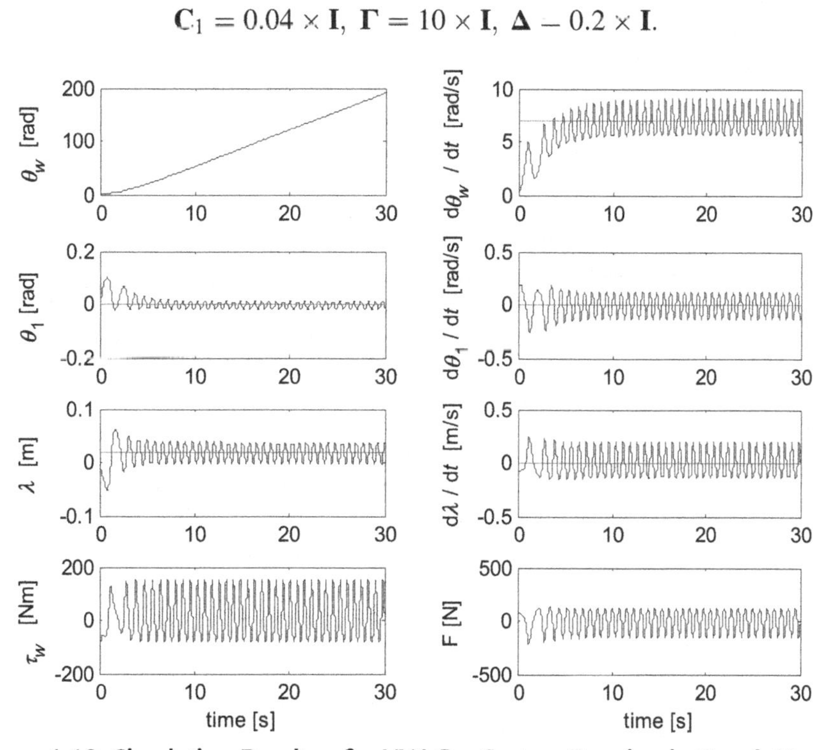

Figure 1.12: Simulation Results of a UW-Car System Running in Rough Terrain Using LQR Controller.
The horizontal lines show the desired value of equilibrium.

For the sliding surfaces, $T_f = 1$ s was used. In order to avoid chattering associated with the sliding mode control law, we have approximated the discontinuous sign function (sgn(s)) with continuous saturation functions of small boundary layers. Assuming the initial condition and the rough terrain are the same as used by the LQR controller, the simulation results are shown in Figure 1.13.

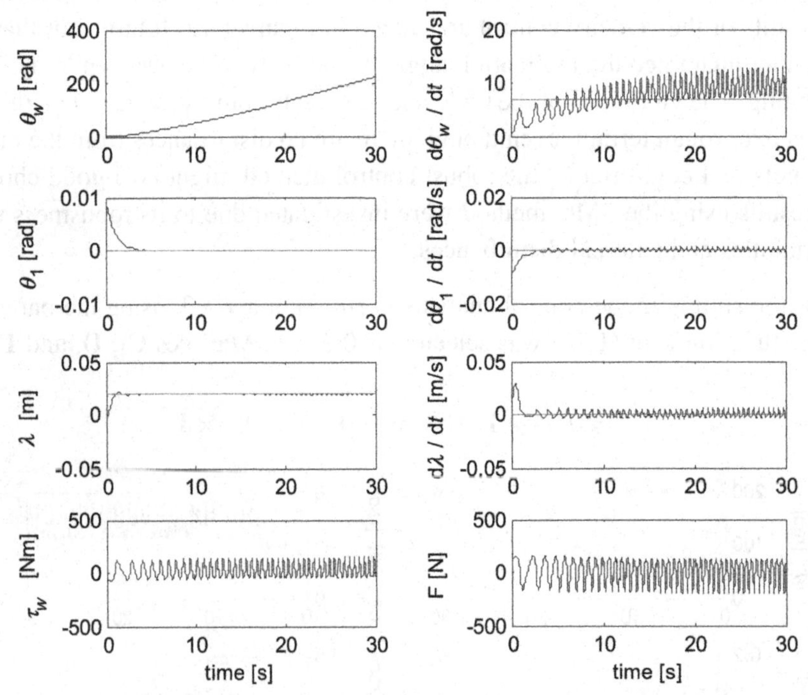

Figure 1.13: Simulation Results of a UW-Car System Running in Rough Terrain Using SMC Controller.
The horizontal lines show the desired value of equilibrium.

It turns out that the SMC controller (1.77) guarantees the inclination angle of the body to be approximately zero and the seat to vibrate very slightly near a certain position, even when the UW-Car system is running on rough terrain.

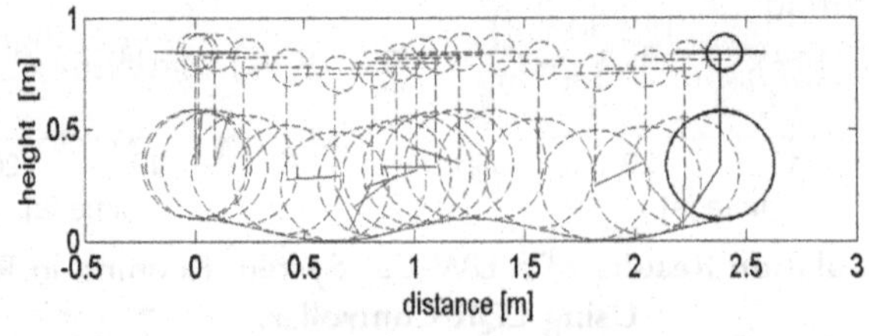

Figure 1.14: Motion Simulation of a UW-Car System Accelerating from its Initial Position.

1.5.4 Consistent Trajectories in Acceleration Behavior

A simulation is conducted in order to confirm the effectiveness of the proposed control method based on the desired trajectory of acceleration in Figure 1.14. A block diagram of the control method used in the simulation is shown in Figure 1.15. Unlike the block diagram shown in Figure 1.7, that shown in Figure 1.15 does not have a dynamics canceling block. The effectiveness of the dynamics canceling algorithm is discussed below. The total simulation time is 15 seconds. The simulation begins with all of the variables set to zero as the initial condition. During the simulation, the driver requests $u_h = r_w \dot{\theta}_w^{f*}$ at time zero. In addition, the gain of state feedback **K** is calculated with an optimum regulator and takes the following values:

$$\mathbf{K} = \begin{bmatrix} -1960 & -2060 & -70.0 & -614.0 & -594.0 \\ 827.0 & 913.0 & 63.9 & 220.0 & 369.0 \end{bmatrix}.$$

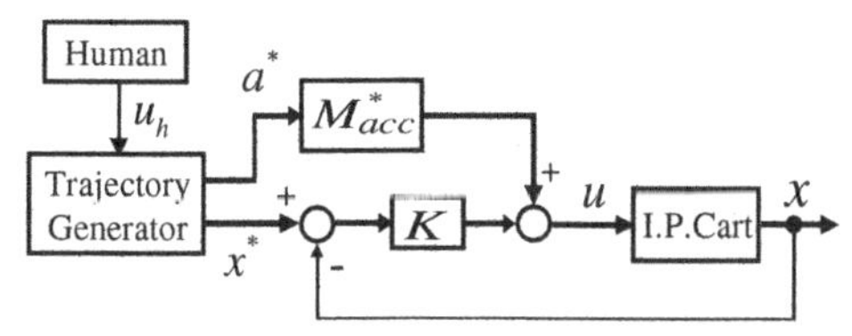

Figure 1.15: Block Diagram of Acceleration Control.

The parameters of the trajectories generation algorithm are shown in Table 1.2.

The trajectory of the main variables and their desired trajectory are shown in Figure 1.16. One problem with this method was that each variable and desired variable did not match at the end of the simulation. After a certain period of time, the control inputs generated by the desired trajectory of acceleration became zero, and consequently the method considered in this simulation cannot resolve the error between the variables and the desired variables. On the other hand, the angular velocity of the wheel and the displacement of the seat balancer were almost able to track the desired trajectory while

Table 1.2: Parameters for the Control Algorithm Used in Simulation

$\dot{\theta}_w^{f*}$	4.90	rad/s	ρ_1	14.4	No unit
α_1	2.0	1/s	ρ_2	90.5	No unit
α_2	3.0	1/s	χ_1	3.30	No unit
α_3	4.0	1/s	χ_2	11.9	No unit
β_1	−45.6	rad/s	γ_0	118	No unit
β_2	71.7	rad/s	γ_1	−32.5	No unit
β_3	−30.9	rad/s			

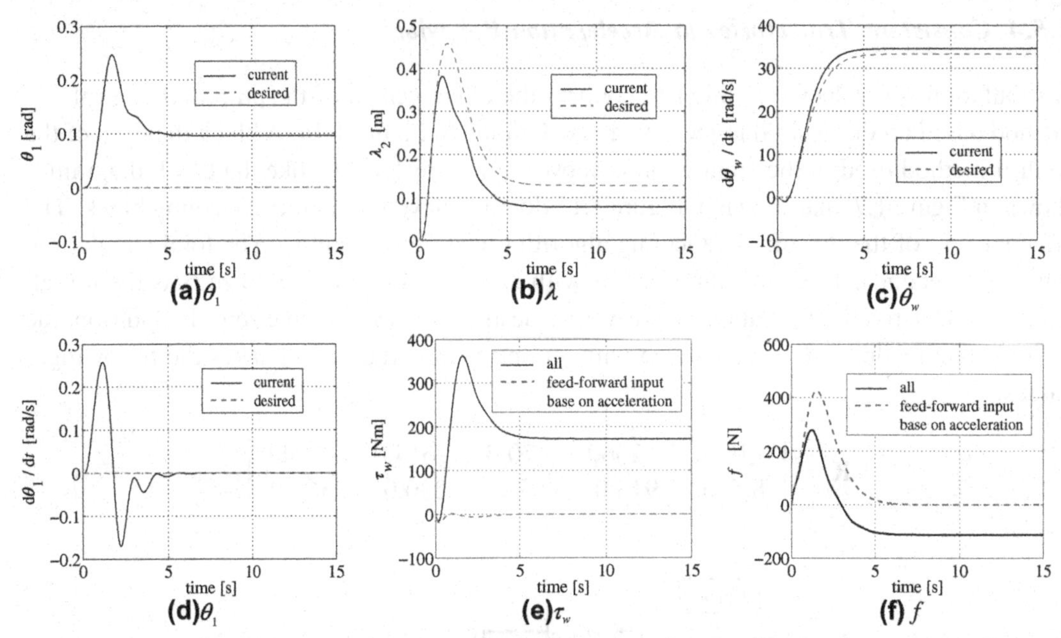

Figure 1.16: Simulation Results of a UW-Car System Using Feedforward Input Based on Trajectory Generation in Acceleration Behavior.

suppressing the angle of the pendulum of the cart. Thus, a quick transition to goal states was realized. This is not possible with simple state feedback control. Next, as control inputs, τ_W and f are also shown in Figure 1.16. The feedforward inputs generated by the desired trajectory of acceleration are also shown in Figure 1.16. The element of the feedforward input of τ_W is small in Figure 1.16(e). On the other hand, the element of the feedforward input of f is greater than the total input of f in Figure 1.16(f). The magnitude of the elements of input depends on the parameters of viscous friction. However, the main reason for this phenomenon in the case of f is that the feedforward method generates a required control input in order to force the seat balancer to move forward against the inertial resistance force of the seat balancer generated by the acceleration of the cart in the vertical direction. Large acceleration of the cart can be obtained while maintaining a perpendicular angle of the inverted pendulum, because the seat balancer moves forward, which is planned in advance.

1.5.5 Dynamics Canceling Inputs

In order to confirm an improvement in control performance by combining the feedforward inputs generated by the desired trajectory of acceleration and the dynamics canceling inputs, a further simulation was conducted. The calculation scheme of the dynamics canceler block diagram in Figure 1.7 is described in the following. From Eq. (1.9), we obtain

$$
\begin{aligned}
V_1(x^*) &= (D_\mathrm{w} + D_1)\dot{\theta}_\mathrm{w}^* - D_1\dot{\theta}_1^* - 2m_2 r_\mathrm{w}\sin(\theta_1^*)\dot{\theta}_1^*\lambda^* \\
&\quad - r_\mathrm{w}\left\{(m_1 l_1 + m_2 l_2)\sin\theta_1^* + m_2\lambda^*\cos\theta_1\right\}\dot{\theta}_1^{*2} \\
G_1(x^*) &= 0 \\
V_2(x^*) &= -D_1\dot{\theta}_\mathrm{w}^* + D_1\dot{\theta}_1^* + 2m_2\lambda^*\dot{\theta}_1^{**}\dot{\lambda}^* \\
G_2(x^*) &= -(m_1 l_1 + m_2 l_2)g\sin\theta_1^* - m_2 g\lambda^*\cos\theta_1^* \\
V_3(x^*) &= -D_2\dot{\lambda}^* - m_2\lambda^*\dot{\theta}_1^{*2} \\
G_3(x^*) &= -m_2 g\sin\theta_1^*.
\end{aligned}
$$

These inputs cancel both viscous force/torque and gravity force/torque.

The trajectories of the variables and the desired trajectories of the variables are shown in Figure 1.17. All of the variables conformed with the desired trajectories. These results are ideal for the situation in which all parameter values are known. In addition, as control inputs, τ_w and f are shown in Figure 1.17. The feedforward inputs generated by the desired trajectory of acceleration are also shown in these figures. All of the variables approximately matched

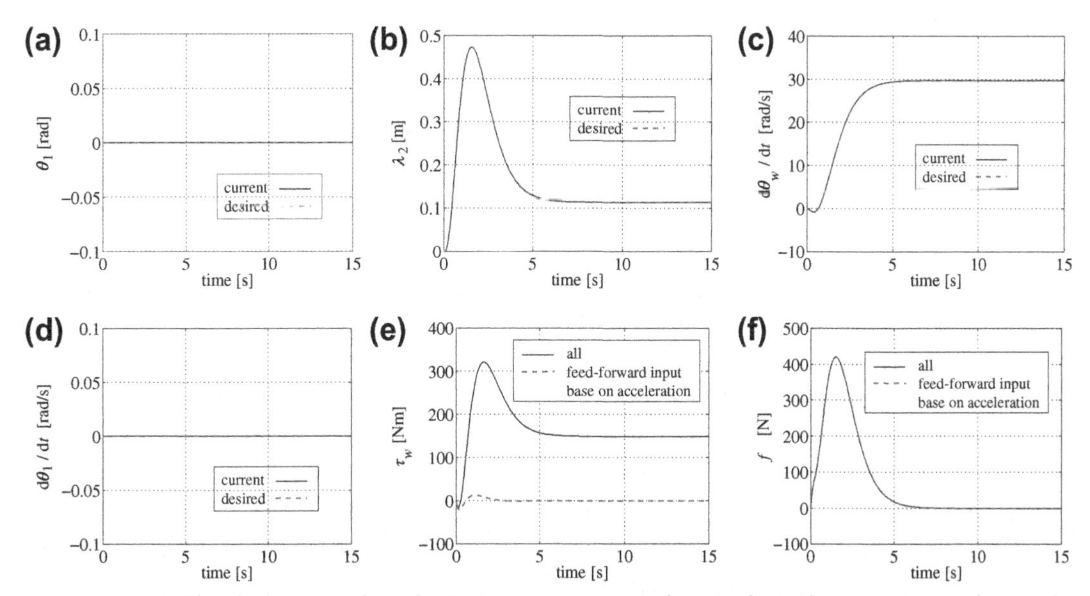

Figure 1.17: Simulation Results of a UW-Car System Using Both Trajectory Generation and Dynamics Canceling Input.

(a) θ_1. (b) λ. (c) $\dot{\theta}_\mathrm{w}$. (d) $\dot{\theta}_1$. (e) τ_w. (f) f.

those of the desired trajectory. Thus, the inputs obtained by feedback control became approximately zero. The main element of τ_W is dynamics canceling inputs, whereas the main element of f is feedforward inputs by the desired acceleration trajectory. Ideal control inputs realized both the feedforward inputs with desired acceleration trajectory and the dynamics canceling inputs. Finally, accelerations acting on the driver during vertical direction are illustrated in Figure 1.18. The acceleration of the cart changes in positive and negative values. However, the total acceleration acting on the driver is always positive. Therefore, the driver does not feel a vibration phenomenon during acceleration.

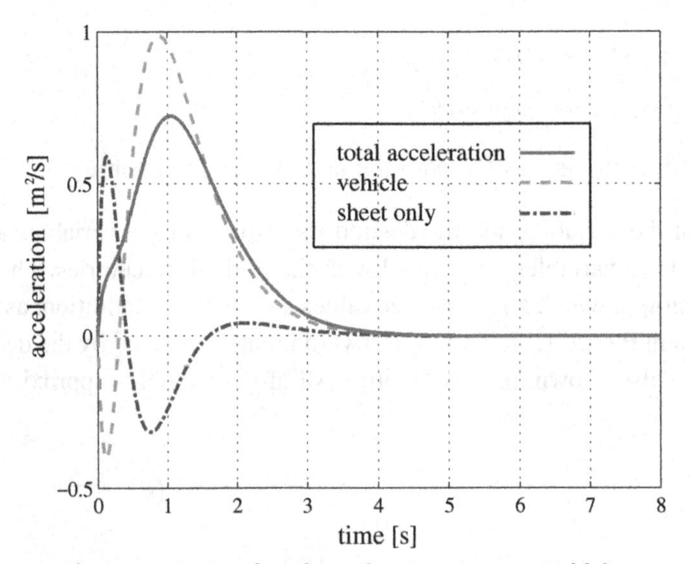

Figure 1.18: Acceleration of a Person on a Vehicle.

1.6 Conclusion

A novel transportation system, the UW-Car, which is composed of an MWIP base and a movable seat, is proposed in this chapter. The dynamic model of a UW-Car system is obtained by applying Lagrange's motion equations, for both the case of running on flat ground and on rough terrain. Based on the model, velocity control and optimal braking strategy are discussed. A terminal sliding mode control method is proposed for velocity control, which guarantees that the UW-Car system can always keep the body upright and the seat in a desired position. Compared to the LQR method, the TSMC method presents better performance. A strategy of optimal braking with optimal parameters computed by a GA is proposed, which guarantees that the UW-Car system stops in a short distance by implementing online switching three terminal sliding mode controllers. We also proposed a control method for realizing quick acceleration while maintaining the pendulum angle perpendicular for the UW-Car. The control inputs calculated from desired acceleration trajectories and the dynamics canceling inputs were combined to generate feasible feedforward inputs. In order to prove the correctness of the model and the effectiveness of the methods, simulations using MATLAB were performed and confirmed the theoretical results.

Appendix

$$M_{11} = (m_{\mathrm{w}} + m_1 + m_2)r_{\mathrm{w}}^2 + I_{\mathrm{w}} \qquad M_{22} = m_1 l_1^2 + m_2 l_2^2 + I_{\mathrm{b}} + I_{\mathrm{s}}$$

$$M_{12} = (m_1 l_1 + m_2 l_2) r_{\mathrm{w}} \qquad G_1 = (m_1 l_1 + m_2 l_2) g$$

$$\mathbf{M}^{(1)}\left(\mathbf{q}^{(1)}\right) = \begin{bmatrix} M_{11} & * & * \\ (M_{12}C_1 - m_2 r_{\mathrm{w}} \lambda S_1) & M_{22} + m_2\lambda^2 & * \\ m_2 r_{\mathrm{w}} C_1 & m_2 l_2 & m_2 \end{bmatrix}$$

$$\mathbf{N}^{(1)}\left(\mathbf{q}^{(1)}, \dot{\mathbf{q}}^{(1)}\right) = \begin{bmatrix} -D_{\mathrm{w}}\dot{\theta}_{\mathrm{w}} + \left(M_{12}S_1 + m_2 r_{\mathrm{w}} \lambda C_1\right)\dot{\theta}_1^2 + m_2 r_{\mathrm{w}} S_1 \dot{\theta}_1 \dot{\lambda} \\ -D_1\dot{\theta}_1 + G_1 S_1 + m_2 g \lambda C_1 - m_2 \lambda \dot{\theta}_1 \dot{\lambda} \\ -D_2\dot{\lambda} + m_2 \lambda \dot{\theta}_1^2 + m_2 g S_1 \end{bmatrix}$$

$$\mathbf{B}^{(1)} = \begin{bmatrix} 1 & -1 & 0 \\ 0 & 0 & 1 \end{bmatrix}^{\mathrm{T}}$$

$$\mathbf{R}(\mathbf{x}) = \begin{bmatrix} 1 & * & * & * & * \\ 0 & 1 & * & * & * \\ 0 & 0 & M_{11} & * & * \\ 0 & 0 & (M_{12}C_1 - m_2 r_{\mathrm{w}} x_2 S_1) & M_{22} + m_2 x_2^2 & * \\ 0 & 0 & m_2 r_{\mathrm{w}} C_1 & m_2 l_2 & m_2 \end{bmatrix}$$

$$\mathbf{H}(\mathbf{x}) = \begin{bmatrix} x_4 \\ x_5 \\ -D_{\mathrm{w}} x_3 + \left(M_{12}S_1 + m_2 r_{\mathrm{w}} x_2 C_1\right)\dot{\theta}_1^2 + 2m_2 r_{\mathrm{w}} S_1 x_4 x_5 \\ -D_1 x_3 + G_1 S_1 + m_2 g x_2 C_1 - 2m_2 x_2 x_4 x_5 \\ -D_2 x_5 + m_2 x_2 x_4^2 + m_2 g S_1 \end{bmatrix}$$

$$
\mathbf{M}_{\mathrm{w}} = \begin{bmatrix} m_{\mathrm{w}} & * & * & * & * & * \\ 0 & m_{\mathrm{w}} & * & * & * & * \\ -m_{\mathrm{w}} r_{\mathrm{w}} C_{\xi} & -m_{\mathrm{w}} r_{\mathrm{w}} S_{\xi} & m_{\mathrm{w}} r_{\mathrm{w}}^{2} & * & * & * \\ 0 & 0 & 0 & I_{\mathrm{w}} & * & * \\ 0 & 0 & 0 & 0 & 0 & * \\ 0 & 0 & 0 & 0 & 0 & 0 \end{bmatrix}
$$

$$
\mathbf{M}_{1} = \begin{bmatrix} m_{1} & * & * & * & * & * \\ 0 & m_{1} & * & * & * & * \\ -m_{1} r_{\mathrm{w}} C_{\xi} & -m_{1} r_{\mathrm{w}} S_{\xi} & m_{1} r_{\mathrm{w}}^{2} & * & * & * \\ 0 & 0 & 0 & 0 & * & * \\ m_{1} l_{1} C_{1} & -m_{1} l_{1} S_{1} & m_{1} l_{1} r_{\mathrm{w}} \left(S_{1} S_{\xi} - C_{1} C_{\xi} \right) & 0 & m_{1} l_{1}^{2} + I_{1} & * \\ 0 & 0 & 0 & 0 & 0 & 0 \end{bmatrix}
$$

$$
\mathbf{M}_{2} = \begin{bmatrix} m_{2} & * & * & * & * & * \\ 0 & m_{2} & * & * & * & * \\ -m_{2} r_{\mathrm{w}} C_{\xi} & -m_{2} r_{\mathrm{w}} S_{\xi} & m_{2} r_{\mathrm{w}}^{2} & * & * & * \\ 0 & 0 & 0 & 0 & * & * \\ m_{2} l_{2} C_{1} - m_{2} \lambda S_{1} & -m_{2} l_{2} S_{1} - m_{2} \lambda C_{1} & \Omega_{1} & 0 & m_{2} \left(l_{2}^{2} + \lambda^{2} \right) + I_{2} & * \\ m_{2} C_{1} & -m_{2} S_{1} & m_{2} r_{\mathrm{w}} \left(S_{1} S_{\xi} - C_{1} C_{\xi} \right) & 0 & m_{2} l_{2} & m_{2} \end{bmatrix}
$$

$$
\Omega_{1} = m_{2} r_{\mathrm{w}} \left[l_{2} \left(S_{1} S_{\xi} - C_{1} C_{\xi} \right) + \lambda \left(S_{1} C_{\xi} + C_{1} S_{\xi} \right) \right]
$$

$$
\mathbf{dM}_{\mathrm{w}} = \dot{\mathbf{M}}_{\mathrm{w}} = \begin{bmatrix} 0 & * & * & * & * & * \\ 0 & 0 & * & * & * & * \\ m_{\mathrm{w}} r_{\mathrm{w}} S_{\xi} \cdot \dot{\xi} & -m_{\mathrm{w}} r_{\mathrm{w}} C_{\xi} \cdot \dot{\xi} & 0 & * & * & * \\ 0 & 0 & 0 & 0 & * & * \\ 0 & 0 & 0 & 0 & 0 & * \\ 0 & 0 & 0 & 0 & 0 & 0 \end{bmatrix}
$$

$$
\mathbf{dM}_{1} = \dot{\mathbf{M}}_{1} = \begin{bmatrix} 0 & * & * & * & * & * \\ 0 & 0 & * & * & * & * \\ m_{1} r_{\mathrm{w}} S_{\xi} \cdot \dot{\xi} & -m_{1} r_{\mathrm{w}} C_{\xi} \cdot \dot{\xi} & 0 & * & * & * \\ 0 & 0 & 0 & 0 & * & * \\ -m_{1} l_{1} S_{1} \cdot \dot{\theta}_{1} & -m_{1} l_{1} C_{1} \cdot \dot{\theta}_{1} & m_{1} r_{\mathrm{w}} l_{1} \left(C_{1} S_{\xi} + S_{1} C_{\xi} \right) \left(\dot{\xi} + \dot{\theta}_{1} \right) & 0 & 0 & * \\ 0 & 0 & 0 & 0 & 0 & 0 \end{bmatrix}
$$

$$\mathbf{dM}_2 = \dot{\mathbf{M}}_2 = \begin{bmatrix} 0 & * & * & * & * & * \\ 0 & 0 & * & * & * & * \\ m_2 r_w S_\xi \cdot \dot{\xi} & -m_2 r_w C_\xi \cdot \dot{\xi} & 0 & * & * & * \\ 0 & 0 & 0 & 0 & * & * \\ -m_2\left[(l_2 S_1 + \lambda C_1)\cdot\dot{\theta}_1 + S_1\dot{\lambda}\right] & m_2\left[(-l_2 C_1 + \lambda S_1)\cdot\dot{\theta}_1 - C_1\dot{\lambda}\right] & \Omega_2 & 0 & 2m_2\lambda\dot{\lambda} & * \\ -m_2 S_1 \cdot \dot{\theta}_1 & -m_2 C_1 \cdot \dot{\theta}_1 & \Omega_3 & 0 & 0 & 0 \end{bmatrix}$$

$$\Omega_2 = m_2 r_w \left\{ \left[l_2 (S_1 C_\xi + C_1 S_\xi) + \lambda (C_1 C_\xi - S_1 S_\xi) \right] \cdot (\dot{\xi} + \dot{\theta}_1) + (S_1 C_\xi + C_1 S_\xi) \dot{\lambda} \right\}$$

$$\Omega_3 = m_2 r_w \left[(S_1 C_\xi + C_1 S_\xi) \right] \cdot (\dot{\xi} + \dot{\theta}_1)$$

$$\mathbf{N}_w\left(\mathbf{q}^{(2)}, \dot{\mathbf{q}}^{(2)}\right) = \begin{bmatrix} 0 & 0 & a_{w1} & 0 & 0 & 0 \end{bmatrix}^{\mathrm{T}}$$

$$\mathbf{N}_1\left(\mathbf{q}^{(2)}, \dot{\mathbf{q}}^{(2)}\right) = \begin{bmatrix} 0 & 0 & a_{11} & 0 & a_{12} & 0 \end{bmatrix}^{\mathrm{T}}$$

$$\mathbf{N}_2\left(\mathbf{q}^{(2)}, \dot{\mathbf{q}}^{(2)}\right) = \begin{bmatrix} 0 & 0 & a_{21} & 0 & a_{22} & a_{23} \end{bmatrix}^{\mathrm{T}}$$

$$a_{w1} = -\frac{1}{2}\dot{\mathbf{q}}^{\mathrm{T}} \frac{\partial \mathbf{M}_w}{\partial \xi} \dot{\mathbf{q}} = -m_w r_w \left(S_\xi \cdot \dot{x}\dot{\xi} - C_\xi \cdot \dot{y}\dot{\xi} \right)$$

$$a_{11} = -\frac{1}{2}\dot{\mathbf{q}}^{\mathrm{T}} \frac{\partial \mathbf{M}_1}{\partial \xi} \dot{\mathbf{q}} = m_1 r_w \cdot \left[-S_\xi \cdot \dot{x}\dot{\xi} + C_\xi \cdot \dot{y}\dot{\xi} - l_1 (C_1 S_\xi + S_1 C_\xi) \cdot \dot{\theta}_1 \dot{\xi} \right]$$

$$\begin{aligned} a_{12} &= -\frac{1}{2}\dot{\mathbf{q}}^{\mathrm{T}} \frac{\partial \mathbf{M}_1}{\partial \theta_1} \dot{\mathbf{q}} \\ &= m_1 l_1 \cdot \left[S_1 \cdot \dot{x}\dot{\theta}_1 + C_1 \cdot \dot{y}\dot{\theta}_1 - r_w (C_1 S_\xi + S_1 C_\xi) \cdot \dot{\theta}_1 \dot{\xi} \right] \end{aligned}$$

$$\begin{aligned} a_{21} &= -\frac{1}{2}\dot{\mathbf{q}}^{\mathrm{T}} \frac{\partial \mathbf{M}_2}{\partial \xi} \dot{\mathbf{q}} \\ &= m_2 r_w \cdot \left\{ -S_\xi \cdot \dot{x}\dot{\xi} + C_\xi \cdot \dot{y}\dot{\xi} - \left[l_2 (S_1 C_\xi + C_1 S_\xi) + \lambda (C_1 C_\xi - S_1 S_\xi) \right] \cdot \dot{\xi}\dot{\theta}_1 \right. \\ &\quad \left. + (S_1 C_\xi + C_1 S_\xi) \cdot \dot{\xi}\dot{\lambda} \right\} \end{aligned}$$

$$\begin{aligned} a_{22} &= -\frac{1}{2}\dot{\mathbf{q}}^{\mathrm{T}} \frac{\partial \mathbf{M}_2}{\partial \theta_1} \dot{\mathbf{q}} \\ &= m_2 \cdot \left\{ (l_2 S_1 + \lambda C_1) \cdot \dot{x}\dot{\theta}_1 + (l_2 C_1 - \lambda S_1) \cdot \dot{y}\dot{\theta}_1 \right. \\ &\quad - r_w \left[l_2 (S_1 C_\xi + C_1 S_\xi) + \lambda (C_1 C_\xi - S_1 S_\xi) \right] \cdot \dot{\xi}\dot{\theta}_1 \\ &\quad \left. + S_1 \cdot \dot{x}\dot{\lambda} + C_1 \cdot \dot{y}\dot{\lambda} - r_w (S_1 C_\xi + C_1 S_\xi) \dot{\xi}\dot{\lambda} \right\} \end{aligned}$$

$$
\begin{aligned}
a_{23} &= -\frac{1}{2}\dot{\mathbf{q}}^{\mathrm{T}}\frac{\partial \mathbf{M}_2}{\partial \lambda}\dot{\mathbf{q}} \\
&= m_2 \cdot \left[S_1 \cdot \dot{x}\dot{\theta}_1 + C_1 \cdot \dot{y}\dot{\theta}_1 - r_{\mathrm{w}}\left(S_1 C_\xi + C_1 S_\xi\right) \cdot \dot{\xi}\dot{\theta}_1 - \lambda\dot{\theta}_1\right].
\end{aligned}
$$

References

[1] M. Sasaki, N. Yanagihara, O. Matsumoto, K. Komoriya, Steering control of the personal riding-type wheeled mobile platform (PMP), IEEE International Conference on Intelligent Robots and Systems (2005) 1697–1702.

[2] Available from: <http://www.ibotnow.com/>

[3] Available from: <http://www.segway.com/>

[4] A. Salerno, J. Angeles, A new family of two-wheeled mobile robots: modeling and controllability, IEEE Transaction on Robotics 23 (1) (2007) 169–173.

[5] F. Grasser, A. D'Arrigo, S. Colombi, A. Rufer, Joe: A mobile, inverted pendulum, IEEE Transactions on Industrial Electronics 49 (1) (2002) 107–114.

[6] S.C. Lin, P.S. Tsai, H.C. Huang, Adaptive robust self-balancing and steering of a two-wheeled human transportation vehicle, Journal of Intelligent and Robotic Systems: Theory and Applications 62 (1) (2011) 103–123.

[7] K. DucDo, G. Seet, Motion control of a two-wheeled mobile vehicle with an inverted pendulum, Journal of Intelligent and Robotic Systems 60 (3–4) (2010) 577–605.

[8] A. Salerno, J. Angeles, The control of semi-autonomous two-wheeled robots undergoing large payload-variations, IEEE International Conference on Robotics and Automation (2004) 1740–1745.

[9] Y.-S. Ha, S. Yuta, Trajectory tracking control for navigation of the inverse pendulum type self-contained mobile robot, Robotic and Autonomation Systems 17 (1996) 65–80.

[10] R. Marino, On the largest feedback linearizable subsystem, Systems Control Letters 6 (1986) 345–351.

[11] K. Pathak, J. Franch, S.K. Agrawal, Velocity and position control of a wheeled inverted pendulum by partial feedback linearization, IEEE Transaction on Robotics 21 (3) (2005) 505–513.

[12] M. Karkoub, M. Parent, Modeling and non-linear feedback stabilization of a two-wheel vehicle, Journal of Systems and Control Engineering 218 (8) (2004) 675–686.

[13] M.W. Spong, The swing up control problem for the acrobat, IEEE Control Systems Magazine 15 (5) (1995) 49–55.

[14] Z.J. Li, J. Luo, Adaptive robust dynamic valance and motion controls of mobile wheeled inverted pendulums, IEEE Transaction on Control System Technology 17 (1) (2009) 233–241.

[15] S. Jung, S.S. Kim, Control experiment of a wheel-driven mobile inverted pendulum using neural network, IEEE Transactions on Control Systems Technology 16 (2) (2008) 297–303.

[16] V. Sankaranaryanan, A.D. Mahindrakar, Control of a class of underactuated mechanical systems using sliding modes, IEEE Transaction on Robotics 25 (2) (2009) 459–467.

[17] B.S. Park, S.J. Yoo, J.B. Park, Y.H. Choi, Adaptive neural sliding mode control of nonholonomic wheeled mobile robots with model uncertainty, IEEE Transactions on Control System Technology 17 (1) (2009).

[18] J. Huang, Z.H. Guan, et al., Sliding-mode velocity control of mobile-wheeled inverted-pendulum systems, IEEE Transaction on Robotics 26 (4) (2010) 750–758.

[19] H. Ashrafiuon, R.S. Erwin, Sliding control approach to underactuated multibody systems, American Control Conference (2004) 1283–1288.

[20] P.S. Tsai, L.S. Wang, F.R. Chang, Modeling and hierarchical tracking control of tri-wheeled mobile robots, IEEE Transaction on Robotics 22 (5) (2006) 1055–1062.

[21] X.H. Yu, J.X. Xu, Variable structure systems: Toward the 21st Century, Springer, Berlin, 2002.

[22] Z.H. Man et al. A robust MIMO terminal sliding mode control scheme for rigid robotic manipulators, IEEE Transactions on Automatic Control 39(12), 2464–2470

[23] S.-Y. Chen, F.-J. Lin, Robust nonsingular terminal sliding model control for nonlinear magnetic bearing system, IEEE Transaction on Control Systems Technology 19 (3) (2011) 636–643.
[24] H. Liu, J.F. Li, Terminal sliding mode control for spacecraft formation flying, IEEE Transaction on Aerospace and Electronic Systems 45 (3) (2009) 835–846.
[25] Y.S. Guo, C. Li, Terminal sliding mode control for coordinated motion of a space rigid manipulator with external disturbance, Applied Mathematics and Mechanics 29 (5) (2008) 583–590.
[26] K.-B. Park, T. Tsuji, Terminal sliding mode control of second-order nonlinear uncertain systems, International Journal of Robust and Nonlinear Control 9 (1999) 769–780.
[27] S. Kidane, R. Rajamani, L. Alexander, P.J. Starr, M. Donath, Development and experimental evaluation of a tilt stability control system for narrow commuter vehicles, IEEE Transactions on Control System Technology 18 (6) (2010) 1266–1279.
[28] A. Shimada, N. Hatakeyama, High speed movement control making use of zero dynamics on inverted pendulum, Journal of SICE 42 (9) (2006) 1035–1041.
[29] A. Alasty, H. Pendar, Equations of motion of a single-wheel robot in a rough terrain, in: Proc. IEEE International Conference on Robotics and Automation, Barcelona, Spain, 2005, pp. 18–22.

CHAPTER 2

Testing of Intelligent Vehicles Using Virtual Environments and Staged Scenarios

Arda Kurt, Michael Vernier, Scott Biddlestone, Keith Redmill, Ümit Özgüner

The Ohio State University, OH, USA

Chapter Outline

2.1 Introduction

Sufficient testing is an essential part of engineering research and development. For our main focus, Intelligent Transportation Systems (ITS), the testing phase often includes preliminary testing in numerous simulation environments as well as physical tests involving one or more vehicles.

Incorporating virtual equipment in the testing and evaluation of controlled engineering systems is now commonplace, and the availability of powerful computation and data acquisition and control equipment has made hardware-in-the-loop something of a standard practice. These approaches are also being employed in the design and testing of intelligent and automated vehicle technologies. There are a number of factors that justify this approach, including:

- The cost or availability of equipment
- The time and effort required to manage and staff field test activities
- Risk and safety considerations, especially with unproven technologies.

Advances in Intelligent Vehicles. http://dx.doi.org/10.1016/B978-0-12-397199-9.00002-1

To the best of our knowledge, the earliest uses of virtual hardware in the area of intelligent transportation systems appears in the work of Robert Fenton, beginning in the late 1960s. Fenton was one of the earliest active researchers in the area of automated vehicles and driver–vehicle interactions.

In Ref. [1], he describes a longitudinal control strategy for car following, what we would call platooning in an automated highway system, the implementation of this controller on a drive-by-wire experimental vehicle, and experiments conducted at highway speeds. Because there was no practical method or sensor available to measure the separation distance (headway) to the lead vehicle and the velocity or relative velocity of the lead vehicle, he created a simulated "phantom car", whose behavior could be varied according to the desired test, to produce the required sensor readings for his control experiments.

In Ref. [2], he describes an early driver-assist technology for vehicle following. He designed a haptic interface to provide information about lead car headway and relative velocity to the human driver and experimentally demonstrates the resulting improvement in car-following performance versus an unaided driver, especially in short-headway situations. In addition to the practical sensor issues faced in Ref. [1], these experiments involved human subjects not affiliated with the research program, so the use of a simulated lead vehicle during initial trials and while training test subjects was an important safety factor and training aid.

One major limitation of tests involving multi-vehicle scenarios, as we observed throughout our more recent experience in this particular field [3–5], is that outdoor testing can have a high cost in terms of logistics and scheduling, as mentioned above. Since adequately equipped testing areas are not always readily accessible, scheduling for transportation of the multiple parties involved and for favorable weather conditions can be a problem in and of itself.

In order to mitigate some of the logistics, cost, and safety problems listed above, this chapter describes a low-cost, flexible supplement to outdoor testing for intelligent transportation research and applications. Starting with fully virtual computer simulations and gradually increasing the real-life components of the tested scenarios, complex cyber-physical systems can be studied without many of the associated costs of actual outdoor testing.

For realistic urban traffic, where autonomous vehicles interact with human-driven vehicles, the aforementioned complexities of outdoor tests can be even more daunting; therefore, an indoor testbed that emulates the focused aspects of an outdoor environment is often effective for tests involving multiple vehicles, higher-level decision-making, and situation assessment. The generally lower speeds of urban traffic scenarios, as opposed to automated highway systems, makes indoor testing a particularly attractive option for consistent, repeatable tests.

The indoor testbed at The Ohio State University Control and Intelligent Transportation Research Laboratory [6], named SimVille and seen in Figure 2.1, has served as the

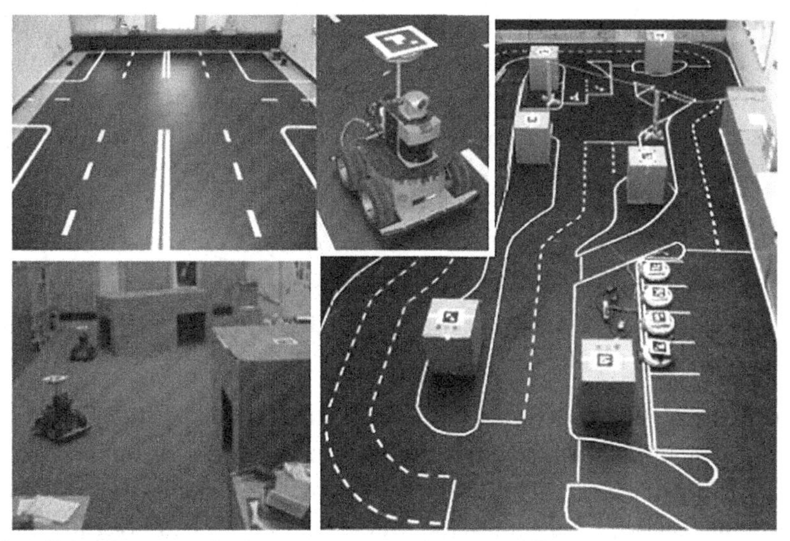

Figure 2.1: SimVille Indoor Testbed at OSU Control and Intelligent Transportation Research Lab, Through Various Incarnations and with Different Mobile Robots.

intermediate, semi-virtual test environment between computer simulations and fully developed outdoor tests for a number of projects over the years.

In the following subsections, the concepts behind the semi-virtualized iterations of the testing procedure, including simulations, small-scale testbeds, and full-scale outdoor testing, will be described and examples will be used as illustrations of these.

2.2 Procedure

The testing procedure described in this chapter starts with purely virtual simulations after the initial design process, as seen in Figure 2.2, and gradually moves to full-scale, purely physical tests as the design is finalized.

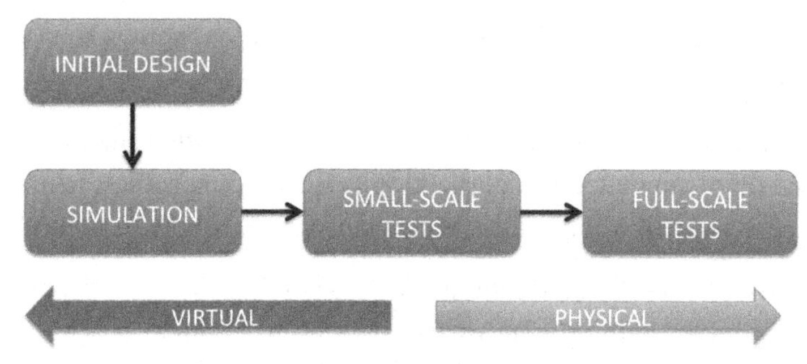

Figure 2.2: Testing Procedure Starting from Purely Virtual and Arriving at a Purely Physical Realization.

The reasons for following a gradual transition from the virtual world to the physical world include reduced testing costs and the rapid development and completion of the more virtualized test scenarios due to the flexibility offered by simulated or small-scale environments in changing aspects, as opposed to a full-scale, real-life testbed.

2.2.1 Simulation

The initial stages of the overall testing procedure take place solely in the virtual world, as high-fidelity computer simulations are very useful in detecting and correcting initial design and implementation errors rapidly.

Although there are many simulators available for ITS simulations depending on the specific issues studied, the examples in the remainder of this chapter will focus on the Gazebo [7] and Stage [8] simulators used at CITR. These two simulation environments use the Player [8] interface to communicate with and control the simulated robots. The Player interface is also available for actual robots through a uniform interface so that control and sensing software can be written without knowledge of the specific hardware of the robot or how virtual the robot may be.

The Stage simulator provides a lightweight environment that focuses on the interaction of a vast number of autonomous agents. These vehicles navigate on the stacked two-dimensional planes that create Stage's 2.5-dimensional simulation region. This setup constitutes the first process in our proposed procedure.

One particular vehicular autonomy study [9], which will be visited throughout the narrative in the following sections, is based on a collaborative convoying scenario, where a three-vehicle convoy consisting of two autonomous vehicles led by a manual vehicle continuously circles the figure-8 route shown in the diagram in Figure 2.3. They obey the traffic light, accelerating and decelerating simultaneously to maintain set inter-vehicle distances and to allow the convoy to clear the intersection as smoothly and rapidly as possible. They also obey the stop sign and intersection precedence rules, determining the presence of a manually driven fourth vehicle using information obtained from its DSRC transmissions.

This scenario, seen in Figure 2.4 as the initial Stage simulation, shows four vehicles interacting with one another at an intersection.

In contrast, Gazebo is a fully three-dimensional simulator with higher fidelity physics corresponding to roll and tilt of simulated vehicles and fully 3D terrain profiles, neither of which are available in Stage. The same Player interface is provided for the robots simulated in the Gazebo.

An example simulation for a fully autonomous vehicle coming to a stop at an intersection and detecting the other vehicles with an LIDAR can be seen in Figure 2.5. This particular simulation example was based on the DARPA Urban Challenge 2007 [10] autonomous vehicle competition.

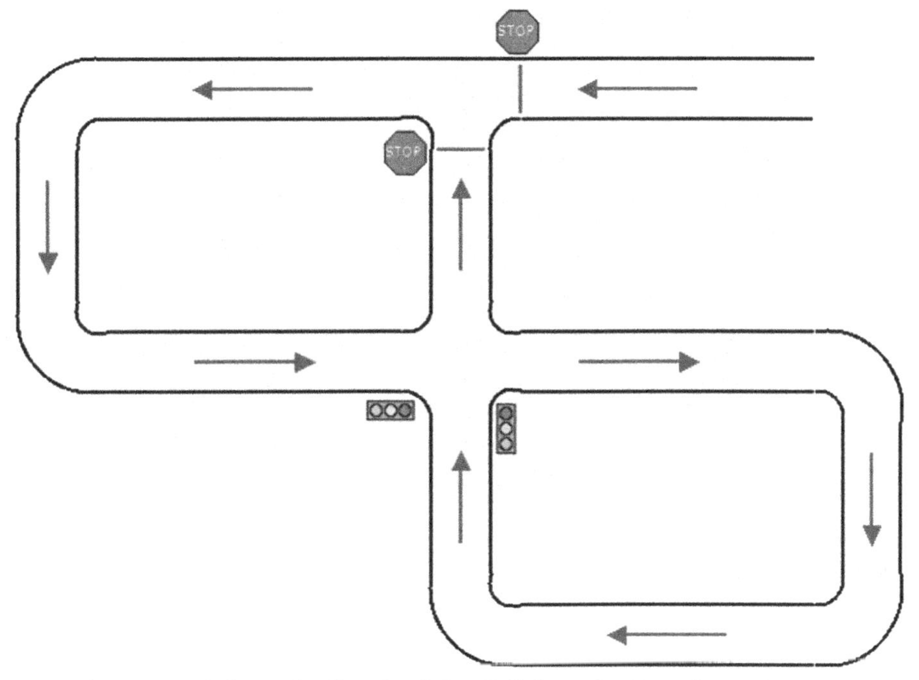

Figure 2.3: Schematic Sketch of the Collaborative Experiment Layout.

2.2.2 Small-Scale Testing

Indoor testbeds can be used both as a lower cost tool for Intelligent Vehicle research as well as a valuable educational tool that, compared to pure simulation, greatly benefits the teaching environment.

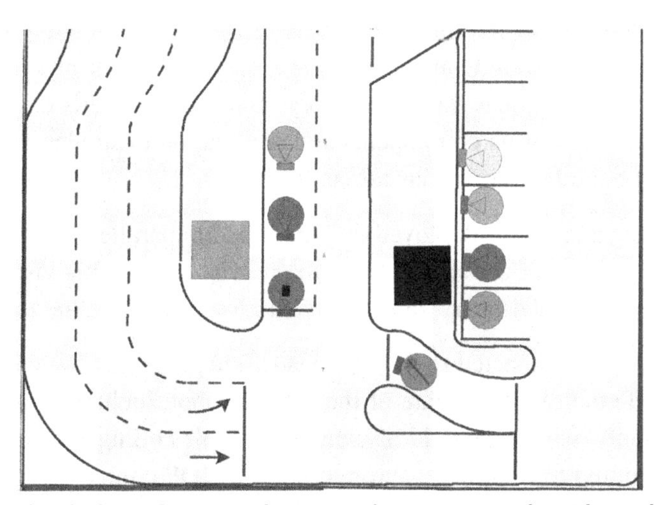

Figure 2.4: Stage Simulation of Four-Robot Experiment Interacting Through an Intersection.

Figure 2.5: Simulation of an Intersection Scenario with Simulated LIDAR Rays Visible. © *2008 IEEE. Reprinted with permission from "Simulation and testing environments for the DARPA Urban Challenge", IEEE International Conference on Vehicular Electronics and Safety, 2008. ICVES 2008, 22–24 September 2008, pp. 222–226.*

As part of the continuing Cyber-Physical Systems (CPS) research program at OSU, the CITR testbed discussed in the following paragraphs was used as a successful teaching tool for high school students. In a series of Summer Institute programs lasting for 2 weeks each summer, groups of students learned the basics of intelligent vehicle research, including how to control a small-scale robot, how to read data from sensors, and how to make decisions based on the sensory data [11]. A simple obstacle-avoidance scenario was used to test and demonstrate their understanding of the concepts on the CITR testbed.

The testbed architecture, as the middle ground between fully virtual and completely real hardware, consists of several mobile agents, a scale network of roads, and multiple hardware and software modules that emulate various sensory modalities. Details on the proposed architecture itself, and each of its components and application examples, can be found in Ref. [6]. The overall module descriptions are as follows.

A number of mobile agents can be deployed on the testbed in parallel. Each mobile agent has a unique identification number and a matching unique visual tag. This visual tag can be identified and tracked via a stationary camera system, located above the testbed.

The Virtual GPS module, connected to the overhead camera system, uses image-processing techniques to generate the real-time state of the mobile robot, including its position, yaw orientation, and velocity, within the field of view of multiple ceiling-mounted cameras. This information is transmitted to each robot through unicast UDP communication via wireless 802.11b interfaces.

Each robot fuses the Virtual GPS data with on-board sensor data from wheel encoders or inertial sensors to create more accurate state estimates. Each robot broadcasts their own state estimates through Vehicle-to-Vehicle (V2V) communications, which are available to other robots and to the central monitoring module. The state estimates, combined with scenario-specific reference trajectories or missions and information about the infrastructure received by each robot on a Vehicle-to-Infrastructure (V2I) channel, are used by the on-board computer and control software to navigate the mobile robot within the area.

The simulation scenario of four vehicles interacting through two intersections described in the previous subsection and shown in the simulator, when implemented with mobile robots on the indoor testbed, can be seen in Figure 2.6.

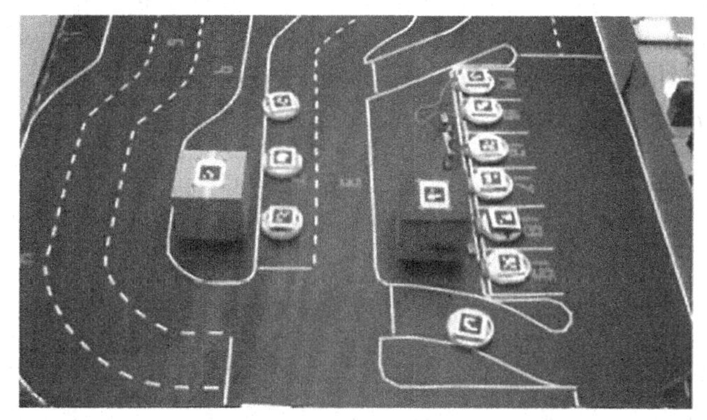

Figure 2.6: Four Mobile Robots Interacting Through a Stop-Sign Controlled Intersection.

Another version of the OSU-CITR indoor testbed and a mid-sized mobile robot as they were used for the DARPA Urban Challenge can be seen in Figure 2.7.

Figure 2.7: OSU-CITR Testbed for Intersection Scenarios and Pioneer 3-AT Mobile Robot Used on the Testbed. *© 2008 IEEE. Reprinted with permission from "Simulation and testing environments for the DARPA Urban Challenge", IEEE International Conference on Vehicular Electronics and Safety, 2008. ICVES 2008, 22–24 September 2008, pp. 222–226.*

This version of the testbed was used, in parallel with simulations such as the one described in the previous subsection and illustrated in Figure 2.5, to test various concepts and algorithms for the DARPA Urban Challenge, such as intersection precedence, obstacle avoidance for partially blocked intersections and U-turns, and dynamic re-planning for blocked intersections.

2.2.3 Full-Scale Testing

After successful testing of scenarios in simulation and indoor testbed environments, the development procedure was moved to semi-virtualized outdoor tests.

The example convoying scenario, which was visited as a simulation and an indoor test and is illustrated in Figures 2.4 and 2.6, was finally realized as an outdoor test with four actual vehicles. The satellite image of the test site as it corresponds to the figure-8 layout of the scenario can be seen in Figure 2.8.

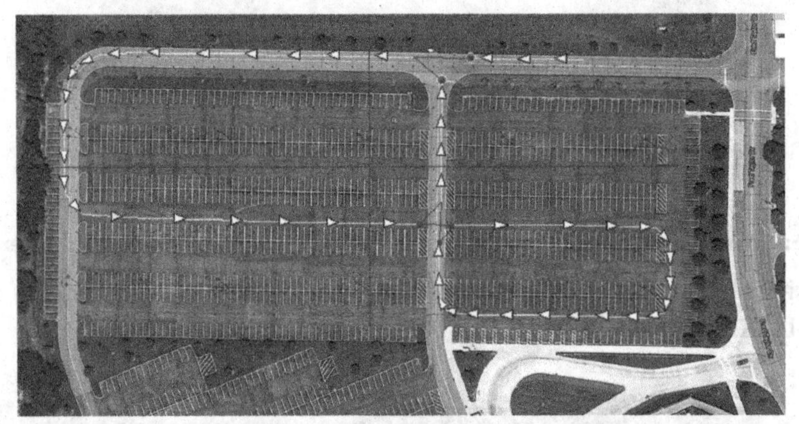

Figure 2.8: Satellite Picture of the Test Site and the Correspondence to the Schematic Layout.

The four vehicles, as they drive through the intersections in the outdoor test site, can be seen in Figure 2.9.

Figure 2.9: Experimental Vehicles Driving Through One Intersection at the Outdoor Test Site at OSU.

The tests on autonomous decision-making at intersections for the DARPA Urban Challenge 2007, as discussed above for the case of simulations and small-scale tests, culminated with the full-scale experiments using fully autonomous vehicles as seen in Figure 2.10.

Figure 2.10: Outdoor Tests for Intersection Decision-Making at DARPA Urban Challenge.

2.3 Virtual Sensors and Scenarios

At each stage of the described testing procedure there are a number of interacting agents and modules. In a pure simulation environment, each module and agent is virtual, while a later-stage full-scale outdoor test benefits from higher-fidelity physical systems.

In between the fully virtual and fully physical worlds, such as the one inhabited by small-scale indoor testbeds described above, it is possible to implement a number of modules and agents virtually. These virtual entities can range from virtual sensors that emulate various real-life sensors to either cut costs or test experimental sensor configurations, to virtual agents or obstacles that physical vehicles interact with for safety or logistics reasons.

In the following subsections, virtual sensors and agents will be discussed through specific examples.

2.3.1 Virtual Sensors

In this subsection, a suite of virtual sensors named the Virtual Sensor Systems, developed to be used in conjunction with physical robots, will be used as demonstrative examples.

The Virtual Sensor System was implemented as a software agent using Stage [12]. This system received input from the various software and hardware components already in place in the Control and Intelligent Transportation Research Laboratory. Vehicle-to-vehicle messages were used to place the virtual robots into the Stage simulation environment. Sensors could then be mounted on the virtual robots in this environment. As the physical robots navigated through the testbed, sensor data would be simulated in real time and transmitted to the robot for use by the control software. Vehicle-to-infrastructure messages were also received in order to update a virtual traffic light so that simulated camera images of the entire combined environment were possible.

In order to reconstruct the other static objects in the environment that needed to be recognized by the virtual sensors, markers were placed on them so that the Virtual GPS module could measure their position state. In the SimVille testbed, several boxes were used in the physical testbed as buildings surrounding the roadways, and virtual counterparts reflected the physical state of these boxes; however, because of the flexibility of this virtual component, the virtual model does not have to exactly represent the physical object. As an example, a marker could be placed on the ground within the testbed and assigned to represent a tree by the robots' virtual sensors, as shown in Figure 2.11. This figure also shows the virtual representation of the Create robot platform and the buildings of SimVille.

In order to demonstrate the flexibility and ease of use within the Virtual Sensor System, three sensors, an LIDAR, magnetometer, and a lane edge sensor, were initially implemented. Several parameters of each of the sensors were modeled so that they can be configured to satisfy a specific need of the test scenario or mimic the capabilities of an off-the-shelf sensor that is either physically installed on another robot or is being evaluated for such purposes. For the LIDAR, the sensor's field of view, minimum and maximum detectable distance along each ray, and the number of rays within the field of view can be modified through Stage's world definition file. Figure 2.12 shows a forward-pointing, front-mounted laser rangefinder on a Create robot detecting objects in front of it.

Figure 2.11: Objects in Both the Physical Testbed and in the Virtual Sensor System. *© 2011 IEEE. Reprinted with permission from "Virtual sensor system: Merging the real world with a simulation environment", 14th International IEEE Conference on Intelligent Transportation Systems (ITSC), 5–7 October 2011, pp. 1904–1909.*

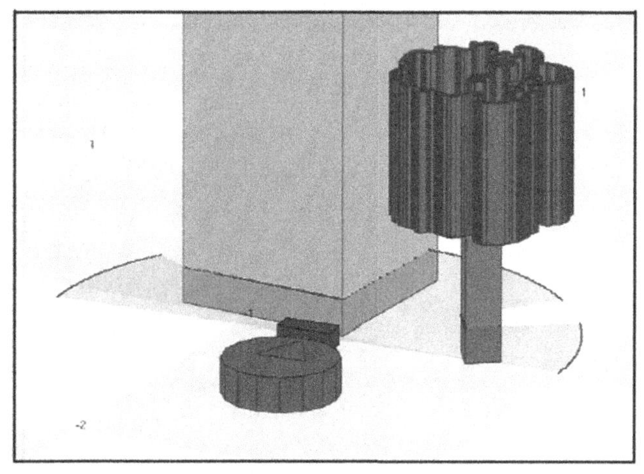

Figure 2.12: Mobile Robot with Laser Rangefinder. © *2011 IEEE. Reprinted with permission from "Virtual sensor system: Merging the real world with a simulation environment", 14th International IEEE Conference on Intelligent Transportation Systems (ITSC), 5–7 October 2011, pp. 1904–1909.*

A magnetometer has the ability to measure the strength and direction of a magnetic field. This sensor has been used to detect the field of magnets embedded into the lanes of a road [13]. Several magnetometers would be placed on the front of the robot to measure the magnetic field of the embedded magnets. These measurements can be used to determine the offset of the vehicle from the center of the lane and can be used as input to a lateral steering controller. Figure 2.13 shows a five-sensor array virtually mounted to the front of a Create robot.

A lane edge sensor is a fictitious sensor that has been used in teaching basic intelligent vehicle control techniques in a class at OSU to detect the distance between the vehicle and the edge of a road lane. In a physical environment, this sensor could be implemented by processing an image of the lane edge provided by a downward-pointing camera mounted on a vehicle, but in the Virtual Sensor System a sensor's output is not needed to be modeled or implemented in a realistic or physically practical manner.

The implementation of this sensor was based upon the output of a laser rangefinder mounted close to the virtual ground. This technique exploited a design feature in the Stage simulator where each object displayed in the environment has a solid volume in order to reduce the computation time needed to determine the measurement. Lane markers were solid rectangles placed on top of the ground rather than a simple image textured on the ground polygons. A Create robot with a left-pointing, virtually mounted lane edge sensor is shown in Figure 2.14.

These sensors are useful in various aspects of Intelligent Transportation System testing, and the use of the Virtual Sensor System allows different tests to be swapped out without any physical modification of the testbed. One such application used a single robot in an environment with a virtually mounted LIDAR to navigate through an unknown environment.

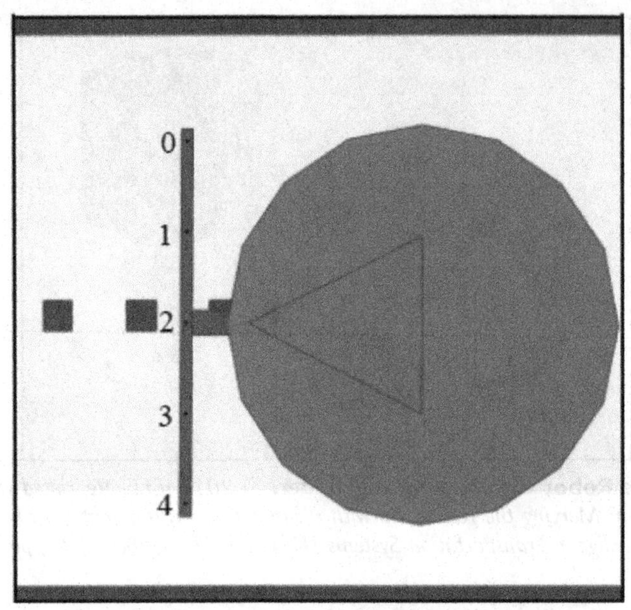

Figure 2.13: Mobile Robot with Front-Mounted Magnetometer. *© 2011 IEEE. Reprinted with permission from "Virtual sensor system: Merging the real world with a simulation environment", 14th International IEEE Conference on Intelligent Transportation Systems (ITSC), 5–7 October 2011, pp. 1904–1909.*

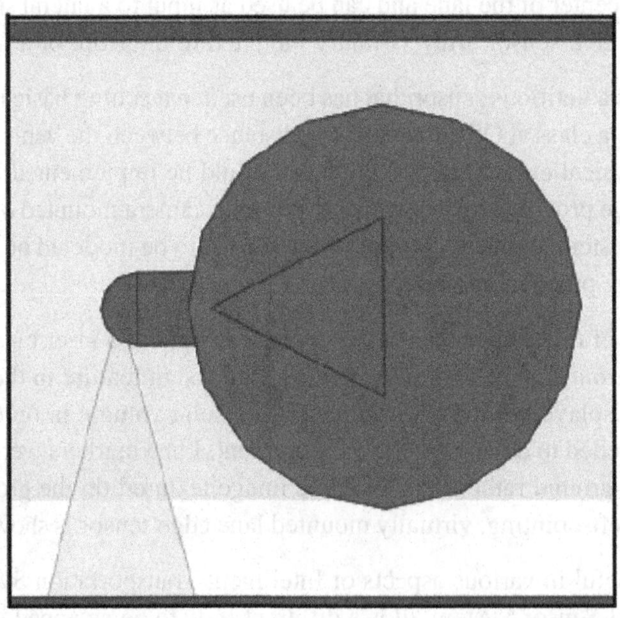

Figure 2.14: Mobile Robot with Lane-Edge Sensor. *© 2011 IEEE. Reprinted with permission from "Virtual sensor system: Merging the real world with a simulation environment", 14th International IEEE Conference on Intelligent Transportation Systems (ITSC), 5–7 October 2011, pp. 1904–1909.*

This test is shown in Figure 2.15. The states of the physical robot and the physical obstacles were tracked within the system. Virtual walls were added to keep the robot within the specified test area. The robot interprets the virtual LIDAR hits in order to navigate through the environment without collision. This information is also used to generate a map as shown in Figure 2.16.

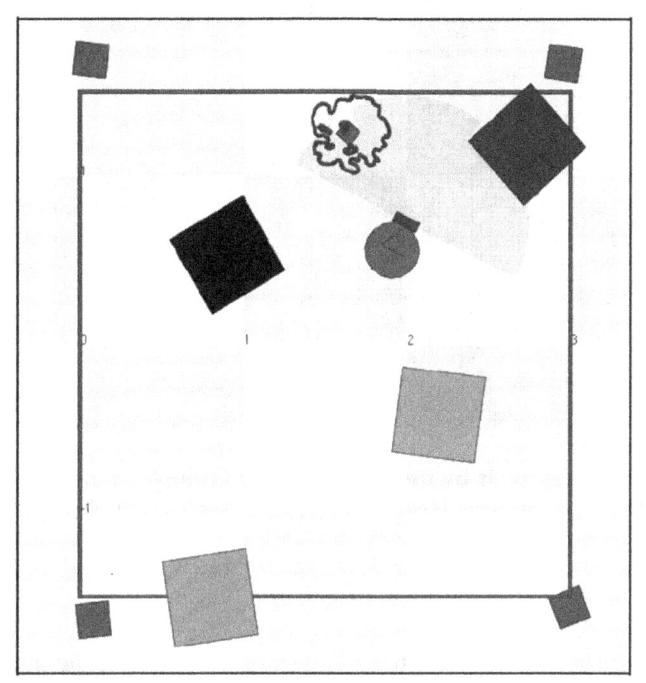

Figure 2.15: Stage Representation of an Actual Mobile Robot Interacting with Real-Life Objects Through a Virtual Sensor. *© 2011 IEEE. Reprinted with permission from "Virtual sensor system: Merging the real world with a simulation environment", 14th International IEEE Conference on Intelligent Transportation Systems (ITSC), 5–7 October 2011, pp. 1904–1909.*

2.3.2 Virtual Agents and Scenarios

Testing intelligent vehicles in interaction with virtual agents and different scenarios can be traced back to the studies conducted by Fenton at OSU [1,2], as discussed in the introduction.

The main advantage of testing various aspects of an autonomous vehicle in interaction with a computer model of another vehicle, as opposed to an actual car, is that safety-related potential design errors will not cause accidents. If the developed autonomous vehicle is following a virtual car, be it a pure mathematical construct in the vehicle software or a separate simulation suite generating the necessary signals, there is no risk of actually hitting the lead car through an error in the following code. Similarly, intersection precedence scenarios can be tested

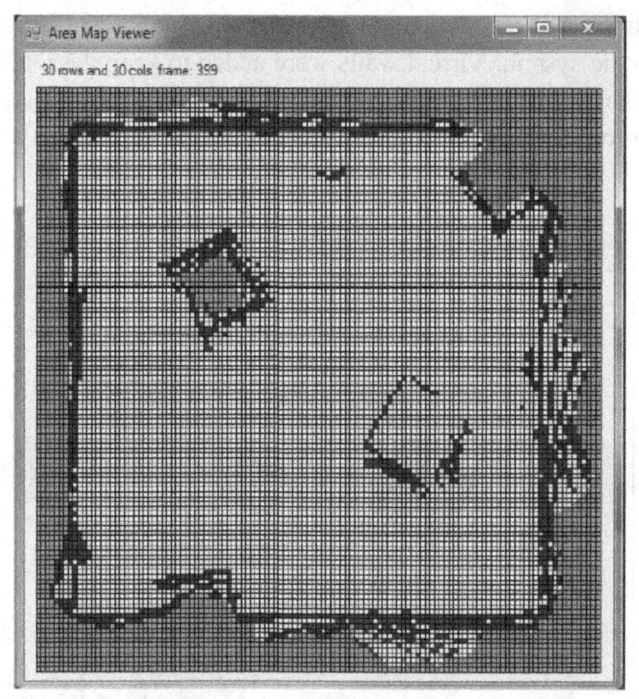

Figure 2.16: Occupancy Map Built by the Mobile Robot Using Virtual Sensor Returns. *© 2011 IEEE. Reprinted with permission from "Virtual sensor system: Merging the real world with a simulation environment", 14th International IEEE Conference on Intelligent Transportation Systems (ITSC), 5–7 October 2011, pp. 1904–1909.*

safely, at least in the earlier stages, by setting up virtual vehicles for the intelligent vehicle to interact with.

In the earliest studies benefitting from this idea, Fenton et al. called these constructs "phantom cars". As we have developed a series of intelligent vehicles over the years and started using modeled, simulated, and recorded computer constructs, the term evolved into "virtual cars" and "virtual agents". The following paragraphs describe one application of "virtual car" use with specific focus on scenario safety.

This example focuses on the testing stages of a "Driver Intention Estimator" developed at OSU CITR in collaboration with Honda R&D of Americas, Inc. The actual estimator software uses Vehicle-to-Vehicle (V2V) communications to detect, track, and predict the behavior of other vehicles as the ego vehicle approaches an intersection. The details of estimator implementation and underlying ideas can be found in Refs [14,15].

The input to the estimator suite is the measurement vector **z** containing information on the tracked vehicle and populated using V2V communications based on Dedicated Short Range Communication (DSRC) standards. DSRC radios on each vehicle connected to GPS receivers and the Controller-Area Network (CAN) bus of the vehicle both collect and record and also broadcast real-time information about the vehicle. Transmitted at a 10-Hz rate, the "safety message" generated by the DSRC radio includes the following pieces of information:

- Radio timestamp
- GPS timestamp
- Vehicle width
- Vehicle length
- GPS position – latitude
- GPS position – longitude
- GPS position – elevation
- Velocity
- Longitudinal acceleration
- Lateral acceleration
- Yaw
- Yaw rate
- Steering-wheel position
- Accelerator (gas pedal) position
- Left turn-signal On indicator
- Right turn-signal On indicator
- Brake activated indicator.

Using a set of measurements from the ego vehicle (directly through the CAN bus) and one set of measurements from the tracked vehicle (through DSRC), the estimator software was designed to analyze and predict the safety of the overall scenario as two vehicles approach an intersection.

Since collecting data or conducting real-time tests with two vehicles in a combination of safe and unsafe scenarios is both dangerous and time consuming, the concept of virtual agents was utilized by manipulating the data recorded from a single vehicle to represent two separate vehicles approaching the same intersection at the same time.

Given two separate approaches of the same experimental vehicle to the same intersection at two different times, we were able to shift the time of one data log to align with the other. $z1 = z[k0, k1]$ and $z2 = z[k2, k3]$ being recorded from the same vehicle, shifting the second data set to $z3(k) = z2(k + (k2 - k0))$, we were able to use real-life data with potentially unsafe conditions, without actually using and risking two vehicles.

For example, Honda Experimental Vehicle (Vehicle 1) approaches the intersection of 10 Mile Rd and Amanda Lane, marked A in Figure 2.17, heading west on 10 Mile Rd (approaching from east). At time 10:50:52 a.m. (39,052 seconds after midnight), the vehicle reaches the intersection and continues straight on 10 Mile Rd.

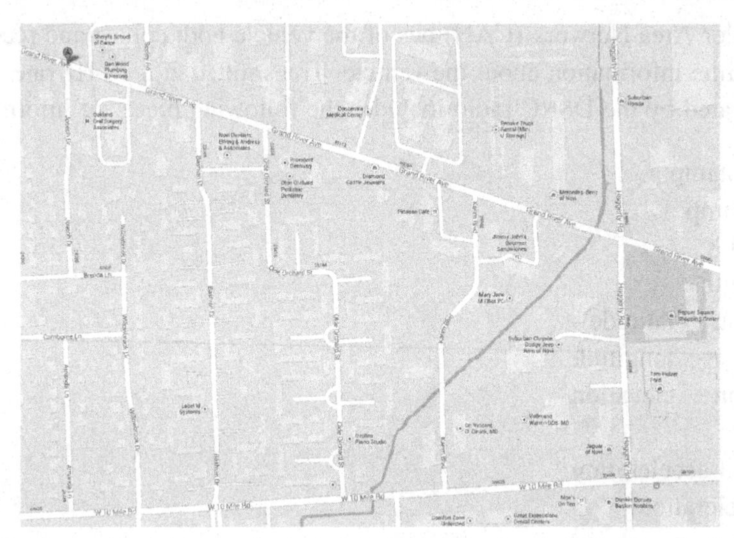

Figure 2.17: Map of the Area at Detroit, MI, Used for Data Collection by Honda R&D. The intersection marked with A is used in the example. *Image from Google Maps. © 2012 Google.*

Later in the day, the same vehicle (Vehicle 1) approaches the same intersection (intersection A) from north on Amanda Ln. It arrives at the intersection at 2:45:20 p.m. (53,120 seconds after midnight), slows down, and makes a left turn on to Ten Mile Road, leaving the intersection heading east.

At this point, we have access to the logged data from the entire day of collection. By taking log snippets starting from 10 seconds before the intersection and ending at 5 seconds after the intersection, we can generate two 15-second log pieces, which capture the relevant intersection approach behavior. The first log trace starts at timestamp 39,042.00 and ends at 39,057.00, the second trace starts at 53,110.00 and ends at 53,125.00. Normally, these 150-sample-long log traces (collected at 10 Hz) are temporally separated by hours (14,068 seconds). However, we can modify the timestamps of the second log trace by subtracting 14,068 from the timestamps of each sample, therefore aligning the two logs in time.

The two-vehicle scenario, with recorded data from a single vehicle, can be seen as it runs on the driver intention and safety estimator software in Figure 2.18. As far as the software is concerned, there are two separate vehicles arriving at the same intersection on a potentially fatal collision course. To test or record the same scenario with two vehicles would not be possible due to safety concerns, yet the virtualization of the scenario made it possible with only a single experimental car.

Similar time-shift manipulations can also be expanded to modify the actual rate of change of variables in the measurement vector. Instead of shifting the timescale, we could also dilute or compact the same time progression (by decimating or interpolating the signals) so that a

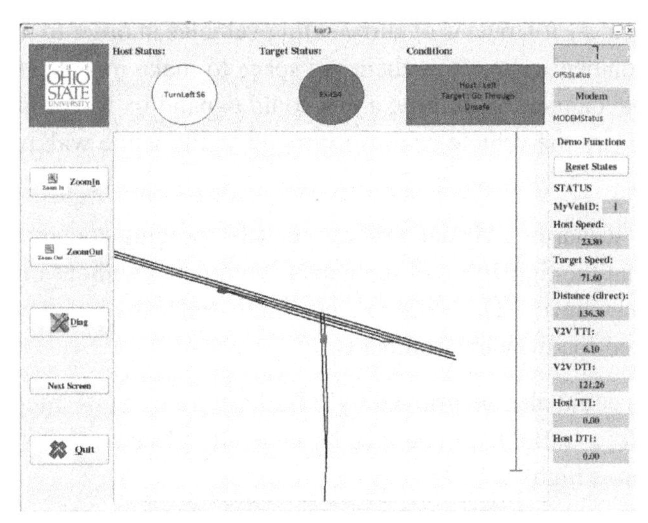

Figure 2.18: Two Vehicles, One of Which is Virtual, Approaching the Intersection of Grand River Ave and Joseph Dr in Michigan, as Seen on the Map Screen of the Estimator Software.

slower or faster intersection approach can be emulated. However, the more the recorded data are modified, the more unrealistic thc measurements become. At some point, the benefits of highly manipulated real-life data versus pure simulation become debatable.

2.3.3 Virtual Environments

The virtual environment in conjunction with a physical testing setup can significantly increase the set of possible testing scenarios, in addition to making them easier to implement. Specifically, in an indoor testbed, primary objects can be added to the environment without physically creating them in the laboratory. Roadways can be quickly mapped out and changed if necessary. Virtual barriers could temporarily block these roadways or various parking-lot configurations could be analyzed.

The flexibility of a partially virtual environment also allows for augmentation of existing physical objects. A simple box placed physically next to the robot could be seen as an intricate three-dimensional model of an office building by virtual sensors. Curbs or other roadway defects can be added to an existing path in the testbed without consuming testing time with physical fabrication practices. Cars could be placed randomly in designated parking spots as well.

A common virtual obstacle interface could be used outside of the indoor testbed as well. In mixed-mode traffic environments, it may be beneficial for some human-driven cars to artificially modify its shape or size as observed by other surrounding cars using partially virtual sensors. Vehicles driven by students or novice drivers could be made to seem larger in order to increase the safety tolerances of surrounding vehicles in order to remove stress from this unfamiliar environment or to allow them the space to make mistakes without causing otherwise unavoidable damage. Similarly, a bus could expand its footprint to protect pedestrians as they board the vehicle or move through traffic while walking to their destinations.

In a convoy-forming scenario, a virtual car between two merging convoy groups could be helpful in insuring that the tightening of the gaps is completed without interference from vehicles outside of the convoy. This might be especially helpful when these maneuvers are attempted in congested roadway environments.

Virtual environments could also be utilized by infrastructure systems integrated into a roadway. For example, a traffic light could be dynamically placed at a two-way intersection only when congestion is high.

Virtual barriers could be used to block a road or other road entrance if it would be beneficial to proper traffic flow. This technique could be used to direct traffic around an accident or other disaster scenarios, allowing emergency response crews the space they need to operate successfully.

These barriers could also be used selectively so that some vehicles are allowed to drive straight through while others must be redirected, as seen in Figure 2.19. The virtual block, shown blocking the lower outgoing lane, could be applied selectively to only a subgroup of vehicles, so the first intelligent vehicle (IV1) detects the obstacle and reroutes itself while the second one (IV2) detects no obstruction and continues straight. This could be useful in zone areas that require permits for parking. Empty spots would be filled with virtually parked cars that are only visible to non-pass holders. Another example would be barriers that act as gatekeepers where only specific vehicles are permitted in the roadway such as emergency vehicles near a hospital or employee vehicles when on a company premises.

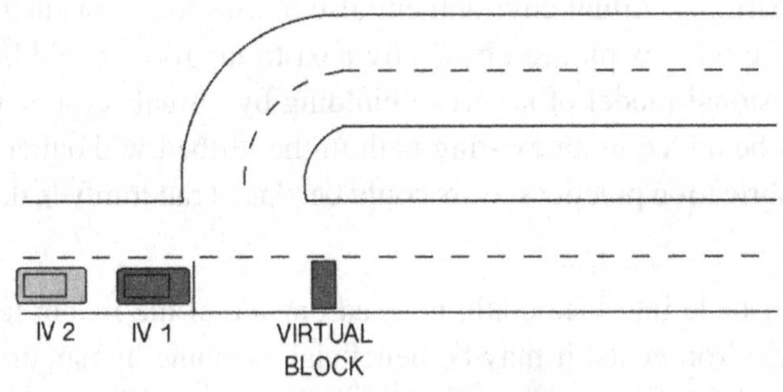

Figure 2.19: Traffic-Flow Shaping Through Virtual Obstacles.
Two IVs approaching an intersection; the first vehicle is rerouted while the second vehicle can pass.

Through the use of virtual roadway design, individual lanes could be restructured to reduce congestion. Lanes could switch direction in uneven congestion. Turn lanes could also be introduced if the overall traffic patterns would benefit.

2.4 Conclusions

In this chapter, various aspects of testing for Intelligent Vehicle applications were discussed. Since ITS vehicles and systems are generally designed and built for outdoor use, the tests for these systems are costly, time consuming, and sometimes even unsafe.

Using the developed procedure, which starts from simulation and gradually moves to full-scale outdoor tests as the design matures, it is possible to cut the development time and costs and improve the safety conditions, especially in the earlier stages of development. Each stage of this iterative testing process has been successfully utilized at the Ohio State University Control and Intelligent Transportation Research Laboratory (OSU-CITR).

As the tests start purely in the simulated, virtual domain and moves towards the physical realm, modules and agents used in the tests can be virtualized for a variety of reasons. Virtual sensors, agents, scenarios, and environments are described, discussed in terms of utility and demonstrated with examples as they are used in interaction with the physical systems that are tested. Using these virtual subsystems as parts of ITS tests can be traced almost five decades back to the earliest days of intelligent-vehicle research.

In summary, this chapter provides guidelines and demonstrative examples for streamlining the intelligent-vehicle testing process and discusses the benefits of virtualized subsystems for the entire series of tests.

Acknowledgments

The studies presented in this chapter were supported partially by the National Science Foundation under the Cyber-Physical Systems (CPS) Program (ECCS-0931669) and partially by Honda R&D of Americas, Inc.

References

[1] J.G. Bender, R.E. Fenton, A study of automatic car following, IEEE Transactions on Vehicular Technology 18 (3) (1969) 134–140.

[2] R.E. Fenton, W.B. Montano, An intervehicular spacing distance for improved car-following performance, IEEE Transactions on Man–Machine Systems 9 (2) (1968) 29–35.

[3] Q. Chen, U. Ozguner, Intelligent off-road navigation algorithms and strategies of team Desert Buckeyes in the DARPA Grand Challenge 2005, in: M. Buehler, K. Iagnemma, S. Singh (Eds.), The 2005 DARPA Grand Challenge, Springer Tracts in Advanced Robotics, vol. 36, Springer, Berlin, 2007, pp. 183–203.

[4] Q. Chen, U. Ozguner, K.A. Redmill, Ohio State University at the 2004 DARPA Grand Challenge: Developing a completely autonomous vehicle, Intelligent Systems, IEEE 19 (5) (2004) 8–11.

[5] K.A. Redmill, U. Ozguner, The Ohio State University automated highway system demonstration vehicle, SAE Transactions 1997: Journal of Passenger Cars (1999) 1332.

[6] S. Biddlestone, A. Kurt, M. Vernier, K.A. Redmill, U. Ozguner, An indoor intelligent transportation testbed for urban traffic scenarios, in: 12th International IEEE Conference on Intelligent Transportation Systems, 2009, ITSC 09, 2009, pp. 1–6.

[7] N. Koenig, A. Howard, Design and use paradigms for Gazebo, an open-source multi-robot simulator, in: IEEE/RSJ International Conference on Intelligent Robots and Systems (IROS 2004), Proceedings (2004) 2149–2154.

[8] The Player Project. <http://playerstage.sourceforge.net>.

[9] U. Ozguner, K.A. Redmill, Cyber-physical systems cooperative vehicle demonstration: Phase I, in: IEEE Vehicular Networking Conference (VNC) Demo Summaries, Amsterdam, 15–16 November 2011, pp. 7–8.

[10] U. Ozguner, K.A. Redmill, S. Biddlestone, M.F. Hsieh, A. Yazici, C. Toth, Simulation and testing environments for the DARPA Urban Challenge, in: IEEE International Conference on Vehicular Electronics and Safety (ICVES), Columbus, OH, 2008, pp. 222–226.

[11] V. Gadepally, A. Krishnamurthy, U. Ozguner, A hands-on education program on cyber physical systems for high school students, Journal of Computational Science Education 3 (2) (2012) 11–17.

[12] M.A. Vernier, U. Ozguner, Virtual sensor system: Merging the real world with a simulation environment, in: 14th International IEEE Conference on Intelligent Transportation Systems (ITSC), 2011, pp. 1904–1909.

[13] D.Y. Im, Y.J. Ryoo, Y.Y. Jung, J. Lee, Y.H. Chang, Development of steering control system for autonomous vehicle using array magnetic sensors, in: Proceedings of the International Conference on Control, Automation, and Systems, Seoul, Korea, 2007, pp. 690–693.

[14] A. Kurt, U. Ozguner, A probabilistic model of a set of driving decisions, Proceedings of the 14th IEEE Conference on Intelligent Transportation Systems, ITSC 2011 (October 2011).

[15] A. Kurt, J. Yester, Y. Mochizuki, U. Ozguner, Hybrid-state driver/vehicle modelling, estimation and prediction, in: 13th International IEEE Conference on Intelligent Transportation Systems (ITSC), 2010, pp. 806–811.

CHAPTER 3

Mathematical Modeling of Connected Vehicles

Daiheng Ni

University of Massachusetts, Amherst, MA, USA

Chapter Outline

Advances in Intelligent Vehicles. http://dx.doi.org/10.1016/B978-0-12-397199-9.00003-3

3.1 Connected Vehicle Technology

The US Department of Transportation (USDOT) has announced its connected vehicle initiative, previously known as Vehicle Infrastructure Integration (VII) and later IntelliDrive, and commonly referred to as Vehicular Ad-hoc Networks (VANET) outside of the transportation community. The initiative proposes to equip future vehicles and the roadside with wireless communication, computing, and positioning capabilities (see Figure 3.1). Utilizing Vehicle-to-Vehicle (V-V) and Vehicle-to-Infrastructure (V-I) Dedicated Short-Range Communications (DSRC), this initiative envisions making driving safer and transportation systems more efficient by making vehicles aware of other vehicles around them. Vehicles are also able to communicate with roadside infrastructure so that drivers have access to enough functional information to streamline traffic and make roadways safer. As such, future highways and streets will become an environment that encompasses ubiquitous computing and communication. Consequently, innovative applications can be deployed to dramatically increase safety, throughput, and energy efficiency.

The following is typical in the design of such connected vehicle-enabled systems. In a virtual environment, vehicles are flowing on a freeway. Suddenly, two vehicles crash and an analyst

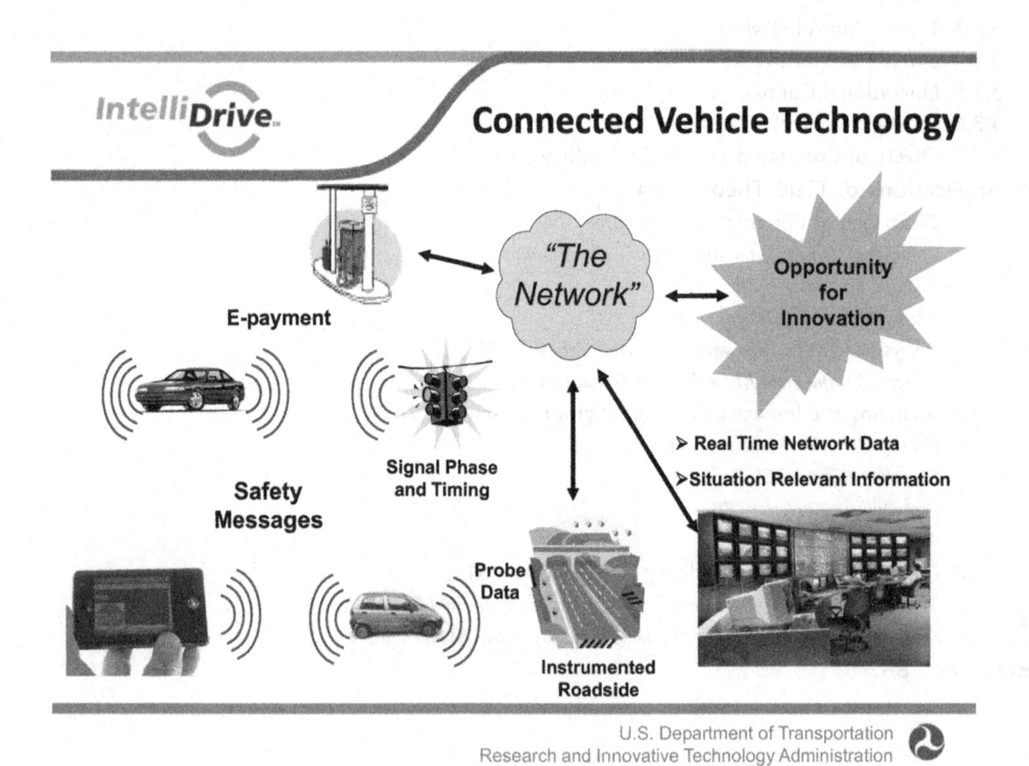

Figure 3.1: The USDOT Connected Vehicle Initiative.

says: "Wait a second." He then rewinds the simulation to a point before the crash. After evaluation, the analyst identifies three potential causes of the crash: (1) the two vehicles were too close, (2) the following driver was distracted, and (3) the leading driver applied emergency braking. "Okay, let's see what happens if these vehicles were connected", says the analyst as he turns on a special equipment on board each vehicle, called on-board equipment or OBE, and resumes the simulation. Surprisingly, the potential crash becomes to a near miss and a life has just been saved! Due to connected vehicle technology, three reasons contributed to this success: (1) the leader was warned by her OBE against an emergency vehicle entering from the on-ramp ahead, so she was able to brake smoothly; (2) when she applied the brake, a "braking ahead" notice was transmitted to her follower who was alerted to a potential collision; and (3) even before this, an "observe safe distance" warning from the follower's OBE has motivated him to stop accelerating, so the subsequent event became easier to handle.

The objective of this chapter is to provide a theoretical foundation for the modeling of connected vehicles that underlies the above virtual environment to assist the design and test of connected vehicle applications.

3.2 The Modeling Approach

To explain the proposed approach, it is necessary to understand how a transportation system with connected vehicles differs from an ordinary one. The top part of Figure 3.2 illustrates a transportation system. Such a physical world is internalized by drivers as a perceived world that serves as the basis for drivers to make driving decisions. For example, the bottom part of Figure 3.2 illustrates what is perceived by the driver in the vehicle at the lower left corner. Each driver carries his or her own goals (e.g. safely make the destination in one hour) and exhibits different characteristics (e.g. reaction time, desired speed, and route preference). The driver's reasoning intelligence, together with his or her goals and characteristics, are combined with the perceived world to make driving decisions such as target route, lane keeping, and driving speed. Hence, these decisions are carried out as actions and executed on the vehicle. Consequently, the vehicle moves dynamically as the driver directs. Meanwhile, the motion of this vehicle and other vehicles modifies the physical world that is dynamically reflected in the driver's perceived world, which triggers the next iteration of the above process. Therefore, the above process can be summarized as a driver–vehicle–environment closed-loop approach.

The following is typical without connected vehicles: (1) the perceived world is sampled from the physical world (a continuous physical process) and the sampling rate depends on how often the driver scans the physical world; (2) the perceived world is a vague figure of the physical world and the vagueness depends on the driver's perception accuracy; (3) the perceived world is a partial picture of the physical world and the scope depends on how far

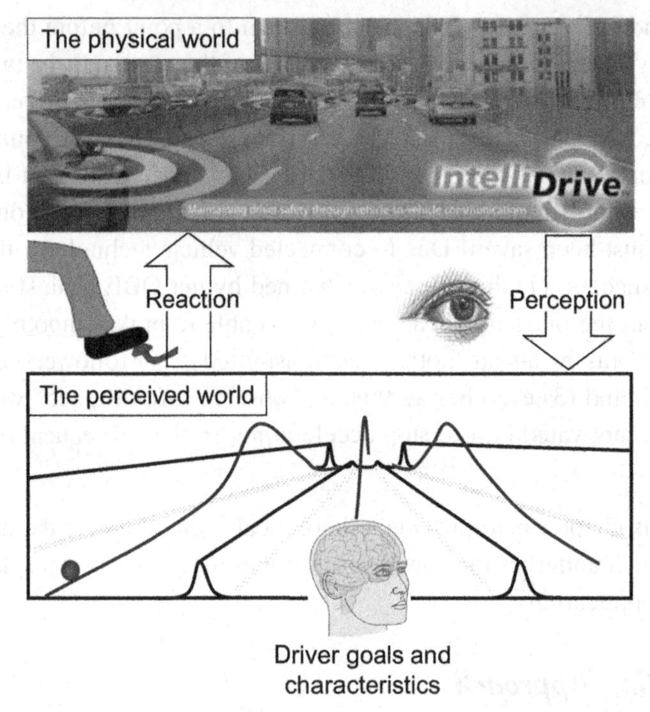

Figure 3.2: The Driver–Vehicle–Environment Closed-Loop Approach.

the driver is able to perceive; (4) there is considerable uncertainty in the driver's decision-making due to the above limitations; (5) the driver requires substantial lead time to account for the uncertainty.

In contrast, in a transportation system with connected vehicles, the driver's capabilities are significantly enhanced due to V-V and V-I DSRC: (1) the perceived world becomes a more frequent sample of the physical world, especially at critical moments, since the connected vehicle technology can assist the driver to constantly monitor the physical world and draw his or her attention as the need arises; (2) the perceived world becomes a clearer figure of the physical world since the connected vehicle technology provides the driver with more accurate information (such as accurate measurement of distance, speed, and acceleration) about the physical world; (3) the perceived world covers a larger scope of the physical world since V-V and V-I DSRC extend the driver's situational awareness beyond his or her perception limit; (4) there is less uncertainty in the driver's decision-making since his or her perceived world approximates the physical world closer; (5) accurate information necessitates less time for the driver to respond and reduced uncertainty decreases perception-reaction variance.

Therefore, the core issue is how to mathematically represent the driver's perceived world in which the connected vehicle's effects on the driver are captured. This is the purpose of the next section.

3.3 Field Theory

The Field Theory of traffic flow [1] is ideal to serve the above needs because it is mathematically amenable to representing the natural way of human thinking and reasoning, and readily admits the modeling of connected vehicle effects. Basically, in this theory, physical world objects (e.g. roadways, vehicles, and traffic control devices) are perceived by the subject driver as component fields. The driver interacts with an object "at a distance" (as opposed to "by contact") and the interaction is mediated by the field associated with the object. In addition, the field may vary when perceived by different drivers, especially those assisted by connected vehicle technology. The superposition of these component fields represents the overall hazard encountered by the subject driver. Hence, the objective of the driver is to seek the least hazardous route by navigating his or her vehicle through the overall field along its valley.

3.3.1 Roadways

Roadways can be represented as a gravity field in the longitudinal x direction so that vehicles are accelerated forward just as free objects fall to the ground. The gravity G_i acting on a driver–vehicle unit i can be expressed as $G_i = m_i \times g_i$, where m_i is the vehicle's mass and g_i is the acceleration of roadway gravity perceived by driver i (see Figure 3.3). Meanwhile, this gravity is counteracted by a resistance R_i perceived by the driver due to his or her willingness

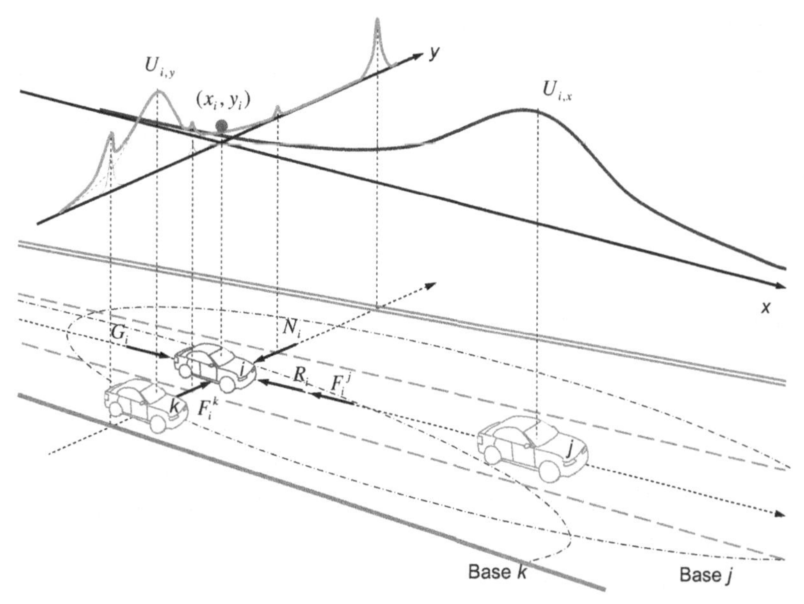

Figure 3.3: A Perceived Field.

to observe traffic rules (e.g. speed limit). As such, the net force $m_i\ddot{x}_i = G_i - R_i$ explains the driver's unsatisfied desire for mobility, which vanishes when her desired speed v_i is achieved. Note that drivers' acceleration efforts are related to their characteristics Θ, vehicle properties Ω, and roadway conditions Υ. For example, an aggressive driver typically accelerates faster than a normal driver. As such, g_i can be modeled as a random variable whose distribution is a function of (Θ,Ω, Υ). The specific form of this distribution is open and can be tailored depending on the nature of each application. Similar treatment applies to R_i. In the lateral y direction, cross-section elements (e.g. lane lines, edge lines, and center lines) are perceived by the driver as a roadway potential field u_i^R. When the vehicle deviates from its lane, the vehicle is subject to a correction force N_i, which can be interpreted as the stress on the driver keeping to his or her lane, as illustrated in Figures 3.2 and 3.3. The effect of such a force is to push the vehicle back to the center of the current lane. Such a force can be derived from the roadway field as $N_i = \partial u_i^R / \partial y_i$.

3.3.2 Vehicles

Vehicles can each be represented as a potential field. Figure 3.3 illustrates two such fields perceived by driver i: one for vehicle j and the other for vehicle k, both of which are shown with hills in the air and associated bases on the ground. Vehicle i interacts with j and k at a distance mediated by their associated fields, e.g. vehicle j slows down i by a repelling force F_i^j, while k motivates i to shy away by another force F_i^k, where F_i^j and F_i^k can each be derived from their corresponding fields. Note that driver characteristics and the connected vehicle effects can be introduced into the formulation of these fields. More specifically, the shapes and sizes of such fields may vary when perceived by different drivers, or even by the same driver, at different running conditions and with different situational awareness. For example, an aggressive driver is willing to accept shorter following distances and, hence, his or her perceived field covers a smaller base; a driver assisted by connected vehicle technology knows exactly the condition (e.g. location and speed) of his or her neighbor and can potentially respond quickly, so the perceived field may be smaller than that of a driver without such assistance.

3.3.3 Traffic Control Devices

A red light can be represented as a potential field that appears periodically at a fixed location. When it appears, an approaching vehicle will decelerate to a stop. When the signal turns green, its field disappears and the vehicle will accelerate by the roadway gravity. A similar technique applies to stop and yield signs with modifications accordingly. Again, the functional form of such fields is open and can be tailored to specific applications. The representation of pavement markings such as lane lines, center lines, and road edges have been discussed above in representing roadways.

3.3.4 Driver's Responsiveness

The above forces may or may not take effect on the subject driver depending on his or her responsiveness, γ. Consequently, a force that actually acts on the driver $\tilde{F}_i$ is the product of his or her responsiveness γ and the force that he or she might have perceived if the driver had paid full attention to it, F_i, i.e. $\tilde{F}_i = F_i \times \gamma$. It is recognized that the driver's responsiveness to the surroundings vary with his or her viewing angle $\alpha \in [-\pi, \pi]$ and scanning frequency ν, as seen in Figure 3.4. For example, the front area typically receives most attention, side areas are noted by the driver's fair or peripheral vision, and the rear area is only scanned occasionally. As such, if one chooses $\gamma(0) = 1$ and $\gamma(\pi) = 0$, the driver responds to F_i in full if it comes from a leading vehicle (i.e. $\alpha = 0$) and the driver ignores F_i when it comes from a trailing vehicle (i.e. $\alpha = \pi$) respectively.

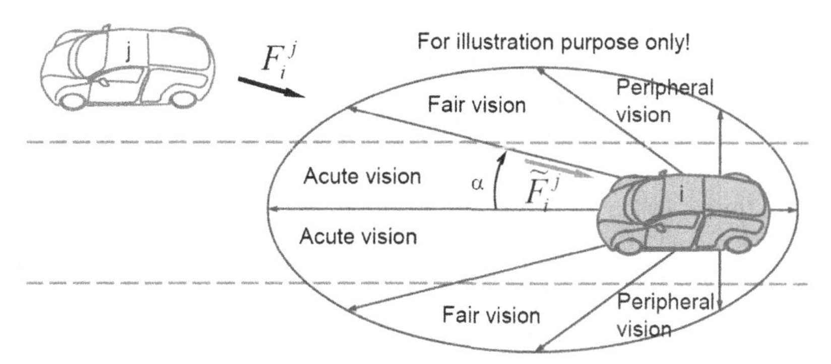

Figure 3.4: Driver's Responsiveness.

3.3.5 Operational Control

A driver's strategy of moving on roadways is to achieve mobility and safety (gains) while avoiding collisions and violation of traffic rules (losses). Such a strategy can be represented using an overall potential field U_i that consists of component fields such as those due to moving bodies U_i^{B}, roadways U_i^{R}, and traffic control devices U_i^{C}, i.e.

$$U_i = U_i^{\mathrm{B}} + U_i^{\mathrm{R}} + U_i^{\mathrm{C}}. \tag{3.1}$$

If U_i is viewed as a mountain range whose elevation denotes the risk of losses, the driver's strategy is to navigate through the mountain range along its valley, i.e. the least stressful route. For example, Figure 3.3 illustrates two sections of such a field. Perceived by driver i, the longitudinal x section of the field, $U_{i,x}$, is dominated by vehicle j since it is the only neighboring vehicle in the center lane. Vehicle i is represented as a ball that rides on the tail of curve $U_{i,x}$ since the vehicle is within vehicle j's field. Therefore, vehicle i is subject to a repelling force F_i^j, which is derived from $U_{i,x}$ as

$$F_i^j = -\frac{\partial U_{i,x}}{\partial x}. \tag{3.2}$$

The effect of F_i^j is to push vehicle i back to keep safe distance. By incorporating the driver's unsatisfied desire for mobility $(G_i - R_i)$, the net force perceived by driver i in the x direction can be determined as

$$m_i\ddot{x}_i = \sum F_{i,x} = G_i - R_i - F_i^j = (m_i g_i - R_i) + \frac{\partial U_{i,x}}{\partial x}. \tag{3.3}$$

The section of U_i in the lateral y direction, $U_{i,y}$ (bold curve), is the sum of two components: the cross-section of the field due to vehicle k (dashed curve) and that due to the roadway field (dotted curve). The former results in a repelling force F_i^k, which makes vehicle i shy away from vehicle k and the latter generates a correction force N_i should vehicle i depart its lane center. Therefore, the net effect can be expressed as

$$m_i\ddot{y}_i = \sum F_{i,y} = F_i^k - N_i = -\frac{\partial U_{i,x}}{\partial x}. \tag{3.4}$$

By incorporating time t, driver i's perception-reaction time τ_i, and driver i's directional response γ_i, Eqs (3.3) and (3.4) can be expressed as

$$m_i\ddot{x}_i(t+\tau_i) = \sum \tilde{F}_{i,x}(t) = \gamma_i^0[G_i(t) - R_i(t)] + \gamma\left(\alpha_i^j\right)\frac{\partial U_{i,x}}{\partial x} \tag{3.5a}$$

$$m_i\ddot{y}_i(t+\tau_i) = \sum \tilde{F}_{i,y}(t) = -\gamma\left(\alpha_i^k\right)\frac{\partial U_{i,y}}{\partial y}. \tag{3.5b}$$

where $\gamma_i^0 \in [0, 1]$ represents driver i's attention to his or her unsatisfied desire for mobility (typically $\gamma_i^0 = 1$), α_i^j and α_i^k are viewing angles that are also functions of time. The above system of equations summarizes Field Theory in generic terms and constitutes the basic law that governs a driver's operational control (i.e. steering and acceleration/deceleration) of his or her vehicle on roadways.

3.3.6 Strategic Control

Compared with operational control, strategic control deals with a larger scope in time and space, e.g. choosing a route among multiple options and the decision to change lane in order to avoid a slow driver or use an exit. Common in modeling lane change and route choice is a discrete choice decision that can be statistically related to the driver's characteristics Θ and his or her situational awareness Ξ (e.g. the attributes of each available option). For example, driver i arriving at a road junction is faced with a set of alternative routes $\{1, 2, \ldots, n\}$. The choice of an appropriate route is influenced by: (1) her characteristics Θ_i, which includes goals (destination), speed preference, and sensitivity to cost/toll, and (2) attributes of each available option $\Xi_n = \{\xi_n^1, \xi_n^2, \ldots\}$ such as traffic speed and cost/toll. A discrete choice model then estimates the probability that driver i chooses a particular alternative n: $P_{in} = \Psi(\Theta_i, \Xi_n, \Xi_m \forall m \neq n)$, where Ξ_m is a vector of attributes of an alternative unknown to the driver $(n \neq m)$ and $S = n + m$ is the total number of options.

3.3.7 Effects of Connected Vehicle Technology

As indicated above, driver assistance enabled by connected vehicle technology can potentially enhance driver capabilities, such as driver characteristics Θ and situational awareness Ξ. Therefore, Θ and Ξ are functions of information I that is brought about by V-V and V-I DSRC, i.e. $\Theta = \Theta(I) = \{\tau(I), \gamma(I), \ldots\}$ and $\Xi = \Xi(I)$. For example, supported by human factor studies [2], the perception-reaction time of a regular driver (without driving assistance), a driver with driving assistance, and a vehicle automated by connected vehicle technology can be represented as samples from different distributions. The differences between regular, assisted, and automated vehicles can be reflected in these underlying distributions. Figure 3.5 illustrates another example. A driver from Boston (point C) to Amberst (D) is traveling on default route I-90. Without connected vehicle technology, the driver has to follow the default route and the network appears dark since there is no information. If traffic information is enabled by connected vehicle technology at some locations, these locations become bright/visible. For instance, if an accident is reported on I-90, the driver may choose an alternative route (Rt. 9), which would have been unknown without connected vehicle technology (such as Rt. 2).

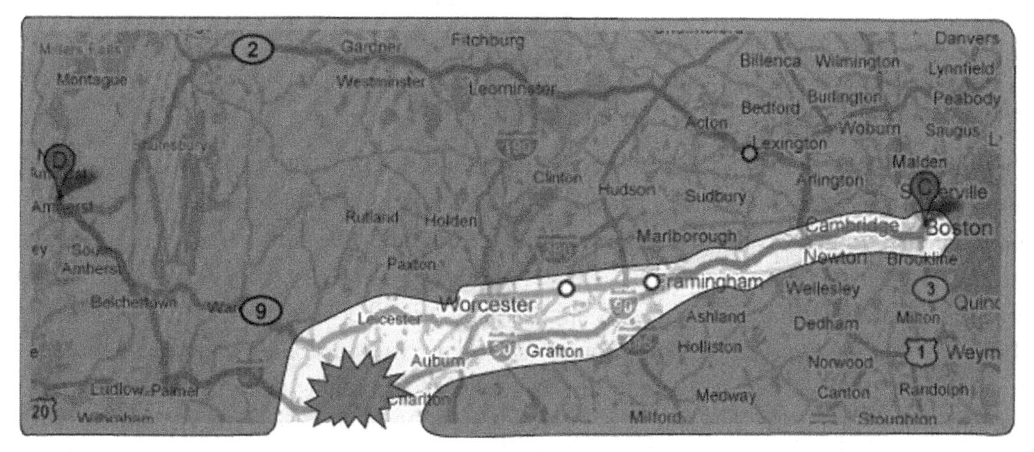

Figure 3.5: Alternative Route.

3.4 Applications of Field Theory

A few potential applications of Field Theory are presented below. Because of space limitations, the discussion only points out application directions with links for further details, if any exist.

3.4.1 Directions of Potential Applications

Connected Vehicle Simulation Virtual Environment

It was noted at the beginning of this chapter that Field Theory was formulated as a theoretical foundation for modeling connected vehicles to assist the design and test of connected vehicle applications. Development of a simulation tool from Field Theory, such as the virtual environment mentioned previously, takes a long time and is not further discussed here.

VANET Simulator

Another application, perhaps a more readily implementable one, is to integrate Field Theory and Network Simulator NS-2 [3]. NS-2 is a discrete event simulator that provides substantial support for simulation of TCP, routing, and multicast protocols over wired and wireless networks with the possibility of using mobile nodes. The motion of these nodes may be tracked internally within the simulation or imported from an external trace file. In either case, Field Theory and its special, simplified cases [1,4,5] can be readily integrated with NS-2. It is worth noting that, outside of the transportation community, many mobility models have been proposed, including SUMO [6], STRAW [7,8], and Eichler [9], containing functions to eliminate collisions [10], VanetMobiSim [11], etc. More literature on this subject can be found in Ref. [12]. Though these mobility models have served their purpose well, realistic traffic flow representation such as Field Theory and its special cases not only transform the validity of VANET simulation but also offer the potential to capture the interdependent dynamics of traffic networks and VANET, as illustrated in Figure 3.6.

Intelligent Vehicle Design

As a third example, Field Theory can be applied to assist the design of intelligent vehicles. For example, Rossetter and Gerdes [13,14] applied a potential field framework for active vehicle lane-keeping assistance. In their approach, the major concern is to correct vehicles from deviating from their lane centers. Hence, roadway potential is considered as a hazard that is translated to virtual control forces applied on the subject vehicle, which are illustrated in Figure 3.7 as the two short springs connected to lane lines. Such an approach can be extended to incorporate forces due not only to deviation from lane centers/road edges/center lines, but also neighboring vehicles, e.g. the two long springs in Figure 3.7. Therefore, Field Theory provides a feasible framework to integrate intelligent vehicle systems such as lane

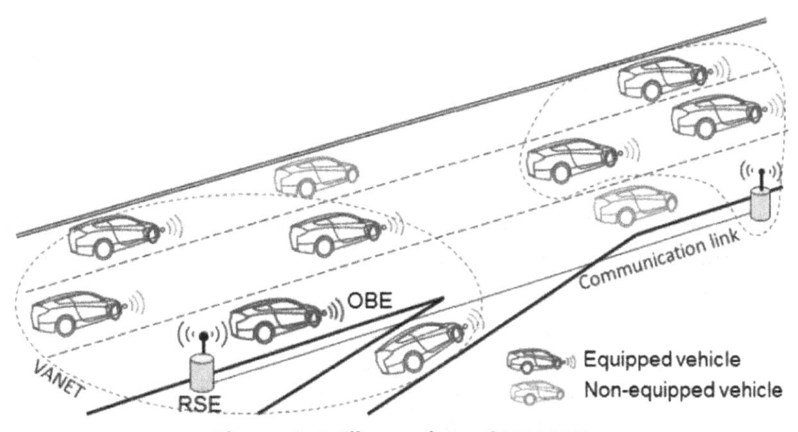

Figure 3.6: Illustration of VANET.

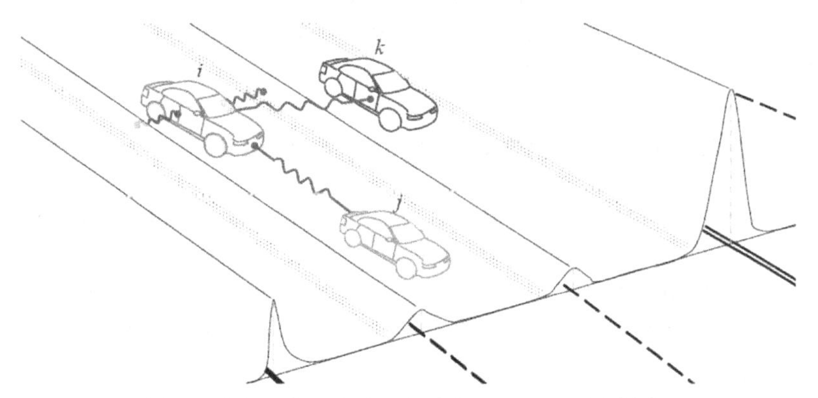

Figure 3.7: Virtual Control Forces on a Vehicle.

keeping, adaptive cruise control, blind spot warning, and cooperative collision avoidance seamlessly into an integrated driving assistance system or automatic driving system.

Analysis of Traffic Flow with Connected Vehicles

As a fourth example, Field Theory can be applied to analyze traffic flow with connected vehicles. Questions that are of major concern to policymakers, system planners, and practitioners are the following:

- What is the infrastructure (equipped vehicles and roadside equipment) needed for success?
- What is the degree of market penetration (i.e. percentage of vehicles equipped with connected vehicle technology) required for effectiveness?

These questions are very difficult to answer mainly because large-scale field experiments are not feasible to deploy and a simulated virtual environment mentioned above has yet to be developed. Nevertheless, it is possible to have a quick peak on highway capacity benefit gained from connected vehicle technology (CVT), though this is a rough, analytical approach. A concrete example of such an analysis is provided below and the reader is referred to Ref. [15] for more information.

General Traffic Flow Modeling and Simulation

As a fifth example, Field Theory can be applied to model and analyze the operation of general traffic flow, which may or may not involve connected vehicles. Issues that are of great interest to both travelers and traffic system operators are:

- When and where will congestion be?
- If there is congestion, what is the impact of the congestion (i.e. how long will it last, how far back it will spill, when will it dissipate)?

Such questions can be addressed in multiple ways using traffic flow models derived from Field Theory. A concrete example of such an analysis is provided below and the reader is referred to Ref. [16] for more information.

3.4.2 Analyzing the Impact of CVT on Highway Capacity

The objective of this example is to provide a rough estimate of the impact of CVT on highway capacity and how the result changes as market penetration varies. The analysis involves two building blocks: (1) incorporating the effects of connected vehicle technology into driving behavior modeling, based on which (2) estimation of highway capacity by conducting a probabilistic analysis.

Driving Behavior Modeling

Driving behavior modeling captures the interaction between vehicles, particularly the response of a vehicle to the vehicle in front, i.e. the F_i^j term in Figure 3.3. This term can be simplified as a safety rule: at any moment, a vehicle should leave enough room in front of it in order to be able to stop safely behind its leading vehicle in the event that the leading vehicle applies emergency braking. Figure 3.8 shows two vehicles following each other. The

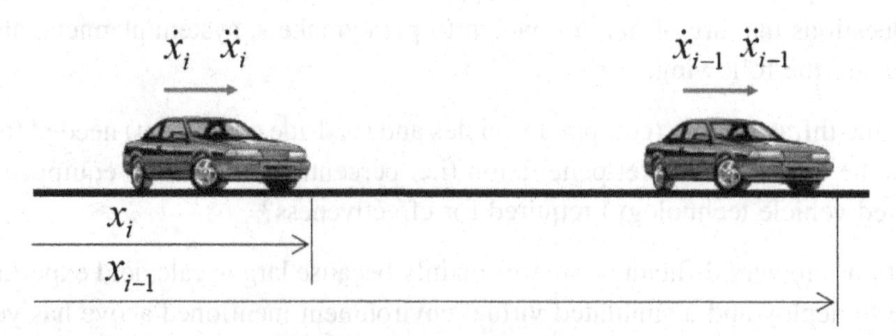

Figure 3.8: Vehicles in Car Following.

leading vehicle with ID $i - 1$ is at position x_{i-1} with speed $\dot{x}_{i-1}$ and acceleration $\ddot{x}_{i-1}$. The subject vehicle i is at position x_i with speed $\dot{x}_i$ and acceleration $\ddot{x}_i$. The minimum safe distance should allow the subject vehicle to stop behind the leading vehicle after a perception-reaction time τ_i and a deceleration process at a tolerable level $b_i = \ddot{x}_i < 0$. After some math, the safety rule can be formulated as

$$S_i(t) = x_{i-1}(t) - x_i(t) \geq l_{i-1} + \dot{x}_i\tau_i - \frac{\dot{x}_i^2(t)}{2b_i} + \frac{\dot{x}_{i-1}^2(t)}{2B_{i-1}}, \tag{3.6}$$

where S_i is the spacing between the two vehicles, B_{i-1} is the maximum deceleration applied by the leading vehicle, and l_{i-1} is the length of the leading vehicle. Under equilibrium conditions, vehicles tend to behave uniformly and thus lose their identities. After suppressing time t, the spacing becomes

$$S \geq l + \dot{x}\tau - \frac{\dot{x}^2}{2b} + \frac{\dot{x}^2}{2B} = \left(\frac{1}{2B} - \frac{1}{2b}\right)\dot{x}^2 + \tau\dot{x} + l. \tag{3.7}$$

To ensure safety, an additional delay θ to τ offers extra protection for the subject vehicle, so the above inequality becomes

$$S = l + \dot{x}(\tau + \theta) - \frac{\dot{x}^2}{2b} + \frac{\dot{x}^2}{2B} = \left(\frac{1}{2B} - \frac{1}{2b}\right)\dot{x}^2 + (\tau + \theta)\dot{x} + l. \tag{3.8}$$

Thus, the safe car-following distance, or equivalently the safe spacing, is explicitly expressed as a function of speed $v \equiv \dot{x}$ (under equilibrium conditions, it is also the traffic speed) with parameters τ, θ, B, b, and l. Among all the parameters, τ and θ characterize the behavior of drivers and are independent of speed v and spacing S, B, b, and l are vehicle properties and can be assumed as constants. Since we also know that density k is related to spacing S:

$$k = \frac{1}{S}, \tag{3.9}$$

the flow q is obtained by substituting k and v into the fundamental relation:

$$q = kv = \frac{v}{Gv^2 + \tau' v + l}, \tag{3.10}$$

where $\tau' = \tau + \theta$ and $G = 1/2B - 1/2b$. In this relation, v can be viewed as the primary input. v and τ' are independent variables. The maximum attainable q is of interest. To find the maximum q (denoted q_m), we solve the equation:

$$\left.\frac{dq}{dv}\right|_{v_m} = -\left.\frac{G - \frac{l}{v^2}}{\left(Gv + \tau' + \frac{l}{v}\right)^2}\right|_{v_m} = 0, \tag{3.11}$$

we get the root:

$$v_m = \sqrt{\frac{l}{G}} \tag{3.12}$$

and correspondingly:

$$q_m = \frac{1}{2\sqrt{Gl} + \tau'}. \tag{3.13}$$

To verify that q_m is indeed a maximum as v varies, one may simply check the second derivative of q at v_m. It turns out that this is true.

Modeling Effect of CVT

The effects of connected vehicle technology are attributed to the change in the distribution of drivers' perception-reaction time. Though there might be other effects, they are not modeled here to keep the analysis feasible and tractable yet able to capture the major effect of connected vehicle technology. The analysis assumes three types of driving modes: non-CVT, CVT assisted, and CVT automated. In the non-CVT mode, drivers operate their vehicles without any assistance from connected vehicle technology, just as a regular driver does. In the CVT-assisted mode, drivers receive connected vehicle technology assistance such as driver advisories (e.g. downstream congestion) and safety warnings (e.g. emergency brake), but these drivers still assume full control of their vehicles. The CVT-automated mode means that a vehicle is operated by CVT-enabled automatic driving features; however, the driver may break the loop and take over at any time as the need arises. The effect of CVT on driver perception-reaction time can be modeled using different underlying distributions (see Figure 3.9).

Probabilistic Analysis

Equation (3.13) above allows the calculation of capacity q_m given perception-reaction time τ if it is uniform across the entire driver population. Such a case is simple but unrealistic. A step forward would be to assume uniform perception-reaction time under each driving mode. Combined with market penetration (analogous to the probability of each driving mode), these perception-reaction times can be used to estimate the average (i.e. mathematical expectation) capacity as the result of the entire driver population. A more realistic scenario would be to assume different distributions of perception-reaction time under different driving modes. For example, we denote by $f_{\text{no}}(t)$ the probability density of perception-reaction time of drivers under non-CVT with mean τ_{no} and variance $\text{Var}(\tau_{\text{no}})$. Similarly, the probability density of the CVT-assisted mode is $f_{\text{as}}(t)$ with mean τ_{as} and variance $\text{Var}(\tau_{\text{as}})$; the probability density of the CVT-automated mode is $f_{\text{au}}(t)$ with mean τ_{au} and variance $\text{Var}(\tau_{\text{au}})$. In addition, market

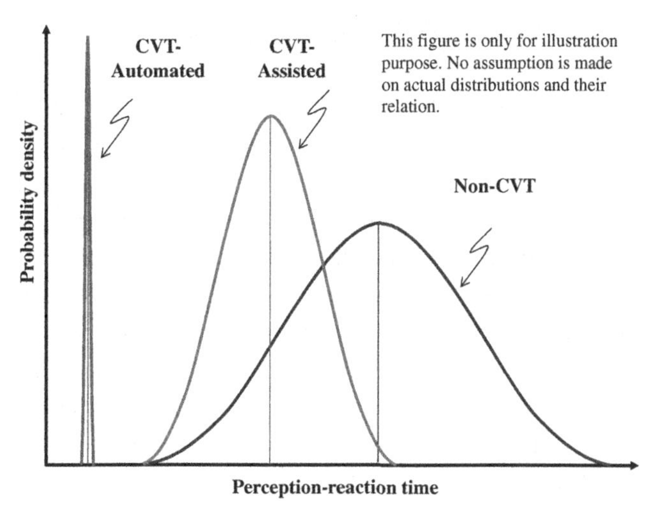

Figure 3.9: Virtual Control Forces on a Vehicle.

penetration rates of road vehicles operating in non-CVT, CVT-assisted, and CVT-automated modes are denoted as p_{no}, p_{as}, and p_{au} respectively. They satisfy the following relationships: $0 \le p_{no}, p_{as}, p_{au} \le 1$ and $p_{no} + p_{as} + p_{au} = 1$. Therefore, the perception-reaction time of an individual driver i, τ_i, is a random variable that can be modeled by drawing first from the percentage/probability of market penetration to determine which driving mode this driver uses and then from the distribution of perception-reaction time of that particular mode. Consequently, the estimation of capacity is to compute the mathematical expectation of q_m based on these underlying distributions and their market penetration rates.

An Illustrative Example

An illustrative example is provided to answer the question regarding degree of market penetration required for effectiveness. This example consists of four cases and in each case the ratio p_{au}/p_{as} is assumed to be constant. In addition, we define the relative change in capacity as

$$\begin{aligned} r(p_{au}/p_{as}, p_{no}) &= \frac{q_m(p_{au}, p_{as}, p_{no})}{q_m(0,0,1)} \\ &= \frac{q_m\left((1-p_{no})\dfrac{p_{au}/p_{as}}{p_{au}/p_{as}+1}, (1-p_{no})\dfrac{1}{p_{au}/p_{as}+1}, p_{no}\right)}{q_m(0,0,1)}, \end{aligned} \tag{3.14}$$

where $q_m(, ,)$ is the capacity corresponding to market penetration (p_{au}, p_{as}, p_{no}), and the second equality is $p_{au} + p_{as} + p_{no} = 1$. This formula can be interpreted as the ratio of increased capacity over the original capacity (i.e. $p_{no} = 100\%$). By employing this definition. we express the change of capacity in relative terms. We obtain the values of r in four cases. i.e. when $p_{au}/p_{as} = 0.1$, 1, 10, and 100. The results are shown in Figure 3.10. It is found that the increase of capacity ranges between 20% and 50% when connected vehicle technology is fully deployed (i.e. $p_{no} = 0$), with the former case corresponding to $p_{au}/p_{as} = 0.1$ and the latter case $p_{au}/p_{as} = 100$. The result seems to suggest that the change of the p_{au}/p_{as} ratio from 1 to 10 has a much stronger effect than that from 10 to 100. A plausible interpretation is that, with a high percentage of CVT-assisted vehicles in the traffic, drivers have more chances to negotiate and hence more room for improvement. As the traffic is dominated by CVT-automated vehicles, they move like a train whose already optimized performance allows little room for improvement. Note that the above example is only a rough estimate under some assumptions and simplifications. Nevertheless, the example does indicate that the benefit of employing connected vehicle technology could be quite significant even when the market penetration of CVT-automated vehicles is small (given full CVT deployment). As more accurate information regarding the involved parameters becomes available, the estimate can be fine-tuned with the expectation of more accurate results. The outcomes can be used to help make the decision on connected vehicle technology deployment in the future.

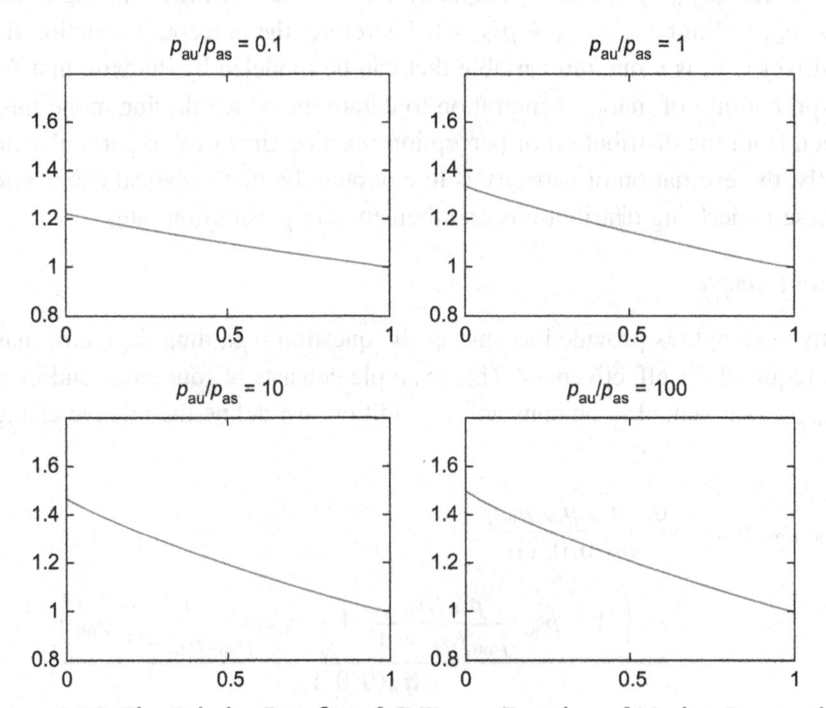

Figure 3.10: The Relative Benefits of CVT as a Function of Market Penetration of CVT in Different Cases.

x-axis: market penetration of non-assisted vehicles; *y*-axis: ratio of increased capacity to the original capacity.

3.4.3 General Traffic Flow Modeling and Analysis

In addition to connected vehicles, Field Theory can be applied to the modeling of traffic consisting of regular vehicles, i.e. those without inter-vehicular communication. Such modeling can be a useful tool to understand traffic phenomena and analyze traffic operations.

The Longitudinal Control Model — A Special Case of Field Theory

If the functional forms of the terms in Eq. (3.5a) are carefully chosen (mainly by experimenting with empirical data), a special case called the Longitudinal Control Model (LCM) can be explicitly derived from Eq. (3.5a) as

$$\ddot{x}_i(t+\tau_i) = A_i\left[1 - \left(\frac{\dot{x}_i(t)}{v_i}\right) - \mathrm{e}^{1-\frac{s_{ij}(t)}{s_{ij}^*(t)}}\right], \tag{3.15}$$

where $\ddot{x}_i(t+\tau_i)$ is the operational control (acceleration or deceleration) of driver i executed after a perception-reaction time τ_i from the current moment t. A_i is the maximum acceleration desired by driver i from a standing start, $\dot{x}_i$ is vehicle i's speed, v_i driver i's desired speed, s_{ij} is the actual spacing between vehicle i and its leading vehicle j, and s_{ij}^* is the desired value of s_{ij}.

No further motivation for this special case is provided other than the following claims: (1) it takes a simple functional form that involves physically meaningful parameters but not arbitrary coefficients; (2) it makes physical and empirical sense; (3) it provides a sound microscopic basis to aggregated behavior, i.e. traffic stream modeling; and (4) it is simple and easy to apply.

The determination of desired spacing $s_{ij}^*(t)$ admits safety rules. Basically, any safety rule that relates spacing to driver's speed choice can be inserted here. Of particular interest is the desired spacing presented earlier that allows vehicle i to stop behind its leading vehicle j after a perception-reaction time τ_i and a deceleration process (at rate $b_i > 0$) should the leading vehicle j apply emergency braking (at rate $B_j > 0$). After some math, the desired spacing can be determined as

$$s_{ij}^*(t) = \frac{\dot{x}_i^2(t)}{2b_i} - \frac{\dot{x}_j^2(t)}{2B_j} + \dot{x}_i\tau_i + l_j, \tag{3.16}$$

where l_j is vehicle j's effective length (i.e. actual vehicle length plus some buffer spaces at both ends). Note that the term $\dot{x}_i^2(t)/2b_i - \dot{x}_j^2(t)/2B_j$ represents the degree of aggressiveness that driver i chooses. For example, when the two vehicles travel at the same speed, this term becomes $\gamma_i\dot{x}_i^2$ with

$$\gamma_i = \frac{1}{2}\left(\frac{1}{b_i} - \frac{1}{B_j}\right), \tag{3.17}$$

where B_j represents driver i's estimate of the emergency deceleration that is most likely to be applied by driver j, while b_i can be interpreted as the deceleration tolerable by driver i. Attention should be drawn to the possibility that b_i might be greater than B_j in magnitude, which translates to the willingness (or aggressive characteristic) of driver i to take the risk of tailgating.

The above constitutes the microscopic version of the Longitudinal Control Model, based on which the macroscopic version is derived.

Under steady-state conditions, vehicles in traffic behave uniformly and, thus, their identities can be dropped. Therefore, the microscopic LCM (Eqs (3.16) and (3.17)) can be aggregated to its macroscopic counterpart (traffic stream model):

$$v = v_{\mathrm{f}}\left[1 - \mathrm{e}^{1-\frac{k^*}{k}}\right], \tag{3.18}$$

where v is traffic space-mean speed, v_{f} free-flow speed, k traffic density, and k^* takes the following form:

$$k^* = \frac{1}{\gamma v^2 + \tau v + l}, \tag{3.19}$$

where γ denotes the aggressiveness that characterizes the driving population, τ average response time that characterizes the driving population, and l average effective vehicle length. Equivalently, the macroscopic LCM can be expressed as

$$k = \frac{1}{(\gamma v^2 + \tau v + l)\left[1 - \ln\left(1 - \frac{v}{v_{\mathrm{f}}}\right)\right]}. \tag{3.20}$$

Note that an earlier version of LCM was proposed in Refs [1,17] that does not explicitly consider the effect of drivers' aggressiveness. To make a distinction, the LCM by default refers to the LCM formulated herein (both microscopic and macroscopic forms), whereas the earlier version of the LCM will be referred to as LCM without aggressiveness.

Analyzing Traffic Flow Using the Longitudinal Control Model

Since the LCM takes a simple mathematical form that involves physically meaningful parameters, the model can be easily applied to help investigate traffic phenomena at both microscopic and macroscopic levels. For illustrative purposes, a concrete example is provided below, in which a moving bottleneck is created by a sluggish truck. Microscopic modeling allows the LCM to generate profiles of vehicle motion so that the cause and effect of vehicles slowing down or speeding up can be analyzed in exhaustive detail; macroscopic modeling may employ the LCM to generate fundamental diagrams that help determine shock paths and develop graphical solutions. Since the LCM is consistent at the microscopic and macroscopic levels, the two sets of solutions not only agree but also complement each other.

A freeway segment contains an on-ramp (which is located 2000 m away from an arbitrary reference point denoting the upstream end of the freeway) followed by an off-ramp 2000 m apart. The freeway was initially operating under condition A (flow 0.3333 veh/s or 1200 veh/h, density 0.1111 veh/m or 17.9 veh/mi, and speed 30 m/s or 67.1 mi/h). At 2:30 p.m., a slow truck enters the freeway traveling at a speed of 5.56 m/s, which forces the traffic to operate under condition B (flow 0.3782 veh/s or 1361 veh/h, density 0.0681 veh/m or 109.6 veh/mi, and speed 5.56 m/s or 12.4 mi/h). After a while, the truck turns off the freeway at the next exit. The impact on the traffic due to the slow truck is illustrated macroscopically and microscopically in the following subsections.

A fundamental diagram (Figure 3.11) is generated using the macroscopic LCM to characterize the freeway with the following parameters: free-flow speed $v_f = 30$ m/s, aggressiveness $\gamma = -0.028$ m/s^2, average response time $\tau = 1$ second, and effective vehicle length $l = 7.5$ m. In addition, the above-mentioned traffic flow conditions, free-flow condition O, and capacity condition C are tabulated in Table 3.1.

To illustrate the application of the LCM, the above problem is addressed in two ways: macroscopic graphical solution and microscopic simulation solution. The microscopic simulation is conducted in deterministic and random fashion.

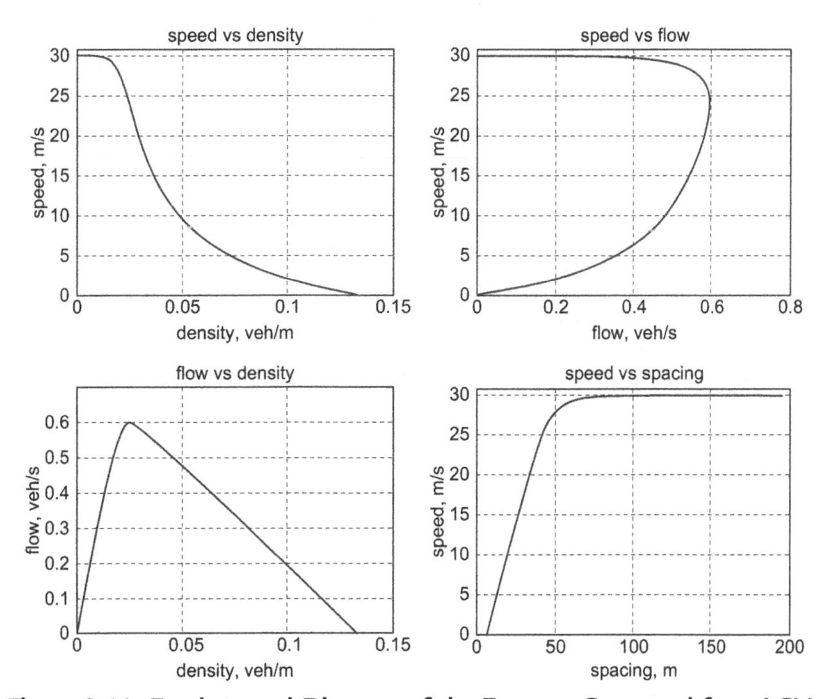

Figure 3.11: Fundamental Diagram of the Freeway Generated from LCM.

Table 3.1: Traffic Flow Conditions

Condition	Flow, q	Density, k	Speed, v
	veh/s (veh/h)	veh/m (veh/mi)	m/s (mi/h)
A	0.3333 (1200.0)	0.0111 (11.9)	30(67.1)
B	0.3782 (1361.6)	0.0681 (109.6)	5.56 (12.4)
C	0.5983 (2154.0)	0.0249 (40.1)	24.03 (53.7)
O	0 (0)	0 (0)	30 (67.1)

Macroscopic approach – graphical solution

The graphical solution to the problem involves finding shock paths that delineate time–space (t–x) regions of different traffic conditions. Figure 3.12 illustrates the time–space plane overlaid with the freeway on the right and a mini-version of the flow–density plot in the top left corner. The point when the slow truck enters the freeway (2:30 p.m.) roughly corresponds to P_1 ($t_1 = 65$, $x_1 = 2000$) on the time–space plane, while the point when the truck turns off

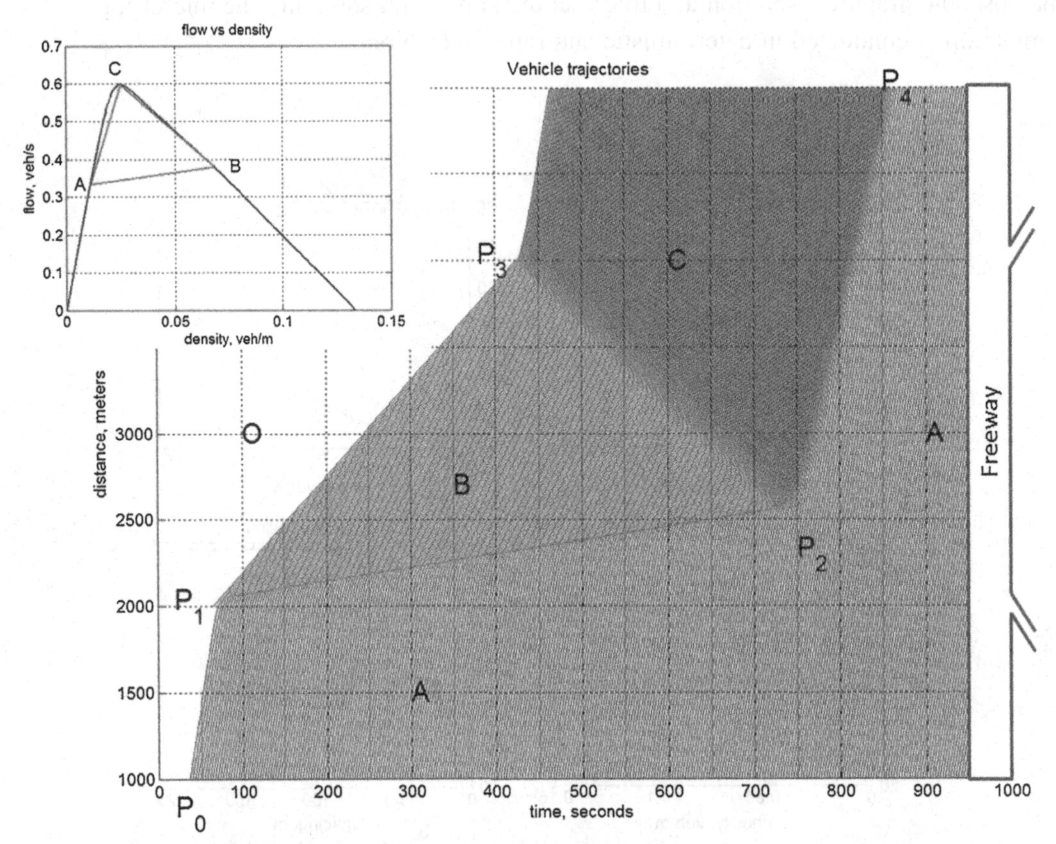

Figure 3.12: A Moving Bottleneck Due to a Slow Truck, Deterministic Simulation.

the freeway is roughly P_3 ($t_3 = 425$, $x_3 = 4000$). Therefore, constrained by the truck, the t–x region under P_1P_3 should contain traffic condition B. On the other hand, the t–x regions before the truck enters and before congestion (i.e. condition B) forms should have condition A. As such, there must be a shock path that delineates the two regions, and such a path should start at P_1 with a slope equal to shock wave speed U_{AB}, which can be determined according to the Rankine–Hugoniot jump condition [18,19]:

$$U_{AB} = \frac{q_B - q_A}{k_B - k_A} = \frac{0.3782 - 0.3333}{0.0681 - 0.0111} = 0.7877 \text{ m/s}. \tag{3.21}$$

Meanwhile, downstream of the off-ramp, congested traffic departs at capacity condition C, which corresponds to a t–x region that starts at P_3 and extends forward in time and space. Hence, a shock path forms between the region with condition C and the region with condition B. Such a shock path starts at P_3 and runs at a slope equal to shock wave speed U_{BC}:

$$U_{BC} = \frac{q_C - q_B}{k_C - k_B} = \frac{0.5983 - 0.3782}{0.0249 - 0.0681} = -5.0949 \text{ m/s}. \tag{3.22}$$

If the flow–density plot is properly scaled, one should be able to construct the above shock paths on the t–x plane. The two shock paths should eventually meet at point $P_2(t_2,x_2)$. Its location can be found by solving the following set of equations:

$$\begin{cases} x_2 - x_1 & = U_{AB} \times (t_2 - t_1) \\ x_2 - x_3 & = U_{BC} \times (t_2 - t_3) \\ (x_2 - x_1) + (x_3 - x_2) & = 2000. \end{cases} \tag{3.23}$$

After some math, P_2 is determined roughly at (716.8,2513.4). After the two shock paths P_1P_2 and P_3P_2 meet, they both terminate and a new shock path forms that delineates regions with conditions C and A. The slope of the shock path should be equal to shock speed U_{AC}:

$$U_{AC} = \frac{q_C - q_A}{k_C - k_A} = \frac{0.5983 - 0.3333}{0.0249 - 0.0111} = 19.2029 \text{ m/s}. \tag{3.24}$$

As such, the shock path can be constructed as P_2P_4. Finally, the blank area in the t–x plane denotes a region with no traffic, i.e. condition O.

Microscopic approach – deterministic simulation

In order to double check the LCM and to verify if its microscopic and macroscopic solutions agree reasonably with each other, the microscopic LCM is implemented in MATLAB, a computational software package. As a manageable starting point, the microscopic simulation is made deterministic with the following parameters: desired speed $v_i = 30$ m/s, maximum acceleration $A_i = 4$ m/s^2, emergency deceleration $B_i = 6$ m/s^2, tolerable deceleration $b_i = 9$ m/s^2, perception-reaction time $\tau_i = 1$ second, and effective vehicle length $l_i = 7.5$ m, where $i \in \{1, 2, 3, \ldots, n\}$ are unique vehicle identifiers. Vehicles arrive at the upstream end of the freeway at a rate of one vehicle every 3 seconds, which corresponds to a flow of $q = 1200$ veh/h. The simulation time increment is 1 second and simulation duration is 1000 seconds.

Figure 3.12 illustrates the simulation result in which vehicle trajectories are plotted on the t–x plane. The varying density of trajectories outlines a few regions with clearly visible boundaries. The motion or trajectory of the first vehicle is predetermined, while those of the remaining vehicles are determined by the LCM. The first vehicle enters the freeway at time $t = 65$ seconds after the simulation starts. This moment is calculated so that the second vehicle is about to arrive at the on-ramp at this particular moment. Hence, the second vehicle and vehicles thereafter have to adopt the speed of the truck, forming a congested region where traffic operates at condition B.

Upstream of this congested region B is a region where traffic arrives according to condition A. The interface of regions B and A, P_1P_2, denotes a shock path in which vehicles in fast platoon A catch up and join slow platoon B ahead. The situation continues and the queue keeps growing until the truck turns of the freeway at $t = 425$ seconds into the simulation. After that, vehicles at the head of the queue begin to accelerate according to the LCM, i.e. traffic begins to discharge at capacity condition C. Therefore, the front of the queue shrinks, leaving a shock path P_3P_2 that separates region C from region B. Since the queue front shrinks faster than the growth of the queue tail, the former eventually catches up with the latter at P_2, at which point both shock paths terminate, denoting the end of congestion. After the congestion disappears, the impact of the slow truck still remains because it leaves a capacity flow C in front followed by a lighter and faster flow with condition A. Hence the trace where faster vehicles in platoon A join platoon C denotes a new shock path P_2P_4.

Comparison of the macroscopic graphical solution and the microscopic deterministic simulation reveals that they agree with each other very well, though the microscopic simulation contains much more information about the motion of each individual vehicle and the temporal–spatial formation and dissipation of congestion.

Microscopic approach – random simulation

Since the microscopic approach allows the luxury of accounting for randomness in drivers and traffic flow, the following simulation may replicate the originally posed problem more realistically. The randomness of the above example is set up as follows, with the choice of distribution forms being rather arbitrary provided that they are convenient and reasonable:

- Traffic arrival follows a Poisson distribution, in which the headway between the arrival of consecutive vehicles is exponentially distributed with mean 3 seconds, i.e. $h_i \sim$ Exponential(3) s, which corresponds to a flow of 1200 veh/h.
- Desired speed follows a normal distribution: $v_i \sim N(30, 2)$ m/s.

- Maximum acceleration follows a triangular distribution: $A_i \sim$ Triangular(3,5,4) m/s^2.
- Emergency deceleration: $B_i \sim$ Triangular(5, 7, 6) m/s^2.
- Tolerable deceleration: $b_i \sim$ Triangular(8, 10, 9) m/s^2.
- Effective vehicle length: $l_i \sim$ Triangular(5.5, 9.5, 7.5) m.

The result of one random simulation run is illustrated in Figure 3.13, where the effect of randomness is clearly observable. Trajectories in region B seem to exhibit the least randomness because vehicles tend to behave uniformly under congestion. Trajectories in region C are somewhat random since the metering effect due to the congestion still remains. In contrast, region A appears to have the most randomness not only because of the Poisson arrival pattern but also the random characteristics of drivers. Consequently, the shock path between regions B and C, P_3P_2, remains almost unaltered, while there are some noticeable changes in shock path P_1P_2. The first is the roughness of the shock path; this is because vehicles in platoon A now join the tail of the queue in a random fashion. The second is that the path might not be linear. As a matter of fact, the initial part of the shock path has a slope roughly equal to U_{AB}, while the remaining part has a slightly steeper slope (due to less intense

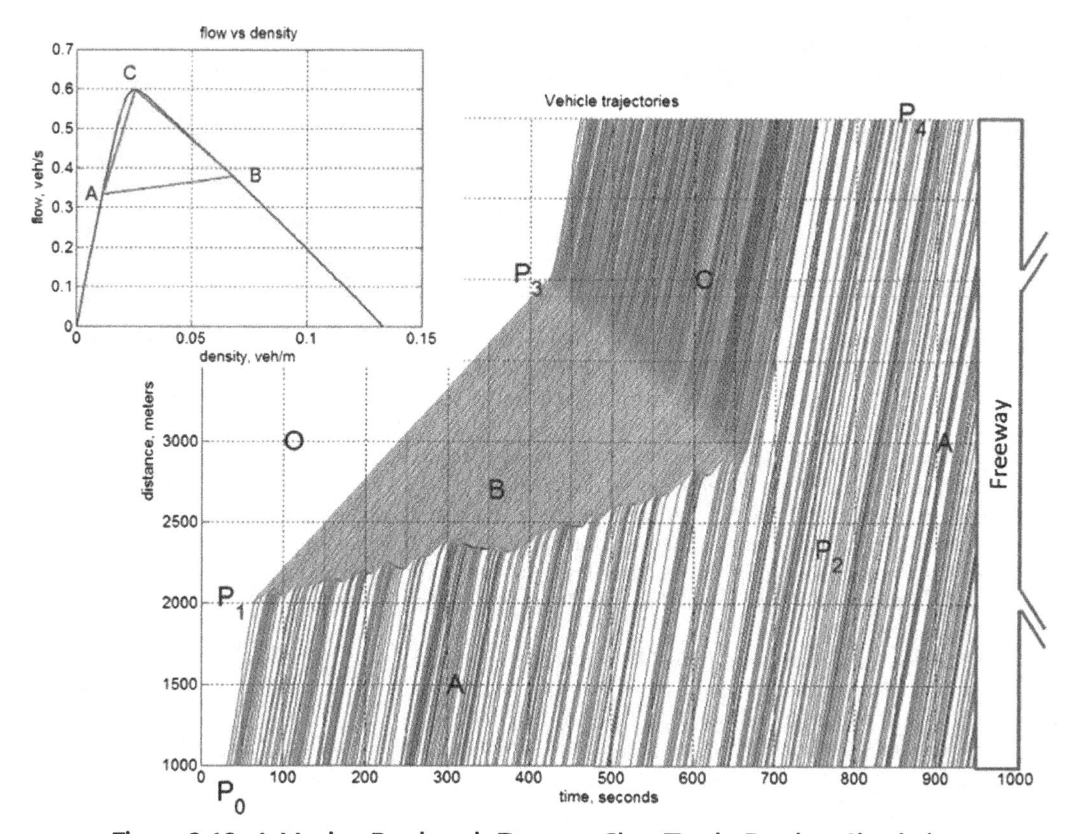

Figure 3.13: A Moving Bottleneck Due to a Slow Truck, Random Simulation.

arrival from upstream during this period), resulting in the termination of congestion earlier than the deterministic case (which is somewhere near P_2). This, in turn, causes the slope of the shock path between regions C and A to shift left.

Note that the slope of this shock path remains nearly the same since this scenario features a fast platoon that is caught up by an even faster platoon.

References

[1] Daiheng Ni, A Unified Perspective on Traffic Flow Theory, Part I: The Field Theory. Applied Mathematical Sciences Vol. 7 (no. 39) (2013) 1929–1946 HIKARI Ltd.

[2] M. Green, How long does it take to stop? Methodological analysis of driver perception-brake times, Transportation Human Factors 2 (3) (2000) 195–216.

[3] K. Fall, K. Varadhan, The VINT project. <http://www.isi.edulnsnarnlnsldocfmdex.html>.

[4] Daiheng Ni, A Unified Perspective on Traffic Flow Theory, Part II: The Unified Diagram. Applied Mathematical Sciences Vol. 7 (no. 40) (2013) 1947–1963 HIKARI Ltd.

[5] Daiheng Ni, Haizhong Wang, A Unified Perspective on Traffic Flow Theory, Part III: Validation and Benchmarking. Applied Mathematical Sciences Vol. 7 (no. 40) (2013) 1965–1982 HIKARI Ltd.

[6] SUMO. <http://sumo.sourceforge.net.>.

[7] D.R. Choffnes, F.E. Bustamante, An Integrated Mobility and Traffic Model for Vehicular Wireless Networks. Cologne, Germany, 2006, 69–78.

[8] A. Kumar Saha, D.B. Johnson, Modeling mobility for vehicular ad-hoc networks, in: VANET '04: Proceedings of the 1st ACM International Workshop on Vehicular Ad Hoc Networks, 2004, pp. 91–92. New York, ACM.

[9] S. Eichler, B. Ostermaier, C. Schroth, T. Kosch, Simulation of car-to-car messaging: Analyzing the impact on road traffic, in: 13th IEEE International Symposium on Modeling, Analysis, and Simulation of Computer and Telecommunication Systems, 2005, pp. 507–510.

[10] S. Krau, Microscopic Modeling of Traffic Flaw: Investigation of Collision Free Vehicle Dynamics. Ph.D. thesis, Universitat zu Koln, 1998.

[11] J. Harri, F. Filali, C. Bonnet, M. Fiore, Vanetmobisim: Generating Realistic Mobility Patterns for Vanets. Los Angeles, CA, 2006, pp. 96–97.

[12] C. Bonnet, J. Harri, F. Filali, Mobility models for vehicular ad hoc networks: A survey and taxonomy, Technical report. Research Report RR-06-168, 2007. <http://www.eurecom.frlutillpublidownload.fr.htm?id=1951>.

[13] J.C. Gerdes, E.J. Rossetter, U. Saur, Combining lanekeeping and vehicle following with hazard maps, Vehicle System Dynamics 36 (4–5) (2001) 391–411.

[14] E.J. Rossetter, A potential field framework for active vehicle lankeeping assistance. Ph.D. thesis, Department of Mechanical Engineering, Stanford University, 2003.

[15] S. Andrews, H. Wang, D. Ni, J. Li, A methodology to estimate capacity impact due to connected vehicle technology, International Journal of Vehicular Technology (2012). Article ID 502432.

[16] D. Ni, J.D. Leonard, G. Leiner, C. Jia, Vehicle longitudinal control and traffic stream modeling, in: Proceedings of the 91st TRB Annual Meeting (Paper #12-0156), Transportation Research Board, Washington, DC, 2012.

[17] D. Ni, Multiscale Modeling of Traffic Flow, Mathematica Aetema, vol. 1(01), Hilaris, pp. 2011, 27–54.

[18] H. Hugoniot, Propagation des Mouvements dans les Corps et specialement dans les Gaz Parfaits, Journal de l'Ecole Poly Technique 57 (3) (1887) (in French).

[19] W.J.M. Rankine, On the thermodynamic theory of waves of finite longitudinal disturbances, Philosophical Transactions of the Royal Society 160 (1870) 277–288.

CHAPTER 4

Guaranteed Collision Avoidance with Discrete Observations and Limited Actuation

Erick J. Rodríguez-Seda*, Dušan M. Stipanović†

United States Naval Academy, Annapolis, MD, USA †University of Illinois, Urbana, IL, USA

Chapter Outline

4.1 Introduction

Avoidance control is one of the foremost fundamental topics of study within autonomous navigation [1–5], intelligent transportation systems [6–8], and differential games research [9–11]. Maintaining a safe distance from other vehicles and obstacles is not only critical for the physical safety of the system and environment, but it is also important when determining solutions for coverage control problems [12,13]. Therefore, several researchers have studied and proposed control policies for the avoidance of obstacles and other vehicles.

Many of the proposed control policies assume that the agent can continuously observe the position of obstacles or assume obstacles to be quasi-static such that previous observations

Advances in Intelligent Vehicles. http://dx.doi.org/10.1016/B978-0-12-397199-9.00004-5

remain valid throughout the entire trajectory (for example, see relevant work on motion planning [14–16], where collision-free trajectories are generated ahead of time). There are, however, multiple scenarios for which continuous obstacle observation is impossible or impractical. For instance, common on-board localization sensors of vehicles, such as sonar radars and computer vision systems, may suffer from slow sampling rates, limited sensing ranges (e.g. short reach and angle of view), interruptions, and failures. Similarly, requiring sensors to continuously operate may extenuate the life of the vehicle's power source [17]. In these situations, it may be preferable to use a *sleep* and *wake-up* type of operation for the localization sensors [18], i.e. to allow sensors to operate at discrete intervals. Furthermore, the idea of observing obstacles at discrete intervals, especially obstacles that do not represent an imminent threat, is intuitive. That is, we may want the vehicle to pay less attention to an obstacle that is far away, and increase the awareness state as the obstacle becomes closer.

Avoidance strategies with discrete observations for pairs of dynamical systems have been studied in Refs [19–22] and for multi-agent systems in Ref. [23]. The instants of observation can be predetermined (i.e. open loop) or updated using the last observation (i.e. closed-loop feedback). Independently of the update approach, the observation rate must be carefully selected: if the obstacle is not observed sufficiently quickly, the vehicle may not be able to prevent a collision. It is worth mentioning that, for some cases, the use of discrete observations with properly chosen intervals can be shown to be equivalent to, or as effective as, the use of continuous observations [19].

In this chapter, we consider the interaction between an autonomous vehicle and an obstacle (which could be another vehicle) and present noncooperative and cooperative avoidance control strategies. The control methodology stems from the work in Ref. [24], where guaranteed avoidance control solutions with continuous observations are formulated taking into consideration errors and uncertainties in the obstacle localization process. Herein, we extend these avoidance strategies to allow the use of discrete observations with time-varying updates. The observation sampling rate is updated in a closed-loop fashion and increases monotonically as the obstacle comes closer to the vehicle. Furthermore, the control policies assume limited sensing range (i.e. reach) and consider the vehicle's dynamic and actuation constraints. That is, the avoidance control inputs are bounded, and therefore can be applied to vehicles with bounded acceleration and control input. The latter consideration is of particular importance since, in practice, autonomous vehicles (especially those with relatively large inertia) cannot stop or change their velocity instantaneously. In this case, the vehicle needs to start the collision avoidance process ahead of time and space to successfully prevent a collision [3,8,25–27]. In addition, we pair the avoidance control policies with a bounded trajectory tracking control and address the existence of unwanted local minima. We then provide a heuristic solution to avoid deadlocks and enforce the convergence of the agent to its desired trajectory.

4.2 Preliminaries

4.2.1 Definitions

First, we describe an interaction between a vehicle and a dynamic obstacle.[1] Let $\mathbf{x}(t) \in \Re^n$ and $\mathbf{y}(t) \in \Re^n$ denote the positions of the vehicle and obstacle respectively, and define $\mathbf{z}(t) = [\mathbf{x}(t)^{\mathrm{T}}, \mathbf{y}(t)^{\mathrm{T}}]$. We assume that the vehicle can detect the obstacle's position whenever the latter lies within its *Detection Region, D,* given by

$$D = \{\mathbf{z} : \mathbf{z} \in \Re^{2n}, \|\mathbf{x} - \mathbf{y}\| \leq R\},$$

where $R > 0$ denotes the vehicle's detection radius. That is, D delimits the area in which the obstacle can be sensed (or observed) by the vehicle. Similarly, we define the vehicle's *Avoidance Region*, Ω, as the zone that the obstacle should not trespass at any given time. Mathematically,

$$\Omega = \{\mathbf{z} : \mathbf{z} \in \Re^{2n}, \|\mathbf{x} - \mathbf{y}\| < r\},$$

where $r \in (0,R)$ is the desired minimum separation distance between the agent and the obstacle. Therefore, any avoidance strategy implemented by the vehicle must prevent $\mathbf{x}(t)$ and $\mathbf{y}(t)$ from entering Ω.

Now, let us assume that the vehicle has limited actuation (i.e. control input and acceleration constraints). This implies that the vehicle cannot stop or achieve a desired velocity instantaneously. Thus, a control policy aimed at avoiding Ω needs to be implemented with enough anticipation, such that the vehicle has sufficient time to decelerate and prevent Ω. Accordingly, we define the vehicle's *Conflict Region, W,* as

$$W = \{\mathbf{z} : \mathbf{z} \in \Re^{2n}, r \leq \|\mathbf{x} - \mathbf{y}\| \leq \bar{r}\},$$

where $\bar{r} \in (r, R)$ is a lower bound on the distance that the vehicle can come from the obstacle that will still allow it to decelerate and avoid Ω. Thus, any avoidance strategy must take effect at least as soon as $\mathbf{x}(t)$ and $\mathbf{y}(t)$ enter W.

Figure 4.1 illustrates the Detection, Conflict, and Avoidance Regions.

4.2.2 System Dynamics

We consider a vehicle whose goal is to approach a desired position or time-varying trajectory in $\Re^n$ while avoiding a dynamic obstacle with unknown control intentions. The motion of the vehicle is governed by

[1] We will refer to the second agent as an obstacle whenever it does not implement an avoidance strategy and as a vehicle whenever it does. In the general case, we will refer to it as an agent.

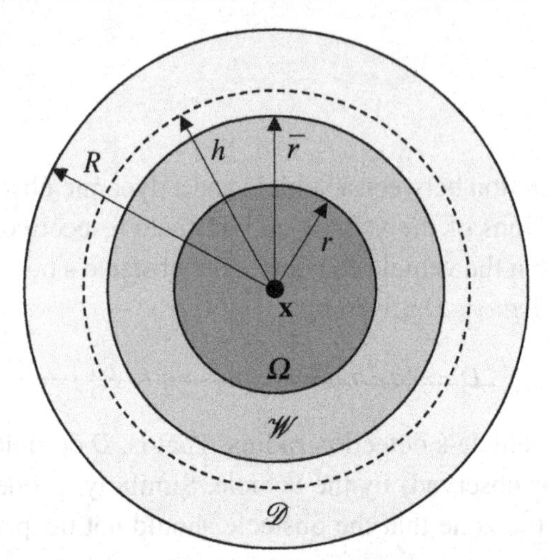

Figure 4.1: Detection (*D*), Conflict (*W*), and Avoidance (*Ω*) Regions.

$$\ddot{\mathbf{x}}(t) = \mathbf{u}(t), \tag{4.1}$$

where $\mathbf{u}(t) \in \Re^n$, the control input, is assumed to be radially bounded. That is, $\exists \mu > 0$ such that:

$$\|\mathbf{u}(t)\| \leq 2\mu, \qquad \forall t \geq 0. \tag{4.2}$$

The equations of motion for the obstacle, on the other hand, are considered to be unknown, but we assume its velocity to be bounded by some non-negative constant η_y, i.e. $\|\dot{\mathbf{y}}(t)\| \leq \eta_y \ \forall t \leq 0$. We further assume that the vehicle can observe or sense the position of the obstacle at time-varying sampling intervals $T_k \geq T_0$, where $T_0 > 0$ is the smallest sampling step. The sampling intervals, which will be designed later in Section 4.3, increase monotonically as a function of the distance between the vehicle and the obstacle. That is, the farther the agent is from the obstacle, the larger the time interval between observations might be. This is a natural choice for the sampling intervals as obstacles that are far away do not need to be sampled as frequently as obstacles that are closer.

4.3 Control Framework

The control objective is to drive the vehicle along a desired trajectory, $\mathbf{x}_d(t) \in \Re^n$, while avoiding a dynamic obstacle with the minimum possible rate of discrete observations. Accordingly, we would like to design a control policy $\mathbf{u}(t)$ such that:

- $\|\mathbf{x}(t) - \mathbf{x}_d(t)\| \to 0$ as $t \to \infty$.
- $[\mathbf{x}^T(t), \mathbf{y}^T(t)]^T \notin \Omega \ \forall t \geq 0$, and
- T_k is maximized $\forall k$, subject to collision avoidance constraints and parameters.

To achieve these objectives, we propose the use of the following control input:[2]

$$\mathbf{u} = \left(1 - \frac{\|\mathbf{u}_\mathrm{a}\|}{\mu}\right)\mathbf{u}_\mathrm{d} + \mathbf{u}_\mathrm{a} - \kappa\dot{\mathbf{x}}, \tag{4.3}$$

where $\mathbf{u}_\mathrm{d} \in \Re^n$ and $\mathbf{u}_\mathrm{a} \in \Re^n$ represent the objective and avoidance control respectively, and κ is a constant given by

$$\kappa = \frac{\mu}{\eta_x}, \qquad \text{for some } \eta_x > \eta_y.$$

As will be shown later in this section, η_x represents the vehicle's desired maximum velocity.

The objective control law is designed such that $\mathbf{x} \to \mathbf{x}_\mathrm{d}$ and is computed as

$$\mathbf{u}_\mathrm{d} = \ddot{\mathbf{x}}_\mathrm{d} + \kappa\dot{\mathbf{x}}_\mathrm{d} - \mathbf{g}(\mathbf{x} - \mathbf{x}_\mathrm{d}),\ \mathbf{g}(\mathbf{x} - \mathbf{x}_\mathrm{d}) = \begin{cases} \kappa_\mathrm{g}(\mathbf{x} - \mathbf{x}_\mathrm{d}), & \text{if } \|\mathbf{x} - \mathbf{x}_\mathrm{d}\| \le \mu_\mathrm{g}/\kappa_\mathrm{g} \\ \mu_\mathrm{g}\dfrac{\mathbf{x} - \mathbf{x}_\mathrm{d}}{\|\mathbf{x} - \mathbf{x}_\mathrm{d}\|}, & \text{otherwise,} \end{cases} \tag{4.4}$$

where $\kappa_\mathrm{g} > 0$ and $\mu_\mathrm{g} \in (0,\mu]$ are design parameters and $\ddot{\mathbf{x}}_\mathrm{d}$ and $\dot{\mathbf{x}}_\mathrm{d}$ are the desired acceleration and velocity respectively. Furthermore, the desired trajectory is assumed to satisfy the following constraint:

$$\|\ddot{\mathbf{x}}_\mathrm{d}(t) + \kappa\dot{\mathbf{x}}_\mathrm{d}(t)\| \le \mu - \mu_\mathrm{g}, \qquad \forall t \ge 0 \tag{4.5}$$

and therefore $\|\mathbf{u}_\mathrm{d}(t)\| \le \|\ddot{\mathbf{x}}_\mathrm{d}(t) + \kappa\dot{\mathbf{x}}_\mathrm{d}(t)\| + \|\mathbf{g}(\mathbf{x}(t) - \mathbf{x}_\mathrm{d}(t))\| \le \mu \forall t \ge 0$.

The avoidance control, on the other hand, is designed to guarantee a safe distance between the agent and the obstacle at all times. It is computed according to:

$$\mathbf{u}_\mathrm{a} = -\frac{\partial V_\mathrm{a}(\mathbf{x}, \mathbf{y}_k)^\mathrm{T}}{\partial \mathbf{x}}, \tag{4.6}$$

where $\mathbf{y}_k \in \Re^n$ represents the observations on the obstacle's position and V_a, called the avoidance function [5,24,28], is given by

$$V_\mathrm{a} = \begin{cases} \Gamma\left(\min\left\{0, \dfrac{\|\mathbf{x} - \mathbf{y}_k\|^2 - R^2}{\|\mathbf{x} - \mathbf{y}_k\|^2 - r^2}\right\}\right)^2, & \text{if } \|\mathbf{x} - \mathbf{y}_k\| \ge h \\ -\mu\|\mathbf{x} - \mathbf{y}_k\| + C, & \text{otherwise} \end{cases}$$

[2] In what follows, we will omit time dependence of signals except when considered necessary.

$$\Gamma = \frac{\mu(h^2 - r^2)^3}{4h(R^2 - h^2)(R^2 - r^2)}, \quad C = \Gamma\frac{(h^2 - R^2)^2}{(h^2 - r^2)^3} + \mu h, \quad h = \bar{r}\left(1 + \frac{\rho\eta_y}{\eta_x + \eta_y}\right),$$

for some $\rho \in (0,1)$. The reader can verify that V_a is non-negative, almost everywhere continuously differentiable, and that $\mathbf{u}_a$ reduces to

$$\mathbf{u}_a = \begin{cases} 0, & \text{if } \|\mathbf{x} - \mathbf{y}_k\| \geq R \\ \kappa_a \dfrac{\left(R^2 - \|\mathbf{x} - \mathbf{y}_k\|^2\right)}{\left(\|\mathbf{x} - \mathbf{y}_k\|^2 - r^2\right)^3}(\mathbf{x} - \mathbf{y}_k), & \text{if } h \leq \|\mathbf{x} - \mathbf{y}_k\| < R \\ \mu\dfrac{\mathbf{x} - \mathbf{y}_k}{\|\mathbf{x} - \mathbf{y}_k\|}, & \text{if } 0 < \|\mathbf{x} - \mathbf{y}_k\| < h \\ \text{not defined}, & \text{if } \|\mathbf{x} - \mathbf{y}_k\| = 0, \end{cases} \tag{4.7}$$

where $\kappa_a = 4\Gamma(R^2 - r_2)$. Note that $\|\mathbf{u}_a(t)\| \leq \mu \forall t \geq 0$.

The sampling vector $\mathbf{y}_k(t)$ is the zero-order-hold function of $\mathbf{y}(t)$ constructed as

$$\mathbf{y}_k(t) = \mathbf{y}(t_k), \text{ for } t \in [t_k, t_{k+1}) \tag{4.8}$$

where $t_0 = 0$ is the initial time and $t_{k+1} = t_k + T_k$ for $k \in \{0, 1, 2, \dots\}$. The sampling steps T_k are chosen such that the minimum possible number of observations (i.e. larger sampling steps) are used, while still avoiding Ω. Therefore, we choose T_k according to

$$T_k = \max\left\{T_0, \frac{\rho}{\eta_x + \eta_y}\min\{\|\mathbf{x}(t_k) - \mathbf{y}(t_k)\|, R\}\right\}. \tag{4.9}$$

Note that the maximum sampling step T_k is limited by the sensing radius R. Recall that the vehicle may not be aware of the obstacle's location whenever the latter is outside of the vehicle's Detection Region.

We now proceed to prove boundedness of the velocity vector and convergence to the desired trajectory when the obstacle is safely away.

Lemma 4.1. Consider the vehicle in (4.1) with control input given by (4.3). If $\|\dot{\mathbf{x}}(0)\| \leq \eta_x$ then $\|\dot{\mathbf{x}}(t)\| \leq \eta_x \forall t \geq 0$.

Proof. Let $V_\eta = 1/2\|\dot{\mathbf{x}}\|^2$. Taking its time derivative and noting that

$$\left\|\left(1 - \frac{\|\mathbf{u}_a(t)\|}{\mu}\right)\mathbf{u}_d(t) + \mathbf{u}_a(t)\right\| \leq \mu, \quad \forall t \geq 0,$$

yields

$$\dot{V}_\eta = \dot{\mathbf{x}}^{\mathrm{T}}\left(\left(1-\frac{\|\mathbf{u}_\mathrm{a}\|}{\mu}\right)\mathbf{u}_\mathrm{d}+\mathbf{u}_\mathrm{a}-\kappa\dot{\mathbf{x}}\right) \le \|\dot{\mathbf{x}}\|\mu-\kappa\|\dot{\mathbf{x}}\|^2 = \|\dot{\mathbf{x}}\|\left(\mu-\frac{\mu}{\eta_x}\|\dot{\mathbf{x}}\|\right).$$

Since $\dot{V}_\eta \le 0$ for all $\|\dot{\mathbf{x}}\| \ge \eta_x$ (i.e. V_η is nonincreasing), we can conclude that the velocity solutions of (4.1) remain bounded by η_x for all $t \ge 0$.

Remark 4.1. Note that (4.3) satisfies (4.2) if $\|\dot{\mathbf{x}}(0)\| \le \eta_x$.

Theorem 4.1. Consider the vehicle in (4.1) with control law (4.3)–(4.5) and let $\tilde{\mathbf{x}} = \mathbf{x} - \mathbf{x}_\mathrm{d}$. Assume that, for some $t_* \ge 0$, $\|\dot{\mathbf{x}}(t_*)\| \le \eta_x$, and $\mathbf{u}_\mathrm{a}(t) \equiv \mathbf{0}\,\forall t \ge t_*$. Then, $\tilde{\mathbf{x}}(t)$, $\dot{\tilde{\mathbf{x}}}(t)$, and $\ddot{\tilde{\mathbf{x}}}(t)$ converge to zero as $t \to 0$.

Proof. Assume that $\exists t_* \ge 0$ such that $\|\dot{\mathbf{x}}(t_*)\| \le \eta_x$ and $\mathbf{u}_a(t) \equiv \mathbf{0}\,\forall t \ge t_*$. Then, (4.1) can be rewritten in terms of the error dynamics as

$$\ddot{\tilde{\mathbf{x}}} = \kappa\dot{\tilde{\mathbf{x}}} - \mathbf{g}(\mathbf{x}). \tag{4.10}$$

Now, consider the following non-negative Lyapunov candidate function:

$$V_x(\tilde{\mathbf{x}}, \dot{\tilde{\mathbf{x}}}) = \frac{1}{2}\|\dot{\tilde{\mathbf{x}}}\|^2 + G(\tilde{\mathbf{x}})$$

for

$$G(\tilde{\mathbf{x}}) = \int_{t_*}^{t} \dot{\tilde{\mathbf{x}}}^{\mathrm{T}}\mathbf{g}(\tilde{\mathbf{x}})\mathrm{d}\tau = \begin{cases} \frac{k_\mathrm{g}}{2}\|\tilde{\mathbf{x}}\|^2, & \text{if } \|\tilde{\mathbf{x}}\| \le \mu_\mathrm{g}/k_\mathrm{g} \\ \mu_\mathrm{g}\|\tilde{\mathbf{x}}\| - \mu_\mathrm{g}/2k_\mathrm{g}, & \text{otherwise.} \end{cases}$$

Note that $V_x \ge 0$ and $V_x \to \infty$ as $\|[\tilde{\mathbf{x}}^{\mathrm{T}}, \dot{\tilde{\mathbf{x}}}^{\mathrm{T}}]^{\mathrm{T}}\| \to \infty$. Taking the time derivative of V_x yields

$$\dot{V}_x = \dot{\tilde{\mathbf{x}}}^{\mathrm{T}}(-\kappa\dot{\tilde{\mathbf{x}}} - \mathbf{g}(\tilde{\mathbf{x}})) + \dot{\tilde{\mathbf{x}}}^{\mathrm{T}}\mathbf{g}(\tilde{\mathbf{x}}) = -\kappa\|\dot{\tilde{\mathbf{x}}}\|^2 \le 0. \tag{4.11}$$

Then, by integrating both sides of (4.11) from t_* to t, we obtain $\dot{V}_x(\tilde{\mathbf{x}}(t), \dot{\tilde{\mathbf{x}}}(t)) \le \dot{V}_x(\tilde{\mathbf{x}}(t_*), \dot{\tilde{\mathbf{x}}}(t_*)) \le 0$, which implies that $\dot{\tilde{\mathbf{x}}} \in L_\infty \cap L_2$. In addition, from (4.10) we obtain $\ddot{\tilde{\mathbf{x}}} \in L_\infty$. Applying Barbalat's Lemma [29], we conclude that $\dot{\tilde{\mathbf{x}}}(t) \to \mathbf{0}$ as $t \to \infty$. Now, returning to (4.10) and differentiating with respect to time yields

$$\dddot{\tilde{\mathbf{x}}} = -\kappa\ddot{\tilde{\mathbf{x}}} - \frac{\partial\mathbf{g}(\tilde{\mathbf{x}})}{\partial\tilde{\mathbf{x}}}\dot{\tilde{\mathbf{x}}},$$

from which we obtain $\dddot{\tilde{\mathbf{x}}} \in L_\infty$. Then, since $\int_{t_*}^{t} \ddot{\tilde{\mathbf{x}}}(\tau)\mathrm{d}\tau \to -\dot{\tilde{\mathbf{x}}}(t_*) < \infty$, we can once again invoke Barbalat's Lemma and conclude that $\ddot{\tilde{\mathbf{x}}}(t) \to \mathbf{0}$ as $t \to \infty$. Finally, using (4.10) and the convergence results for $\dot{\tilde{\mathbf{x}}}(t)$ and $\ddot{\tilde{\mathbf{x}}}(t)$, we have $\mathbf{g}(\tilde{\mathbf{x}}) \to \mathbf{0} \Rightarrow \tilde{\mathbf{x}} \to \mathbf{0}$ as $t \to \infty$, which completes the proof.

4.4 Collision Avoidance Analysis

In this section we derive sufficient conditions for collision avoidance under the proposed control policy. We first evaluate the noncooperative case where the control intentions of the other agent, treated as a dynamic obstacle, are unknown by the vehicle. We assume the worst case scenario in which the obstacle takes an antagonistic role and show that, if the vehicle assumes the proposed control strategy, collisions are guaranteed to never occur. Then, we evaluate the cooperative case where both agents implement the same avoidance strategy. We relax the sufficient conditions for collision avoidance developed for the noncooperative case and show that the vehicles do not enter the Avoidance Region at any given time.

With the objective of formulating the main results of this chapter on collision avoidance, let us first introduce the following lemma. The lemma provides a bound on the velocity of the vehicle with respect to the collision threat.

Lemma 4.2. Consider the vehicle in (4.1) with control input given by (4.3)–(4.9) for $\kappa = \mu/\eta_x$, $\eta_x > 0$. Let $r \in (T_0(\eta_x + \eta_y)\rho^{-1}, R)$, $\bar{r} \in (r, R)$, and $\theta \in (0, \cos^{-1}\rho)$ for $\eta_y \geq 0$ and $\rho \in (0,1)$. Suppose that $\|\dot{\mathbf{x}}(t_0)\| \leq \eta_x$, $\|\dot{\mathbf{y}}(t)\| \leq \eta_y$, and $\|\mathbf{x}(t) - \mathbf{y}(t)\| \in [r, \bar{r}]\ \forall t \in [t_0, t_f]$, where $t_f > 0$, $t_0 \leq t_f + \theta/\omega$, and $\omega = -(\eta_x + \eta_y)/r$. Then, the following inequality holds:

$$\beta(t_f) \geq \|\mathbf{x}(t_f) - \mathbf{y}(t_f)\| f(\mu, \eta_x, \eta_y, \rho, r, \theta), \tag{4.12}$$

where $\beta(t_f) = (\mathbf{x}(t_f) - \mathbf{y}(t_f))^T \dot{\mathbf{x}}(t_f)$ and[3]

$$\begin{aligned} f(\mu, \eta_x, \eta_y, \rho, r, \theta) &= \bar{f}(\mu, \kappa, \omega, \rho, \theta) \\ &= \frac{\mu}{\kappa^2 + \omega^2}\left(\kappa\sqrt{1-\rho^2} + \omega\rho - e^{\kappa\theta/\omega}\left(\frac{\kappa^2 + \omega^2}{\kappa} + \kappa\sqrt{1-\rho^2}\cos\theta\right)\right) \\ &\quad - \frac{\mu e^{\kappa\theta/\omega}}{\kappa^2 + \omega^2}\left(-\kappa\rho\sin\theta + \omega\sqrt{1-\rho^2}\sin\theta + \omega\rho\cos\theta\right). \end{aligned} \tag{4.13}$$

Proof. To simplify the notation, let $\delta = -\theta/\omega$, $t_\delta = t_f - \delta$, $\mathbf{v}(t) = \mathbf{x}(t) - \mathbf{y}(t)$, and $\mathbf{v}_k(t) = \mathbf{x}(t) - \mathbf{y}_k(t)$. From the assumption that $\|\mathbf{v}(t)\| \in [r, \bar{r}]\ \forall t \in [t_0, t_f]$ and $r \in (T_0\eta_y\rho^{-1}, R)$, we have

$$T_k \leq \max\left\{T_0, \frac{\rho}{\eta_x + \eta_y}\min\{\|\mathbf{v}(t_k)\|, R\}\right\} \leq \max\left\{T_0, \frac{\rho}{\eta_x + \eta_y}\bar{r}\right\} = \frac{\rho\bar{r}}{\eta_x + \eta_y}$$

for $t_k \in [t_0, t_f]$. Using the above bound, we can also show that

[3] Note that κ and ω are functions of η_x, η_y, and r.

$$\|\mathbf{v}_k\| \leq \|\mathbf{x}-\mathbf{y}+\mathbf{y}-\mathbf{y}_k\| \leq \overline{r}+\|\mathbf{y}-\mathbf{y}_k\| \leq \overline{r}+T_k\eta_y \leq \overline{r}+\frac{\rho\eta_y}{\eta_x+\eta_y}\overline{r}=h$$

and therefore $\mathbf{u}_{\mathrm{d}}(t) \equiv \mathbf{0}$ for $t \in [t_0,t_{\mathrm{f}}]$, and the velocity solution of (4.1) can be computed as

$$\dot{\mathbf{x}}(t_{\mathrm{f}}) = \mathrm{e}^{-\kappa\delta}\dot{\mathbf{x}}(t_\delta) + \int_{t_\delta}^{t_{\mathrm{f}}} \mathrm{e}^{-\kappa(t_{\mathrm{f}}-\tau)}\mathbf{u}_{\mathrm{a}}(\tau)\mathrm{d}\tau.$$

Accordingly,

$$\begin{aligned}
\beta(t_{\mathrm{f}}) &= \mathbf{v}(t_{\mathrm{f}})^{\mathrm{T}}\mathrm{e}^{-\kappa\delta}\dot{\mathbf{x}}(t_\delta) + \mathbf{v}(t_{\mathrm{f}})^{\mathrm{T}}\int_{t_\delta}^{t_{\mathrm{f}}} \mathrm{e}^{-\kappa(t_{\mathrm{f}}-\tau)}\mu\frac{\mathbf{v}_k(\tau)}{\|\mathbf{v}_k(\tau)\|}\mathrm{d}\tau \\
&\geq -\eta_x\|\mathbf{v}(t_{\mathrm{f}})\|\mathrm{e}^{-\kappa\delta} + \mu\|\mathbf{v}(t_{\mathrm{f}})\|\int_{t_\delta}^{t_{\mathrm{f}}} \mathrm{e}^{-\kappa(t_{\mathrm{f}}-\tau)}\frac{\mathbf{v}(t_{\mathrm{f}})^{\mathrm{T}}\mathbf{v}_k(\tau)}{\|\mathbf{v}(t_{\mathrm{f}})\|\|\mathbf{v}_k(t_{\mathrm{f}})\|}\mathrm{d}\tau \qquad (4.14)\\
&= -\eta_x\|\mathbf{v}(t_{\mathrm{f}})\|\mathrm{e}^{-\kappa\delta} + \mu\|\mathbf{v}(t_{\mathrm{f}})\|\int_{t_\delta}^{t_{\mathrm{f}}} \mathrm{e}^{-\kappa(t_{\mathrm{f}}-\tau)}\cos\phi(\tau)\mathrm{d}\tau,
\end{aligned}$$

where $\phi(\tau)$ defines the angle between $\mathbf{v}(t_{\mathrm{f}})$ and $\mathbf{v}_k(\tau)$ for $\tau \in [t_\delta,t_{\mathrm{f}}]$ and where we used the fact that $\|\dot{\mathbf{x}}(t)\| \leq \eta_x \forall t$ (due to Lemma 4.1).

Our next step is to compute a lower bound on $\int_{t_\delta}^{t_{\mathrm{f}}} \mathrm{e}^{-\kappa(t_{\mathrm{f}}-\tau)}\cos\phi(\tau)\mathrm{d}\tau$. Equivalently, we aim to find an upper bound on $\phi(\tau)$ at every instance of time τ. That is, we would like to define a function $\overline{\phi}(\tau)$ such that $\phi(\tau) \leq \overline{\phi}(\tau)$ for all $\tau \in [t_\delta,t_{\mathrm{f}}]$. Therefore, let us consider Figure 4.2, which illustrates the plane described by the set of vectors $\mathbf{v}$ and $\mathbf{v}_k$ with origin at $\mathbf{x}$. Observe that $\phi(\tau)$ is always upper bounded by the summation of the angle between $\mathbf{v}(t_{\mathrm{f}})$ and $\mathbf{v}(\tau)$ and the angle between $\mathbf{v}(\tau)$ and $\mathbf{v}_k(\tau)$ for $\tau \in [t_\delta,t_{\mathrm{f}}]$, denoted as $\vartheta_{ij}(\tau)$ and $\varphi_{ij}(\tau)$ respectively, i.e. $\|\phi(\tau)\| \leq \|\vartheta(\tau)\| + \|\varphi(\tau)\|$. Hence, a suitable function would be

$$\overline{\phi}(\tau) = \underbrace{\sup_{t\in[t_\delta,\tau]}\|\vartheta(\tau)\|}_{\overline{\vartheta}(\tau)} + \underbrace{\sup_{t\in[t_\delta,\tau]}\|\varphi(t)\|}_{\overline{\phi}(\tau)},$$

where $\overline{\vartheta}(\tau)$ and $\overline{\varphi}(\tau)$ are yet to be determined.

Now, consider $\vartheta(\tau)$. Since $\|\mathbf{v}(\tau)\| \geq r \forall \tau \in [t_\delta, t_{\mathrm{f}}]$ and the velocities of the agents are bounded, we see that $\vartheta(\tau)$ attains its maximum when the vehicle and the obstacle approach each other at maximum speed along the boundary of Ω. Then, using the arc-length formula to compute the maximum length traveled by the agents and invoking its relation with the central angle ϑ, we obtain

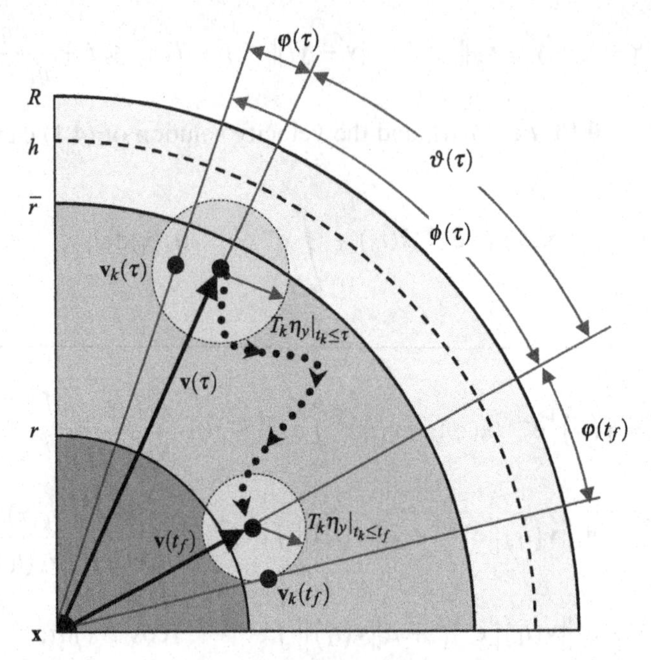

Figure 4.2: Cross-Sectional Plane Containing the Vectors v and v_k, Denoted by the Large Black and Gray Dots Respectively.

$$\vartheta(\tau) \leq \frac{\int_{\tau}^{t_f} \|\dot{\mathbf{v}}(s)\| \mathrm{d}s}{r} \leq \frac{(\eta_x + \eta_y)(t_f - \tau)}{r} = \overline{\vartheta}(\tau), \quad \text{for } \tau \in [t_\delta, t_f].$$

Note that $\overline{\vartheta}(t_\delta) = [(\eta_x + \eta_y)\delta/r] = \theta$ while $\overline{\vartheta}(t_f) = 0$. We are now left to find $\overline{\varphi}(\tau)$. To this end, let us analyze when φ achieves its maximum. We see that the larger admissible value for φ increases as the error between $\mathbf{v}(\tau)$ and $\mathbf{v}_k(\tau)$ increases. Accordingly, $\varphi(\tau)$, $\tau \in [t_k, t_{k+1}]$, can only achieve its maximum when $\tau \to t_{k+1}^- \to t_k + T_k$ (i.e. right before $\mathbf{y}_k$ is updated). That is, φ is upper bounded by the larger possible angle between $\mathbf{x}(t_k) - \mathbf{y}(t_k)$ and $\mathbf{x}(t_{k+1}) - \mathbf{y}(t_k)$. Therefore, let us consider the diagram in Figure 4.3(a), which illustrates not only the angle between $\mathbf{x}(t_k) - \mathbf{y}(t_k)$ and $\mathbf{x}(t_{k+1}) - \mathbf{y}(t_k)$, but also the angle between $\mathbf{x}(t_k) - \mathbf{y}(t_k)$ and $\mathbf{x}(t_{k+1}) - \mathbf{y}(t_{k+1})$. Note that φ can also be upper bounded by the maximum permissible angle between $\mathbf{x}(t_k) - \mathbf{y}(t_k)$ and $\mathbf{x}(t_{k+1}) - \mathbf{y}(t_{k+1})$, represented by ζ in the diagram. Hence, let us compute $\sup_{\tau \in [t_k, t_{k+1}]} \|\zeta(\tau)\|$ instead. We start by noticing that the difference $\mathbf{v}(t_{k+1}) - \mathbf{v}(t_k)$ is upper bounded by

$$\|\mathbf{v}(t_{k+1}) - \mathbf{v}(t_k)\| = \|\mathbf{x}(t_{k+1}) - \mathbf{x}(t_k) + \mathbf{y}(t_{k+1}) - \mathbf{y}(t_k)\| \leq \eta_x T_k + \eta_y T_k, \tag{4.15}$$

which indicates that ζ is maximized when the equality in (4.15) persists, i.e. when $\mathbf{v}(t_{k+1})$ lies on the boundary of the ball centered at $\mathbf{v}(t_k)$ with radius $T_k(\eta_x + \eta_y)$. Therefore, consider the diagram in Figure 4.3(b), which details this case and chooses $\mathbf{e}_1$ and $\mathbf{e}_2$ as an orthonormal basis for the plane containing this cross-sectional area of the ball. Let $\mathbf{e}_2$ be oriented in the

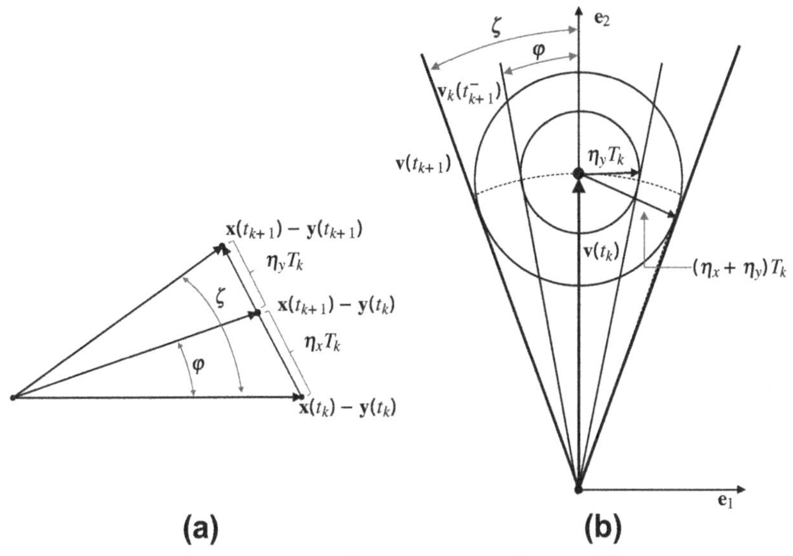

Figure 4.3: Maximum Range of φ and ζ.

same direction of $\mathbf{v}(t_k)$. Then, we can define $\mathbf{v}(t_k) - \|\mathbf{v}(t_k)\|\mathbf{e}_2$. Similarly, $\mathbf{v}(t_{k+1})$ can be written as $\mathbf{v}(t_{k+1}) = c_1\mathbf{e}_1 + c_2\mathbf{e}_2$, where c_1 and c_2 are the new coordinates in the $(\mathbf{e}_1,\mathbf{e}_2)$ system. From (4.15), we see that c_1 and c_2 must satisfy the following equation:

$$c_1^2 + (c_2 - \|\mathbf{v}(t_k)\|)^2 \leq \left(\eta_x + \eta_y\right)^2 T_k^2 = \rho^2 \|\mathbf{v}(t_k)\|^2.$$

Likewise, we observe that ζ is maximized when the ratio $|c_1/c_2|$ attains its maximum. These two conditions yield

$$c_1 = \pm\rho\|\mathbf{v}(t_k)\|\sqrt{1-\rho^2}, \qquad c_2 = \left(1-\rho^2\right)\|\mathbf{v}(t_k)\|.$$

Therefore, $\overline{\varphi}(\tau) = \zeta$ can be computed as

$$\overline{\varphi} = \cos^{-1}\left(\frac{\mathbf{v}(t_k)^{\mathrm{T}}\mathbf{v}(t_{k+1})}{\|\mathbf{v}(t_k)\|\|\mathbf{v}(t_{k+1})\|}\right) = \cos^{-1}\left(\sqrt{1-\rho^2}\right) = \sin^{-1}\rho.$$

Now, let us return to (4.14). Since $\overline{\vartheta}(\tau) \leq \theta_i \leq \cos^{-1}\rho$ and $\overline{\varphi}(\tau) = \sin^{-1}\rho$, we have $\overline{\phi}(\tau) \leq \pi/2$. Therefore, $\cos\phi_{ij} \geq \cos\overline{\phi}(\tau) \geq 0$ for all τ and

$$\begin{aligned}
&\int_{t_\delta}^{t_f} \mathrm{e}^{-\kappa(t_f-\tau)}\cos\phi(\tau)\mathrm{d}\tau \geq \int_{t_\delta}^{t_f} \mathrm{e}^{-\kappa(t_f-\tau)}\cos\overline{\phi}(\tau)\mathrm{d}\tau \\
&= \frac{1}{\kappa^2+\omega^2}\left(\kappa\cos\overline{\phi}(t_f) + \omega\sin\overline{\phi}(t_f) - \mathrm{e}^{-\kappa\delta}\left(\kappa\cos\overline{\phi}(t_\delta) + \omega\sin\overline{\phi}(t_\delta)\right)\right),
\end{aligned} \tag{4.16}$$

where we used the fact that $\dot{\overline{\phi}}(t) = \dot{\overline{\vartheta}}(t) = \omega = -(\eta_x + \eta_y)/r$ is constant. Also note that $\overline{\phi}(t_f) = \overline{\varphi}$ and hence

$$\cos\overline{\phi}_{ij}(t_f) = \cos\overline{\varphi} = \sqrt{1-\rho^2}, \quad \sin\overline{\phi}_{ij}(t_f) = \sin\overline{\varphi} = \rho.$$

Similarly, since $\overline{\phi}(t_\delta) = \theta + \overline{\varphi}$ we have

$$\cos\overline{\phi}(t_\delta) = \sqrt{1-\rho^2}\cos\theta - \rho\sin\theta, \quad \sin\overline{\phi}(t_\delta) = \sqrt{1-\rho^2}\sin\theta + \rho\cos\theta,$$

and returning to (4.16), we obtain

$$\int_{t_\delta}^{t_f} e^{-\kappa(t_f-\tau)}\cos\phi(\tau)d\tau \geq \frac{\kappa\sqrt{1-\rho^2}+\omega\rho}{\kappa^2+\omega^2}$$
$$-e^{-\kappa\delta}\frac{\kappa\sqrt{1-\rho^2}\cos\theta - \kappa\rho\sin\theta + \omega\sqrt{1-\rho^2}\sin\theta + \omega\rho\cos\theta}{\kappa^2+\omega^2}.$$

Therefore, substituting the above equation into (4.14) yields (4.12), and the proof is complete.

Remark 4.2. Mathematically, $\beta/\|\mathbf{v}\|$ represents the scalar projection of the velocity vector $\dot{\mathbf{x}}$ on to the collision threat vector $\mathbf{v}$. Accordingly, we can say that the previous lemma describes the direction of the vehicle's velocity vector with respect to the collision threat. For instance, if for some time t_f, $\beta(t_f) > 0$, then we can conclude that at time t_f the vehicle is moving away from the obstacle.

Remark 4.3. Note that if $\theta = \cos^{-1}\rho$, then (4.13) reduces to

$$\overline{f}(\mu,\kappa,\omega,\rho,\theta) = \mu\left(\frac{\kappa\sqrt{1-\rho^2}+\omega\left(1-e^{\kappa\theta/\omega}\right)}{\kappa^2+\omega^2} - \frac{e^{\kappa\theta/\omega}}{\kappa}\right).$$

Having established Lemma 4.2, we now proceed to state the main results on collision avoidance of this chapter. The following theorem provides a sufficient condition for collision avoidance in a noncooperative case for a vehicle with limited actuation and time-varying discrete observations.

Theorem 4.2 (Noncooperative Collision Avoidance). Consider the vehicle in (4.1) with control input given by (4.3)–(4.9) for $\kappa = \mu/\eta_x > 0$, and assume that the obstacle's velocity is bounded by $\eta_y \geq 0$. Suppose $\exists\, \eta_x > \eta_y$, $\rho \in (0,1)$, $r \geq T_0(\eta_x+\eta_y)\rho^{-1}$ and $\theta \in (0, \cos^{-1}\rho)$ such that:

$$\overline{r} = (\theta+1)r < R\left(1 - \frac{\rho\eta_y}{\eta_x+\eta_y}\right) \tag{4.17}$$

$$f\left(\mu, \eta_x, \eta_y, \rho, r, \theta\right) \geq \eta_y \tag{4.18}$$

hold. Then, if $[\mathbf{x}^{\mathrm{T}}(0), \mathbf{y}^{\mathrm{T}}(0)]^{\mathrm{T}} \notin W \cup \Omega$ and $\|\dot{\mathbf{x}}(0)\| \leq \eta_x$, we have $[\mathbf{x}^{\mathrm{T}}(t), \mathbf{y}^{\mathrm{T}}(t)]^{\mathrm{T}} \notin \Omega \, \forall t \geq 0$.

Proof. Consider the system in (4.1) with control input given by (4.3)–(4.9). Assume that the obstacle's velocity is bounded by some $\eta_y \geq 0$. Let $\kappa = \overline{\mu}/\eta_x$, where $\eta_x > \eta_y$. Suppose that $\|\dot{\mathbf{x}}(0)\| \leq \eta_x$ and that (4.17) and (4.18) hold. Applying Lemma 4.1 we have $\|\dot{\mathbf{x}}(t)\| \leq \eta_x \forall t \geq 0$. Then, we consider the following Lyapunov candidate function:

$$V(t) = \frac{1}{4\left(\|\mathbf{v}(t)\|^2 - r^2\right)^2}. \tag{4.19}$$

Taking its time derivative yields:

$$\dot{V}(t) = \frac{\mathbf{v}(t)^{\mathrm{T}}\dot{\mathbf{y}}(t) - \beta(t)}{\left(\|\mathbf{v}(t)\|^2 - r^2\right)^3} \leq \frac{\|\mathbf{v}(t)\|\eta_x - \beta(t)}{\left(\|\mathbf{v}(t)\|^2 - r^2\right)^3}. \tag{4.20}$$

Now, let $[\mathbf{x}^{\mathrm{T}}(0), \mathbf{y}^{\mathrm{T}}(0)]^{\mathrm{T}} \notin W \cup \Omega$ and suppose that for some time $t > 0$, $\|\mathbf{v}(t)\| \to r$ from above (i.e. the two-agent system approaches the boundary of Ω). Since $\|\mathbf{v}(0)\| > \overline{r}$ and the velocities of the agents are bounded, the agents will require some time Δt to reduce their distance from $\overline{r}$ to r. Therefore, we have $[\mathbf{x}^{\mathrm{T}}(\tau), \mathbf{y}^{\mathrm{T}}(\tau)]^{\mathrm{T}} \in W \forall \tau \in [t - \Delta t, t]$, where it is easy to demonstrate that $\Delta t \geq \delta = (\overline{r} - r)/(\eta_x + \eta_y) = \theta r/(\eta_x + \eta_y)$. Then, applying Lemma 4.2 and using (4.17) and (4.18), we can show that $\beta(t) \geq \|\mathbf{v}(t)\|\eta_y$. Returning to (4.20), we finally obtain $\dot{V}(t) \leq 0$ for $\|\mathbf{v}\| \leq r$. The fact that $\mathbf{v}(t)$ is continuous and $\dot{V}(t)$ is nonpositive for $\|\mathbf{v}(t)\| \leq r$ implies that $V(t) < \infty$. Hence, the solutions of $\mathbf{v}(t)$ are uniformly ultimately bounded by r, which further implies that $[\mathbf{x}^{\mathrm{T}}(t), \mathbf{y}^{\mathrm{T}}(t)]^{\mathrm{T}} \notin \Omega$ for all $t \geq 0$.

We now study the cooperative case, in which both agents (now treated as two autonomous vehicles) implement the proposed avoidance strategy. Without loss of generality, we will assume that the vehicles have the same detection radius R, minimum sampling step T_0, and design parameters ρ and θ. We, however, assume different dynamic constraints.

Theorem 4.3 (Cooperative Collision Avoidance). Assume that both vehicles, namely x and y, have double-integrator dynamics as in (4.1) with control input given by (4.3)–(4.9) and bounded by $2\mu > 0$ and $2\mu_y > 0$ respectively. Suppose $\exists \eta_x > 0,\ \eta_y > 0,\ \rho \in (0,1),\ r \geq T_0(\eta_x + \eta_y)\rho^{-1}$, and $\theta \in (0, \cos^{-1}\rho]$ such that (4.17) and

$$f\left(\mu, \eta_x, \eta_y, \rho, r, \theta\right) \geq 0 \tag{4.21}$$

$$f\left(\mu_y, \eta_y, \eta_x, \rho, r, \theta\right) \geq 0 \tag{4.22}$$

hold. Then, if $[\mathbf{x}^{\mathrm{T}}(0), \mathbf{y}^{\mathrm{T}}(0)]^{\mathrm{T}} \notin W \cup \Omega$, $\|\dot{\mathbf{x}}(0)\| \leq \eta_x$ and $\|\dot{\mathbf{y}}(0)\| \leq \eta_y$, we see that $[\mathbf{x}^{\mathrm{T}}(t), \mathbf{y}^{\mathrm{T}}(t)]^{\mathrm{T}} \notin \Omega \forall t \geq 0$.

Proof. The proof follows similarly to that of Theorem 4.2. It is sufficient to show that, if (4.17), (4.21), and (4.22) are satisfied, then (4.19) is bounded or, alternatively, that if $\|\mathbf{x}(t) - \mathbf{y}(t)\| \to r$, then $(\mathbf{x}(t) - \mathbf{y}(t))^{\mathrm{T}}\dot{\mathbf{x}}(t) \geq 0$ and $(\mathbf{y}(t) - \mathbf{x}(t))^{\mathrm{T}}\dot{\mathbf{y}}(t) \geq 0$. The latter would in turn imply that $\mathbf{v}(t)$ are uniformly ultimately bounded by r.

Theorems 4.2 and 4.3 provide sufficient conditions for r and θ to guarantee collision avoidance. However, we have not discussed how to optimally choose r and θ. Ideally, we would like to minimize the minimum distance at which the vehicle needs to start applying full avoidance control. Accordingly, we have that given μ, T_0, η_x, η_y, and ρ, we would like to choose $r = r^\dagger \in [T_0(\eta_x + \eta_y)\rho^{-1}, R)$, and $\theta = \theta^\dagger \in (0, \cos^{-1}\rho]$ such that

$$\{r^\dagger, \theta^\dagger\} = \arg\min_{r,\theta} \{(\theta + 1)r\} \tag{4.23}$$

subject to (4.17) and (4. 18) or (4.17), (4.21), and (4.22). The optimization problem in Eq. (4.23) is beyond the scope of this chapter, yet it is worth mentioning if there is a need to optimize the design.

4.5 Avoiding Deadlocks

In Section 4.4, we provided sufficient conditions for collision avoidance and showed that the vehicle and obstacle never enter the Avoidance Region. Similarly, in Section 4.3, we showed that the vehicle successfully converges to the desired trajectory if the obstacle remains outside of the vehicle's Detection Region for all future time. However, we have not studied the convergence of the vehicle to the desired trajectory taking into consideration the obstacle's intentions. It turns out that the vehicle, in very particular cases, may not converge to the desired trajectory. That is, the vehicle–obstacle system, under the proposed control law, may suffer from the existence of local minima. Herein, we will develop a control solution to escape local minima for the case in which both agents try to evade each other or, at least, the noncooperative case where the obstacle does not assume an antagonistic role of persistently pursing the vehicle. The scenario where the obstacle relentlessly chases the vehicle does not have a solution since, no matter what the vehicle does next, the obstacle will always prevent the vehicle from stabilizing at the desired trajectory.

The presence of local minima or deadlocks is a fundamental problem in most potential and avoidance field functions methods [30,31]. These situations typically arise as a result of specific symmetries between attractive and repulsive potential field forces. That is, when $\mathbf{u}_{\mathrm{d}} \to -a\mathbf{u}_{\mathrm{a}} \neq \mathbf{0}$ for some $a > 0$. Therefore, a heuristic solution to avoid deadlocks is to perturb the control or trajectory of the agents when a potential deadlock is identified or predicted. A proper perturbation of the control input will break the symmetry between the avoidance and objective control, allowing the vehicle to diverge from the obstacle and converge to the desired trajectory. Based on this principle, we propose the following alteration to the control law in (4.4).

Let $\mathbf{x} - \mathbf{y}_k = \mathbf{v}_k = [\alpha_1, \ldots, \alpha_n]^{\mathrm{T}}$ and assume that $\exists \alpha_i \neq 0$ for some $i \in \{1, \ldots, n\}$ (this is always true for $\|\mathbf{v}_k\| \neq \mathbf{0}$, which was proven in Section 4.4). We define a perpendicular unit vector $\mathbf{v}\perp \in \Re^n$ as

$$\mathbf{v}_\perp = \frac{1}{\left(\sum_{k=1}^{n}\alpha_k^2\right)^{1/2}}\left[\alpha_i, \ldots, \overbrace{-\sum_{k\in\{1,\ldots,n\},k\neq l}\alpha_k}^{i\text{th element}}, \ldots, \alpha_i\right].$$

Then we can redefine (4.4) as

$$\mathbf{u}_\mathrm{d} = \ddot{\mathbf{x}}_\mathrm{d} + \kappa\dot{\mathbf{x}}_\mathrm{d} - (1-\gamma)\mathbf{g}(\tilde{\mathbf{x}}_\mathrm{d}) + \gamma\mu_\mathrm{g}\mathbf{v}_\perp, \tag{4.24}$$

where

$$\gamma = \begin{cases} \varepsilon, & \text{if } h_\gamma(\ddot{\mathbf{x}}_\mathrm{d}, \dot{\mathbf{x}}_\mathrm{d}, \tilde{\mathbf{x}}_\mathrm{d}, \mathbf{u}_\mathrm{a}) \leq \sigma \text{ and } \|\mathbf{u}_\mathrm{a}\| > 0 \\ 0, & \text{otherwise} \end{cases}$$

for $h_\gamma(\ddot{\mathbf{x}}_\mathrm{d}, \dot{\mathbf{x}}_\mathrm{d}, \tilde{\mathbf{x}}_\mathrm{d}, \mathbf{u}_\mathrm{a}) = \|(1 - \|\mathbf{u}_\mathrm{a}\|\mu)(\ddot{\mathbf{x}}_\mathrm{d} + \kappa\dot{\mathbf{x}}_\mathrm{d} - \mathbf{g}(\tilde{\mathbf{x}}_\mathrm{d})) + \mathbf{u}_\mathrm{a}\|$ and small positive constants $\varepsilon \in (0,1)$ and $\sigma > 0$. Note that $\mathbf{v}_\perp$ is perpendicular to $\mathbf{u}_\mathrm{a}$ and, therefore, if $\mathbf{u}_\mathrm{d}|_{\gamma=0} + a\mathbf{u}_\mathrm{a} \rightarrow \mathbf{0}$ for $\mathbf{u}_\mathrm{a} \neq \mathbf{0}$, $\mathbf{v}_\perp$ adds a perturbation such that $\mathbf{u}_\mathrm{d}|_{\gamma=\varepsilon} + a\mathbf{u}_\mathrm{a} \nrightarrow \mathbf{0}$.

Remark 4.4. In the general case, fluctuations and time-varying properties of the agents' desired trajectories might be sufficient to break symmetries between the avoidance and objective control functions, allowing the agents to escape potential deadlocks (for instance, see example in Section 4.6.1). Similarly, in practice, actuation and sensing errors might implicitly perturb the control and trajectories of the agents.

4.6 Examples

To illustrate the performance of the proposed control strategies, we now present two examples. The first example addresses a noncooperative avoidance control scenario, whereas the second evaluates a cooperative case. In the latter, we simulate a symmetric scenario and implement the alternate control policy proposed in (4.24) to avoid a deadlock between both agents.

4.6.1 Noncooperative Example

We consider a vehicle–obstacle system with double-integrator dynamics (4.1). We study the case where the vehicle, denoted by x, implements the proposed trajectory tracking control law (4.4) and avoidance strategy (4.7), while the obstacle, namely y, does not implement an avoidance policy. The vehicle and obstacle are commanded to follow circular trajectories given by

$$\mathbf{x}_{\mathrm{d}}(t) = [A\cos(wt), A\sin(wt)]^{\mathrm{T}}, \quad \mathbf{y}_{\mathrm{d}}(t) = [-A\cos(wt), A\sin(wt)]^{\mathrm{T}}$$

respectively, where $A = 20$ m and $w = 0.1\ \mathrm{s}^{-1}$. The trajectory tracking control parameters are taken to be $\kappa_{\mathrm{g}} = 25\ \mathrm{kg/s}^2$ and $\mu_{\mathrm{g}} = 60$ N. We assume the obstacle's speed to be equal to the desired trajectory's speed, i.e. $\eta_y = 2$ m/s. The vehicle's sensing radius R and minimum observation time step T_0 are assumed to be 10 m and 0.5 s respectively. We take $\mu = 200$ N and $\kappa = 50$ kg/s, which yield $\eta_x = 4$ m/s. Similarly, we choose $\rho = 0.8$, which, according to Theorem 4.2 and (4.23), yields $r = 3.75$ m and $\theta = 0.1$ rad.

The behavior of the two-agent system is illustrated in Figure 4.4. The vehicle and obstacle, which are initialized from rest at $\mathbf{x}(0) = [10\ \mathrm{m}, 0\ \mathrm{m}]^{\mathrm{T}}$ and $\mathbf{y}(0) = [-10\ \mathrm{m}, 0\ \mathrm{m}]^{\mathrm{T}}$, start traveling toward their circular trajectories delineated by the dashed gray line. Note that at time $t \approx 16$ s (see Figure 4.4(b)), their trajectories intersect and the vehicle comes into close proximity with the obstacle. As a reaction, the vehicle deviates from the desired trajectory to avoid the obstacle. Once the obstacle is out of the vehicle's Detection Region, the vehicle retakes its course toward its circular trajectory. Eventually, the vehicle re-encounters the obstacle several times at $t \in \{48\ \mathrm{s}, 80\ \mathrm{s}, 112\ \mathrm{s}\}$ and at every time is able to solve the conflict, successfully avoiding a collision and retaking the desired path. Figures 4.5 and 4.6 illustrate the distance between both agents and the norm of the error between their actual and desired trajectories. Observe that the agents do not enter the Avoidance Regions. Similarly, note that the error in the vehicle's trajectory converges to zero every time it avoids the obstacle.

Finally, Figure 4.7 plots the observation sampling T_k as a function of time. Note that T_k decreases as the vehicle comes close to the obstacle and achieves its maximum when the obstacle is safely away, i.e. when $\|\mathbf{x} - \mathbf{y}\| \geq R$.

4.6.2 Cooperative Example

We now evaluate a cooperative scenario where both agents implement the avoidance strategy. Consider once again the two-agent system with dynamics given by (4.1) and (4.2) for $\mu = \mu_y = 100$ N and $R = 10$ m. We command the agents to stabilize at $\mathbf{x}_{\mathrm{d}} = -\mathbf{y}_{\mathrm{d}} = [20\ \mathrm{m}, 20\ \mathrm{m}]^{\mathrm{T}}$. The control parameters are chosen as[4] $\kappa = 25$ kg/s, $\kappa_y = 33.3$ kg/s, $\kappa_{\mathrm{g}} = 12.5\ \mathrm{kg/s}^2$, $\kappa_{\mathrm{g}y} = 16.7\ \mathrm{kg/s}^2$, and $\mu_{\mathrm{g}} = \mu_{\mathrm{g}y} = \mu$. Therefore, we have $\eta_x = 4$ m/s and $\eta_y = 3$ m/s. Choosing $\rho = 0.95$ and $T_0 = 0.5$ s and applying Theorem 4.3 and (4.23), we obtain $r = 3.7$ m and $\theta = 0.12$ rad. Furthermore, we assume the agents implement the modified control law in (4.24) for $\varepsilon = 0.8$ and $\sigma = 10$ N to avoid the possibility of a deadlock.

The trajectories of the two agents are illustrated in Figure 4.8. The agents start from rest at opposite positions at $\mathbf{x}(0) = -\mathbf{y}(0) = \mathbf{y}_{\mathrm{d}} = -\mathbf{x}_{\mathrm{d}}$. This is a symmetric configuration,

[4] New quantities subscripted with y refer to the y agent.

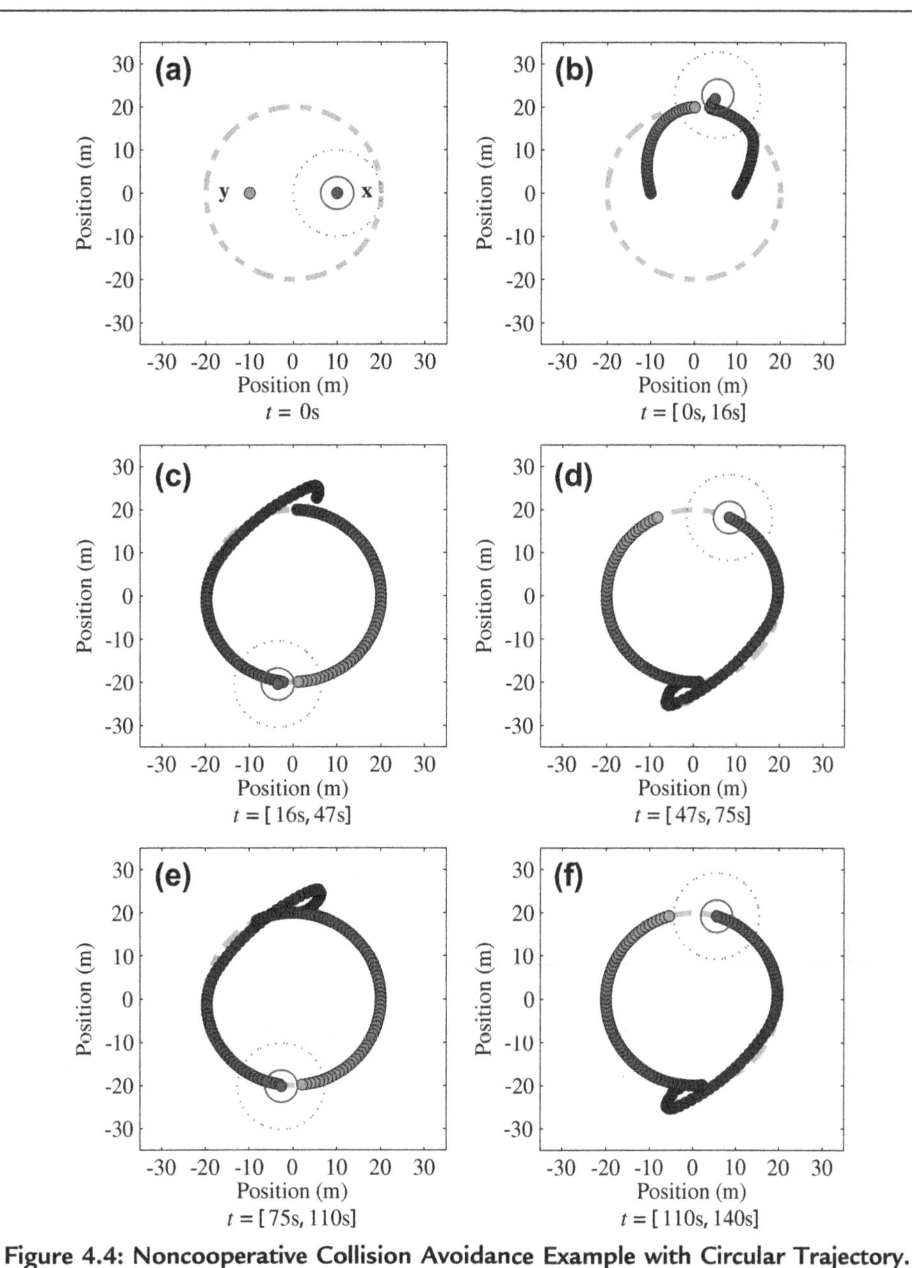

Figure 4.4: Noncooperative Collision Avoidance Example with Circular Trajectory. The positions of the vehicle and the obstacle are traced by the blue and orange dots respectively. The initial positions are illustrated by the darker dots, while newer positions are over-imposed in lighter color and time-spaced by 0.5 s. The desired trajectory is delineated by the dashed line. The vehicle's Avoidance and Detection Regions at the end of each simulation interval are indicated by the circles with solid and dashed lines respectively.

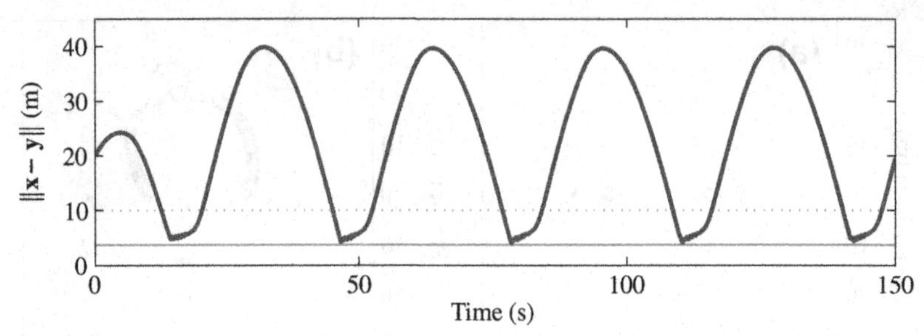

Figure 4.5: Distance Between the Vehicle and the Obstacle for the Noncooperative Case.

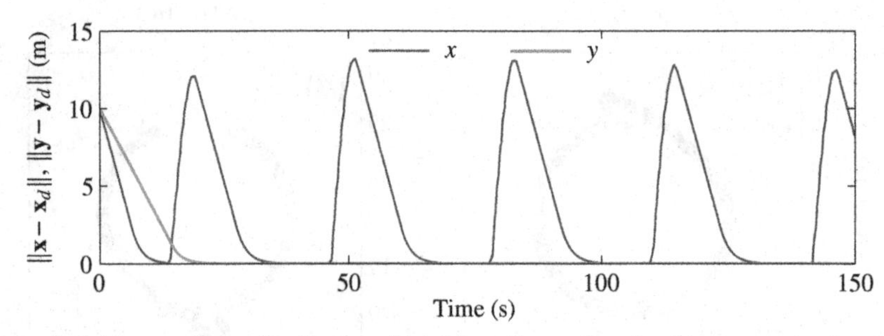

Figure 4.6: Norm of the Trajectory Tracking Error for the Noncooperative Case.

where the agents are commanded to go to each other's initial position. Such a scenario maximizes the risk of collisions and generally leads to a deadlock between both vehicles. Observe that the agents start traveling toward each other according to the trajectory tracking control. Once the agents enter each other's Detection Region (refer to Figure 4.8(b) and (c)), the y agent starts retreating while the x vehicle decreases its velocity. Simultaneously, both agents gradually shift to opposite sides according to (4.24). This scenario continues until the x vehicle reaches its destination and the y agent is able to retake its desired course (see Figure 4.8(e)). Eventually both agents are able to converge to their intended destinations.

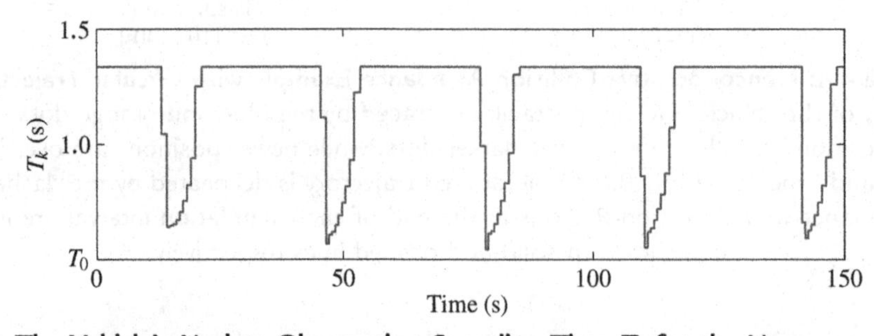

Figure 4.7: The Vehicle's Update Observation Sampling Time T_k for the Noncooperative Case.

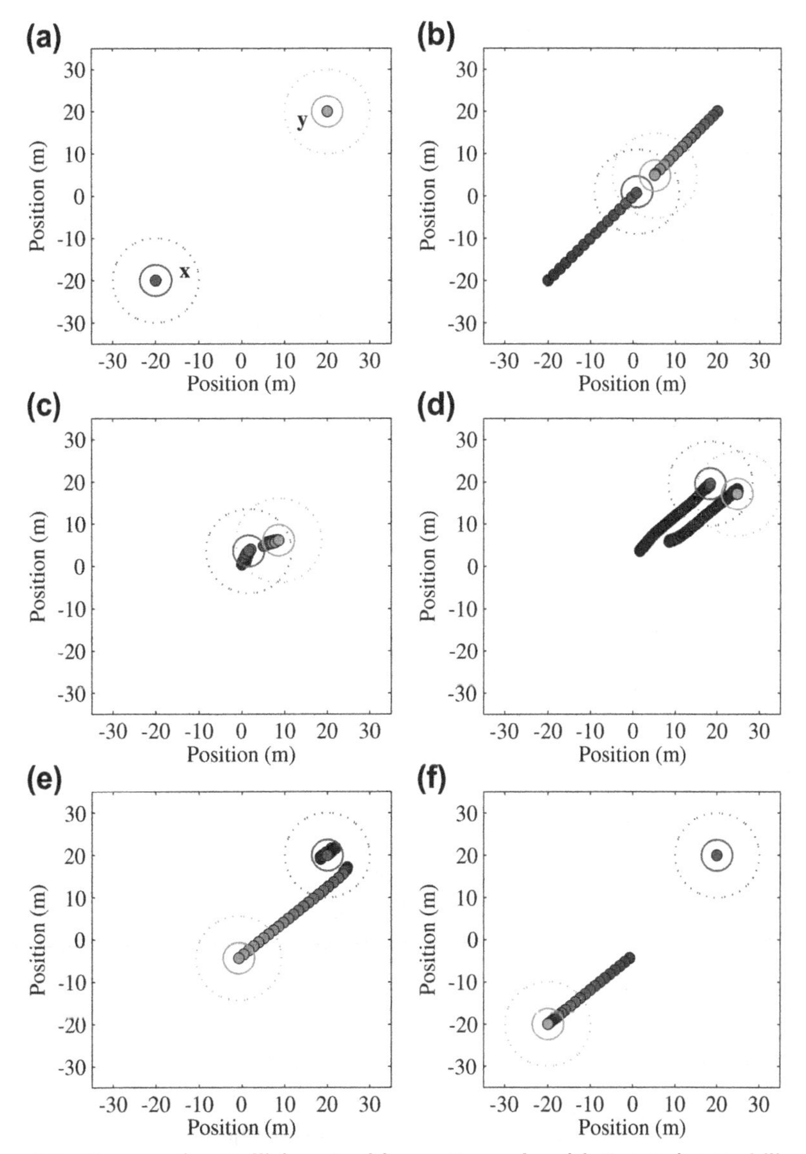

Figure 4.8: Cooperative Collision Avoidance Example with Set-Point Stabilization. The vehicles' Avoidance and Detection Regions at the end of each simulation interval are indicated by the circles with solid and dashed lines respectively. (a) $t = 0$ s. (b) $t = [0\text{ s}, 8\text{ s}]$. (c) $t = [8\text{ s}, 16\text{ s}]$. (d) $t = [16\text{ s}, 70\text{ s}]$. (e) $t = [70\text{ s}, 85\text{ s}]$. (f) $t = [85\text{ s}, 100\text{ s}]$.

Figure 4.9 illustrates the distance between the agents for the entire trajectories. Note that the agents remain outside of each other's Avoidance Region. Similarly, the observation sampling step is plotted in Figure 4.10. The first sampling time for the y vehicle was chosen as $t_1 = T_0/2$ to yield an asynchronous observation update time between both agents. Note that T_k increases when the agents become closer and decreases as they separate.

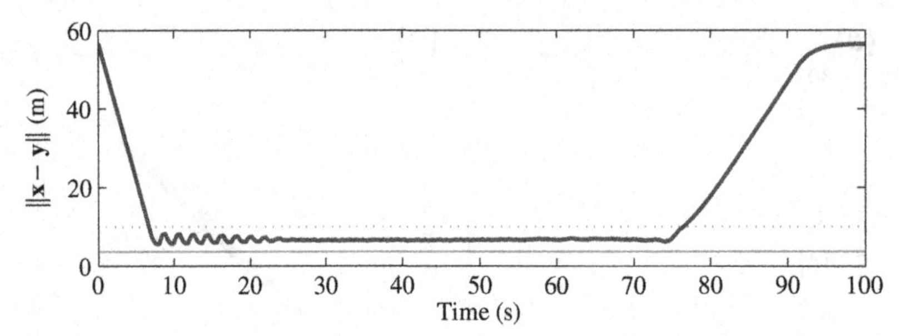

Figure 4.9: Distance Between Both Vehicles for the Cooperative Case.

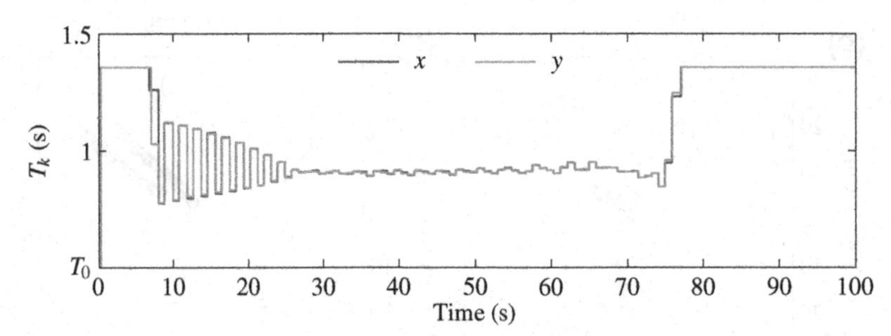

Figure 4.10: The Vehicles' Update Observation Sampling Time T_k for the Cooperative Case.

4.7 Conclusions

In this chapter, we presented noncooperative and cooperative avoidance control strategies with discrete observations that guarantee collision-free trajectories for a two-agent system with actuation constraints. We built the control framework on the concept of avoidance functions and formulated sufficient conditions for guaranteed avoidance using Lyapunov-based analysis. The avoidance strategies admit vehicles with limited sensing range as well as maximum observation sampling rate and are paired with a bounded trajectory tracking control law. We then discussed the existence of local minima and proposed a heuristic solution to escape deadlocks and traps. The solution takes the form of an additional control input, perpendicular to the collision threat, that perturbs the trajectory of the vehicle whenever the system approaches a deadlock. The performance of the control strategies is finally illustrated via two examples. In both examples, it is shown that the vehicles successfully avoid collisions while converging to their desired trajectories.

References

[1] D.V. Dimarogonas, S.G. Loizou, K.I. Kyriakopoulos, M.M. Zavlanos, A feedback stabilization and collision avoidance scheme for multiple independent non-point agents, Automatica 42 (2) (2006) 229–243.

[2] T. Fraichard, J.J. Kuffuer, Guaranteeing motion safety for robots, Autonomous Robots 32 (3) (April 2012) 173–175.

[3] E. Lalish, K.A. Morganseu, Distributed reactive collision avoidance, Autonomous Robots 32 (3) (April 2012) 207–226.
[4] E.J. Rodríguez-Seda, U. Troy, C.A. Erignac, P. Murray, D.M. Stipanović, M.W. Spong, Bilateral teleoperation of multiple mobile agents: Coordinated motion and collision avoidance, IEEE Transactions on Control Systems Technology 18 (4) (2010) 984–992.
[5] D.M. Stipanović, P.F. Hokayem, M.W. Spong, D. Siljak, Cooperative avoidance control for multiagent systems, Journal of Dynamic Systems, Measurement, and Control 129 (2007) 699–707.
[6] I.K. Kuchar, L.C. Yang, A review of conflict detection and resolution modeling methods, Transaction on Intelligent Transportation Systems 1 (4) (2000) 179–189.
[7] A. Vahidi, A. Eskandarian, Research advances in intelligent collision avoidance and adaptive cruise control, Transaction on Intelligent Transportation Systems 4 (3) (2003) 143–153.
[8] Y. Zhai, L. Li, G.R. Widmann, Y. Chen, Design of switching strategy for adaptive cruise control under string stability constraints, In: Proceedings of the American Control Conference, San Francisco, CA, 2011, pp. 3344–3349.
[9] R. Isaacs, Differential Games: A Mathematical Theory with Applications to Warfare and Pursuit, Control and Optimization, John Wiley, New York, 1965.
[10] S.H. Lim, T. Furukawa, G. Dissanayake, H.F. Durrant-Whyte, A time-optimal control strategy for pursuit-evasion games problems, in: Proceedings of the IEEE International Conference on Robotics and Automation, New Orleans, LA, 2004, pp. 3962–3967.
[11] D.M. Stipanović, A. Melikyan, N. Hovakimyan, Some sufficient conditions for multi-player pursuit-evasion games with continuous and discrete observations, in: Advances in Dynamic Games and Their Applications, Annals of the International Society of Dynamic Games, Springer, Berlin, 2009, pp. 133–145.
[12] I.I. Hussein, D.M. Stipanović, Effective coverage control for mobile sensor networks with guaranteed collision avoidance, IEEE Transactions on Control Systems Technology 15 (4) (2007) 642–657.
[13] M. Lindhe, P. Ogreu, K.H. Johansson, Flocking with obstacle avoidance: A new distributed coordination algorithm based on Voronoi partitions, in: Proceedings of the IEEE International Conference on Robotics and Automation, Barcelona, 2005, pp. 1785–1790.
[14] A. Elfes, Using occupancy grids for mobile robot perception and navigation, IEEE Computer 22 (6) (1989) 46–57.
[15] S.M. LaValle, Planning Algorithms, Cambridge University Press, Cambridge, 2006.
[16] H.P. Moravec, Sensor fusion in certainty grids for mobile robots, AI Magazine 9 (2) (1988) 61–74.
[17] F. Kendoul, Survey of advances in guidance, navigation, and control of unmanned rotorcraft systems, Journal of Field Robotics 29 (2) (2012) 315–378.
[18] C. Cassandras, W. Li, Sensor networks and cooperative control, European Journal of Control 11 (4–5) (2005) 436–463.
[19] F.L. Chernousko, A.A. Melikyan, Some differential games with incomplete information, in: Optimization Techniques IFIP Technical Conference Novosibirsk, Lecture Notes in Control and Information Sciences, Springer, Berlin, 1975, 445–450.
[20] A.A. Melikyan, On minimal observations in a game of encounter, Journal of Applied Mathematics and Mechanics 37 (3) (1973) 407–414 (in Russian).
[21] D. Neveu, J.P. Pignon, A. Raimondo, J.M. Nicolas, O. Pourtallier, in: G.J. Olsder (Ed.), New Trends in Dynamic Games and Applications, Annals of the International Society of Dynamic Games, Birkh.
[22] G.J. Olsder, O. Pourtallier, in: G.J. Olsder (Ed.), New Trends in Dynamic Games and Applications, Annals of the International Society of Dynamic Games, Birkh.
[23] D.M. Stipanović, A. Melikyan, N. Hovakimyan, Guaranteed strategies for nonlinear multiplayer pursuit-evasion games, International Game Theory Review 12 (1) (2010) 1–17.
[24] E.J. Rodríguez-Seda, D.M. Stipanović, M.W. Spong, Collision avoidance control with sensing uncertainties, In: Proceedings of the American Control Conference, San Francisco, CA, 2011, pp. 3363–3368.
[25] J. van den Berg, J. Snape, S.J. Guy, D. Manocha, Reciprocal collision avoidance with acceleration–velocity obstacles, in: Proceedings of the IEEE International Conference on Robotics and Automation, Shanghai, China, 2011, pp. 3475–3482.

[26] T. Fraichard, H. Asama, Inevitable collision states: A step towards safer robots?, in: Proceedings of the IEEE/RSJ International Conference on Intelligent Robots and Systems, Las Vegas, NV, 2003, pp. 388–393.
[27] E.J. Rodríguez-Seda, D.M. Stipanović, M.W. Spong, Lyapunov-based cooperative avoidance control for multiple Lagrangian systems with bounded sensing uncertainties, in: Proceedings of the IEEE Conference on Decision and Control, Orlando, FL, 2011, pp. 4207–4213.
[28] G. Leitmann, J. Skowronski, Avoidance control, Journal of Optimization Theory and Applications 23 (4) (1977) 581–591.
[29] H. Khalil, Nonlinear Systems, Prentice Hall, New Jersey, 2002.
[30] O. Khatib, Real-time obstacle avoidance for manipulators and mobile robots, International Journal of Robotics Research 5 (1) (1986) 90–98.
[31] Y. Koren, J. Borenstein, Potential field methods and their inherent limitations for mobile robot navigation, in: Proceedings of the IEEE International Conference on Robotics and Automation, Sacramento, CA, 1991, pp. 1398–1404.

CHAPTER 5

Effect of Human Factors on Driver Behavior

Jianqiang Wang*, Keqiang Li*, Xiao-Yun Lu†

*State Key Laboratory of Automotive Safety and Energy, Tsinghua University, China †California PATH, ITS, University of California, Berkeley, USA

Chapter Outline

Advances in Intelligent Vehicles. http://dx.doi.org/10.1016/B978-0-12-397199-9.00005-7

5.1 Introduction

Road traffic accidents have always been a serious issue in modern society. According to statistical results, more than 90% of accidents have been caused by a driver's mistake and/or fatigue [1]. Therefore, the human driver's behavior has been an important component in Intelligent Transportation System (ITS) research. Some results on driver behavior have been applied to the development of intelligent vehicles [2–4]. There are various aspects of this research field. Some studies focused on specific driving scenarios, including car following [5] and lane changing [6]. The driver's physiological characteristics during driving, such as response time [7], cognition process [8] and fatigue [9], have also been investigated.

Driver behavior depends on many factors. This chapter will focus on the effect of human factors on driver behavior and analyze the differences in subjective evaluation and objective experiments.

5.2 Study Approach

5.2.1 Definitions of Driver Behavior and Characteristics

Driver Behavior

The manner of driver action during driving in real traffic situations with certain vehicles, road and environmental conditions.

The driver's behavior may include many aspects, such as the perception of traffic conditions, decision-making, vehicle operation, using cellphones and navigation systems, talking to other people in the vehicle, eating, drinking, applying cosmetics, looking around, etc.

The vehicle operation includes longitudinal and lateral driving, which reflects the perception of the road, decision-making, and driver's intention and action, such as car following, lane change, lane keeping, acceleration and deceleration, etc. Longitudinal driving behavior mainly focuses on vehicle movement along the driving direction, while lateral driving behavior mainly focuses on vehicle movement perpendicular to the driving direction. These features represent the variation of vehicle states and the relationship with other vehicles when the driver operates on the accelerator/brake pedal, steering wheel, gear, and in-vehicle switches. The vehicle state can be quantified as position, speed, acceleration, and steering angle and rate (single vehicle trajectory), as well as distance headway, time headway, and time to collision (inter-vehicle relationship).

Driver Characteristics

The driving traits, quality, and performance of the driver depend on physiology, psychology, knowledge, culture, traffic laws and regulations, driver's experience and temper, etc.

The driver characteristics may also be classified by skills and styles, such as prudence (aggressive vs. prudent), stability (unstable vs. stable), conflict proneness (risk prone vs. risk avoidance), skillfulness (non-skillful vs. skillful), and self-discipline (law-abiding vs. violation frequent).

5.2.2 Parameters Embodying Driver Behavior and Characteristics

In order to quantitatively analyze driver behavior, several parameters and terms are defined in Table 5.1, which include vehicle parameters and inter-vehicle parameters.

5.2.3 Definitions of Driver Operation and Vehicle Driving Scenarios

Driver operation under different driving scenarios will usually reflect driver behavior characteristics. To compare and comprehend the driver's behavior in the real world, several terms were defined as in Table 5.2. This includes driver actions and driving types. Longitudinal driving is divided into four types, i.e. solo driving, steady-state following, approaching, and non-restricted following.

5.2.4 Relation Diagram of Parameters

Some important parameters such as time headway (THW), time to collision (TTC), and time to lane crossing (TLC) were selected to describe driver behavior. The relation diagram of parameters will reflect driving styles and driver characteristics through behavioral data.

Table 5.1: Nomenclature and Definitions for Vehicle State Parameters

No.	Term	Symbol	Unit	Definition
1	v	Velocity	m/s	The speed of own (host) vehicle
2	v_l	Velocity	m/s	Speed of lead vehicle
3	v_r	Relative speed	m/s	Speed difference between the host vehicle and the relevant vehicle
4	a	Acceleration	m/s^2	Accelerator pedal position more than 5%, and the vehicle speed increasing
5	a_{dmax}	Maximum deceleration	m/s^2	Maximum absolute value of longitudinal deceleration when braking
6	a_{amax}	Maximum deceleration	m/s^2	Maximum value of longitudinal acceleration for every acceleration scenario
7	P_l	Lateral position	m	Distance from the center of vehicle to the lane mark
8	DHW	Distance headway	m	Distance to the lead vehicle
9	THW	Time headway	m	Distance headway divided by the host vehicle speed v
10	TTC	Time to collision	s	Distance headway divided by the relative speed v_r
11	TTCi	The inverse of TTC	s^{-1}	Inverse of TTC
12	TLC	Time to lane crossing	s	Distance from left/right wheel to left/right lane mark divided by the lateral speed
13	TLCi	Inverse of time to lane crossing	s^{-1}	Inverse of TLCi
14	TTCi_1stAccR	The inverse of TTC	s^{-1}	TTCi when the driver first releases the acceleration pedal in approaching.
15	TTCi_1stBra	The inverse of TTC	s^{-1}	TTCi when the driver first activates the brake pedal in approaching.

Scenario A: Steady-State Following

(a) Vehicle Speed vs. DHW

Figure 5.1 shows a sketch map of the relationship between vehicle speed and distance headway, which to a great extent reflects the car-following distance at different vehicle speeds.

(b) Time Headway Histogram

Figure 5.2 shows a sketch map of the distribution of time headway, which to a great extent reflects the THW the driver prefers to keep under a steady-state following scenario.

Table 5.2: Nomenclature and Definitions of Driver Operation and Driving Scenarios

No.	Symbol	Definition
1	Braking	Activation of brake pedal longer than 1 s when the host vehicle speed is decreasing
2	Accelerator pedal release	Accelerator pedal position data becoming lower than a threshold (5% of the allowable deflection)
3	Brake pedal activation	Braking action, and "Brake Pedal Switch" signal becoming 1 (On) from 0 (Off)
4	Solo driving	Host vehicle in motion without a lead vehicle (distance headway set to 0 m)
		Host vehicle speed larger than a set threshold (e.g. 30 km/h in this study for the jammed traffic)
5	Car following	Host vehicle behind a lead vehicle and both in motion; no braking action during the following
6	Steady-state following	Relative speed is very low. Car following with TTCi lower than 0.05 s^{-1} (a speed threshold depending on the road geometry); no braking action
7	Non-restricted following	Car following without any restriction on TTCi
8	Approaching	Gap closing lasts longer than a certain period of time (when the host vehicle speed is greater than the lead vehicle speed)
		Car following under the following conditions: (1) the duration of gap closing is longer than 5 s; (2) TTC is less than 50 s

Scenario B: No Restricted Following

(a) THW vs. TTCi Distribution

Figure 5.3 shows a sketch map of the distributions of THW and TTCi. The contour graph of the distribution is usually shown, the contours being 25%, 50%, 75%, 95%, and 99%. The relation of THW and TTCi to a great extent reflects what safety degree the driver desires under different THW.

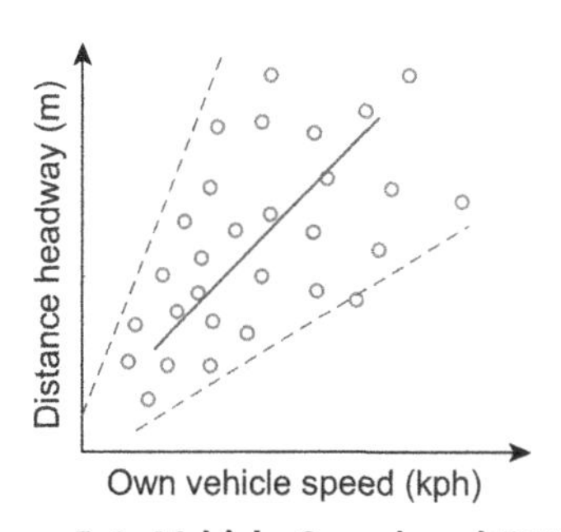

Figure 5.1: Vehicle Speed and DHW.

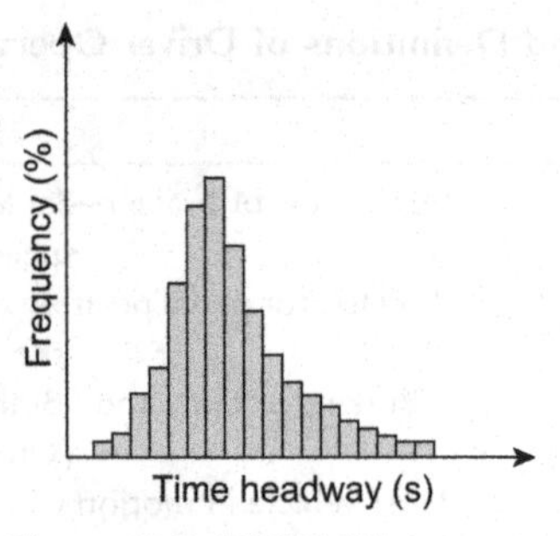

Figure 5.2: THW Histogram.

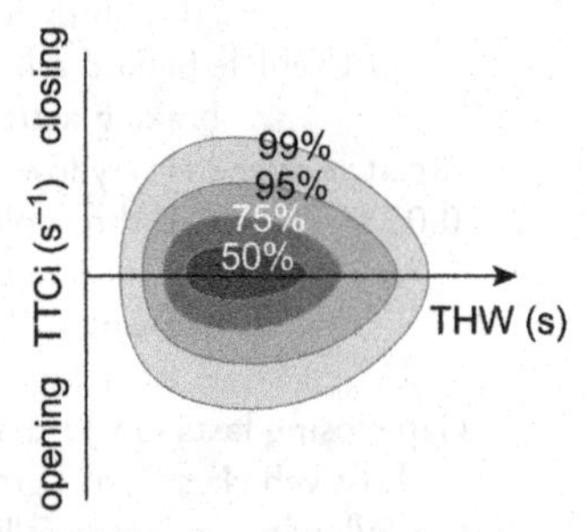

Figure 5.3: Distribution of THW and TTCi.

(b) Longitudinal Acceleration vs. TTCi Distribution

Figure 5.4 shows a sketch map of the distribution of longitudinal acceleration and TTCi. The graph shows the distributions of contours 25%, 50%, 75%, 95%, and 99%. The relationship of longitudinal acceleration and TTCi to a great extent reflects the driver's operating style.

Scenario C: Approaching

(a) Relative Speed and DHW at the Driver's Actions

Figure 5.5 shows a sketch map of the relationship between relative speed and distance headway relative to the driver's actions, which reflects the threshold for the driver to judge

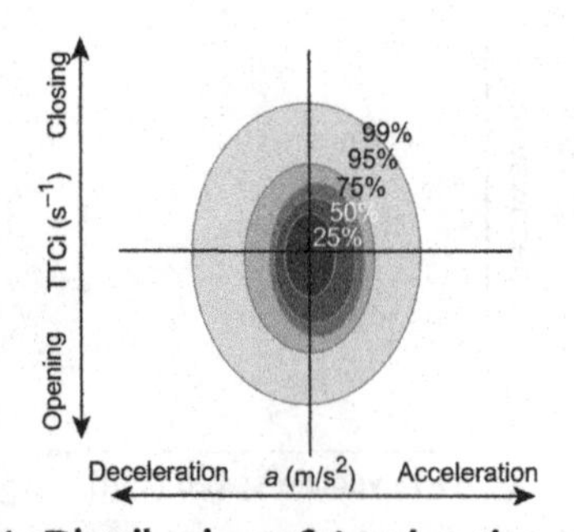

Figure 5.4: Distribution of Acceleration and TTCi.

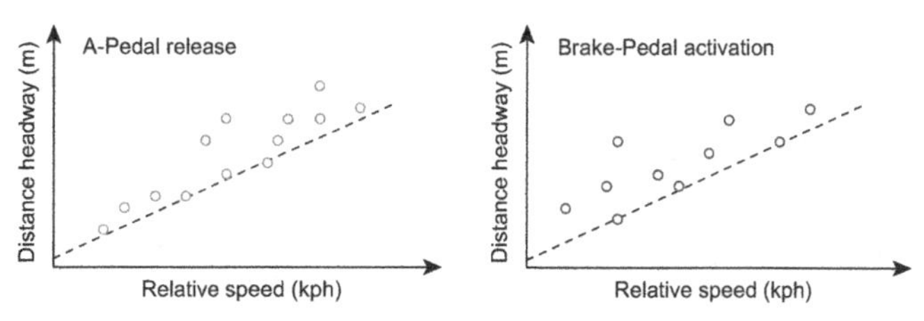

Figure 5.5: Relative Speed and DHW Under Driver Control.

safety and risk. Figure 5.5 represents the acceleration pedal release (left) and brake pedal activation (right).

(b) Subject Vehicle Speed and TTC at the Driver's Actions

Figure 5.6 shows a sketch map of the relationship between own vehicle speed and time to collision for driver actions, which to a great extent reflects the judgmental principle of the risk and decision-making mechanism of operating a vehicle for the driver.

5.2.5 Data Processing and Analysis Approach

The Kalman filtering algorithm is widely applied in signal processing. It is based on the best estimate rule for the estimation of least mean-square error in search of a recursive estimate. It is suitable for real-time processing and computer operation. The Kalman filter was used in Ref. [10] to eliminate noise from the radar data of the longitudinal acceleration in the host vehicle. The detailed mathematical equations of the Kalman filter algorithm are given in Ref. [11]. Only data-processing results with the Kalman filter are presented below.

Figure 5.7(a) shows an example of the distance headway and relative speed data. The data are smoothed with stagnation eliminated. Figure 5.7(b) shows the longitudinal acceleration estimation result, which is much better than the derivative of the vehicle speed.

In this study, the *t*-test method is used for statistical comparison. This method can determine whether there are differences between both samples (such as male group and female group) with an unknown but equal population variance. It is a common method for hypothesis testing. The calculation is simple and the method is suitable for small sample cases [12].

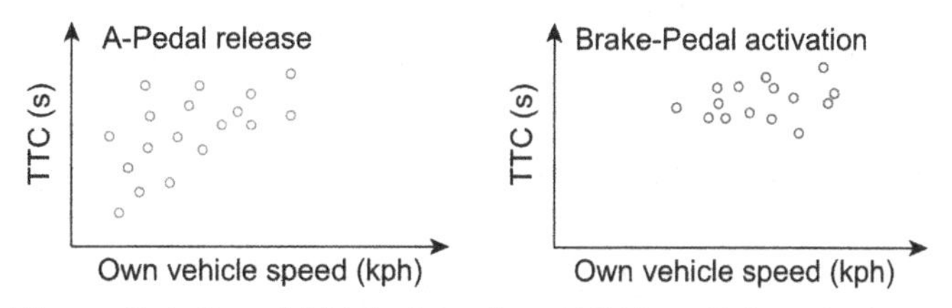

Figure 5.6: Own Vehicle Speed and TTC at Driver Actions.

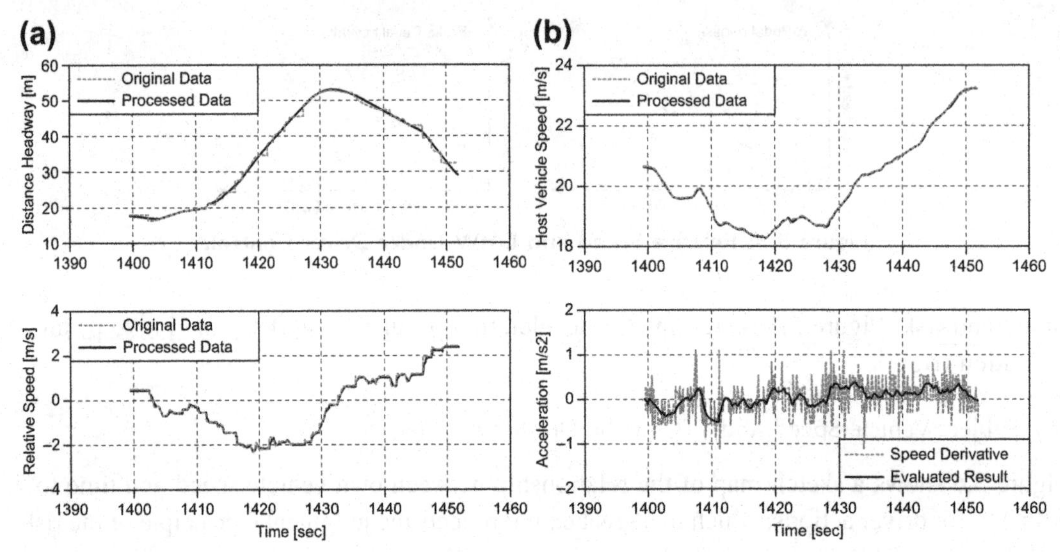

Figure 5.7: (a) Processed DHW and Relative Speed Data. (b) Processed Host Vehicle Speed and Estimated Acceleration.

We formulate a testing variable t as follows:

$$t = \frac{\overline{X} - \overline{Y}}{S_W \sqrt{\frac{1}{n_1} + \frac{1}{n_2}}} \tag{5.1}$$

$$S_W^2 = \frac{(n_1 - 1)S_1^2 + (n_2 - 1)S_2^2}{n_1 + n_2 - 2}, \tag{5.2}$$

where $\overline{X}$ and $\overline{Y}$ are the expectation values of the samples A and B, such as the expectation value of THW for urban access roads and for urban distributor roads. S_1 and S_2 are the standard deviations of samples A and B. n_1 and n_2 are the numbers of the two samples. Assuming that sample A has little difference from sample B, we make a hypothesis H_0 and an alternative hypothesis H_1:

$$\begin{aligned} H_0 &: E_X - E_Y = 0 \\ H_1 &: E_X - E_Y \neq 0, \end{aligned} \tag{5.3}$$

where E_X and E_Y are the expectation values of the two samples. We consider the results for $\alpha = 0.05$, which is the typical level of significance used in the t-test [13].

5.3 Experimental Equipment

Simulations and field tests are important methods for studying driver behavior and developing vehicular on-board electronics such as Advanced Driving Assistance Systems (ADAS) [14].

One simulator and one real vehicular experimental platform have been developed in our research, and are described below.

5.3.1 Simulator

It is generally accepted that the use of a proper driving simulator may shorten the system development cycle and reduce system development costs. A good simulator can be used for studying driver behavior and testing system performance in dangerous conditions and/or situations that cannot or are not easily able to be tested in the real world [15]. In addition, clear advantages of using a driving simulator include its operational flexibility and high level of safety. It can be easily used for recurring experiments, and for testing/evaluating various algorithms.

To comprehensively address driver behavior research, a driving simulation platform was developed in a MATLAB real-time simulation environment [16]. The platform combined two data flow loops, Hardware-in-the-Loop (HIL) and Driver-in-the-Loop (DIL). Its hardware consists of a simulation computer and monitoring computer, a vision computer, ADAS actuators, and a car mock-up. Its main software includes monitor software running in the monitor computer, a vision-rendering software running in the vision computer, and a Simulink-based model running in a simulation computer. These components have been properly integrated through interfaces.

The configuration of the driving simulation platform is shown in Figure 5.8. The car mock-up is a passenger vehicle from Nissan Ltd, with the engine removed. In the DIL simulation, the simulation computer collects driver's manipulating signals through a Control Area Network (CAN) bus connected with the car mock-up. Subsequently, the simulation model calculates the states of the host vehicle and those of its surrounding vehicles, and then sends them to the vision computer through an RS232 serial interface. By utilizing this information, the vision-rendering software in the vision computer generates virtual scenes and projects them on the screen to imitate the driver's view in an actual driving environment.

Monitoring Software

As required by ADAS, monitoring software should have four functions, specifically parameter adjustment, simulation control, real-time display, and data recording. Among the four required functions, real-time display is the most difficult task, because an xPC-based simulation computer does not have enough processing capability for display in real time. Therefore, a monitor was added. To maintain real-time characteristics for displaying data, the sampling frequency of monitor software must be at least 10 Hz. However, experiments have shown that data display would have apparent stagnations if driven by the MATLAB clock, since Graphic User Interface (GUI) was an interpreted programming language that ran

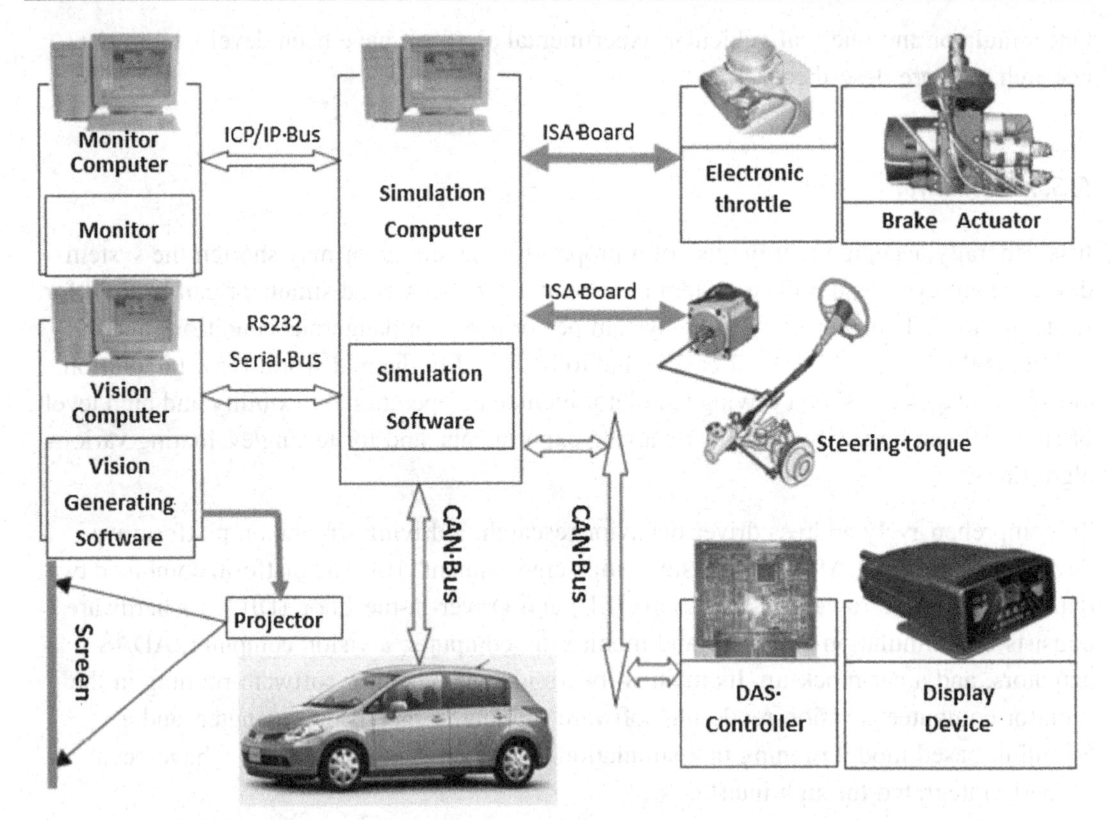

Figure 5.8: System Configuration of Driving Simulation Platform.
CAN, Control Area Network; ISA, Industrial Standard Architecture bus; RS 232, Recommended Standard 232 bus.

slowly. To address this issue, the monitor software was developed by a GUI-Driven-by-S-Function (GUIDSF) method [17]. The appearance and Human–Machine Interface (HMI) of the driving simulator are shown in Figure 5.9.

In the GUI, Active-X controls are used to achieve virtual meters and Real Time (RT) display. From xPC, the Target module in Simulink outputs simulation data to the S-function and then the S-function drives GUI to display them. Experiments indicated that its sampling frequency exceeded 30 Hz, which effectively avoided data-updating delays and display stagnation in a clock-driven method.

Vision-rendering Software

A critical issue of a virtual traffic scene is that it is weak in the sense of "immersion". Apart from its crude approximation to reality, the depth of field, size, and position of objects are also difficult to represent. They are closely related to viewpoint position and view angle

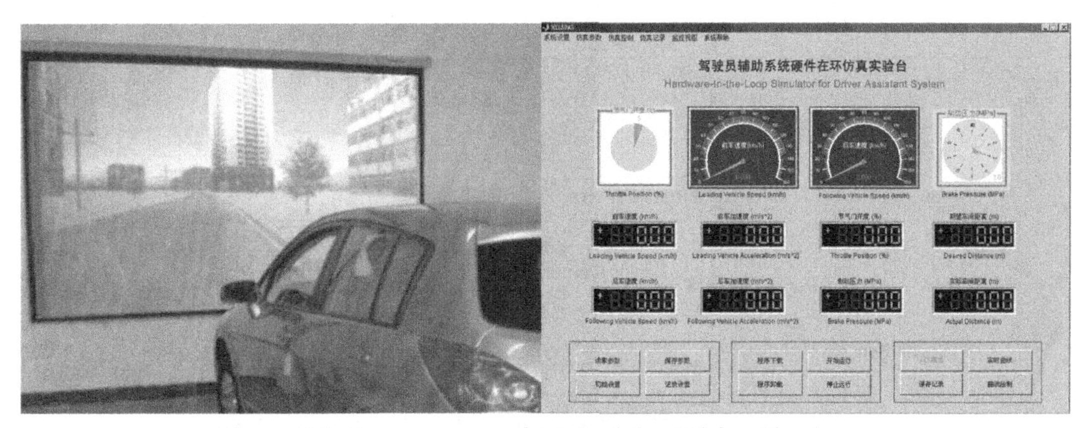

Figure 5.9: Appearance and HMI of the Driving Simulator.

range in the vision-rendering software, positions of car mock-up, screen and driver's eyes. Therefore, to strengthen the "immersion sense", viewpoint position and view angle range must be adjusted according to relative positions of car mock-up, screen, and driver's eyes. An adjustment method based on the principle of optical projection was proposed in Ref. [16].

Compared with existing driving simulators, this configuration possesses a simple structure, strong modularity, and good maintainability. It effectively avoids shortcomings such as excessive complication in structure, difficulties in development, and high cost. A GUI-Driven-by-S-Function method, based on monitor software, eliminates the display stagnation of simulation data. Compared with some effects on the virtual scene, the proposed adjustment method for vision-rendering software further strengthens the driver's feeling of immersion in the virtual traffic environment with improved reliability. It is closer to naturalistic driving and therefore helps in disclosing driver characteristics.

5.3.2 Experimental Vehicle

To investigate driver behavior, an instrumented vehicle testbed was built to measure the information on driver actions and vehicle status, to evaluate driver behavior and different performance, and to obtain the human factor parameters required for the development of driver assistance systems [18].

Hardware

The instrumented vehicle testbed is designed to capture all of the real-time information with sensors and a data collection system. Figures 5.10–5.12 show the system architecture of the testbed and the main equipment.

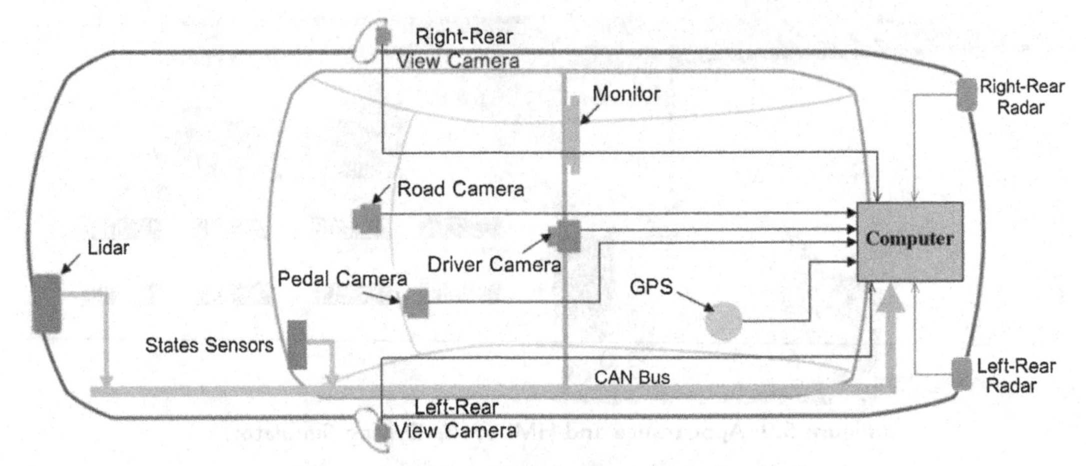

Figure 5.10: Equipment in the Vehicle Testbed.

A Lidar is used to detect the forward vehicle and other objects, and to measure the distance and relative speed. To validate the data of Lidar to meet the research requirement on driver lateral behavior, a CCD camera is used to capture the image of the frontal road environment, which is shown in Figure 5.11(a). The position of the vehicle is obtained with a high-precision GPS. The GPS data based on maps can help to distinguish the driver's

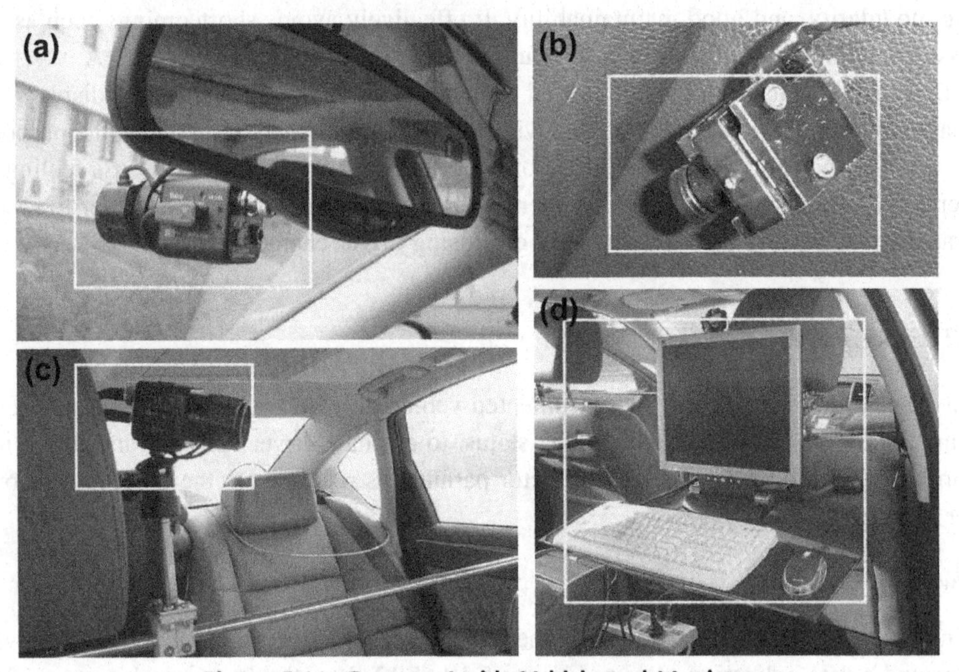

Figure 5.11: Cameras Inside Vehicle and Monitor.
(a) Forward road camera. (b) Pedal and foot camera. (c) Driver hand camera. (d) Monitor.

behavior in different road conditions. The state of the vehicle includes the vehicle speed, longitudinal and lateral acceleration, and yaw rate. There are four CCD cameras. Two of them installed inside the vehicle are shown in Figure 5.11(b) and (c), to capture the images of the driver's hand and foot movements respectively. The other two installed in the rear-view mirrors are shown in Figure 5.12(a) and (b) to capture the images behind the host vehicle.

Figure 5.12: Installation of Rear-View Cameras and Two Radars.
(a) Left: left rear-view camera. (b) Middle: right rear-view camera. (c) Right: rear radars.

The two radars that are installed behind the vehicle mainly collect the rear vehicle status in adjacent lanes for any relative lateral movement (Figure 5.12(c)). The turn light switch, the pedal positions, and steering angle are also recorded to model the driver's behavior.

Data Collection System

The data collection system consists of three parts: CAN bus data, the Lidar and the two radars' data, and the image data of the five cameras and the GPS data. The camera data and the GPS data are from the serial port. The CAN data of FUGA are collected based on the NI-CAN (PCI-CAN Series 2) card and the data collection software of the FUGA's measurement PC.

All of the experimental data are recorded in text file format. The five experimental images and the data display interface are synchronized with respect to the same timeline and are saved as videos.

Data Processing Program

The analysis of the driver experimental data was a complicated task. In order to provide functions for data review and processing, a data analysis program was developed with MATLAB GUI tools.

The program had the following functionalities:

1. Postprocessing the original CAN bus and video data. This step included synchronizing, rearranging and data saving.

2. Overlaying vehicle GPS locations, experimental data, and video image on the city map.
3. Extracting and plotting the intersected data set.
4. Achieving the data statistics and analysis results such as PDF of THW.

5.3.3 Comparison of Simulation and Field Experiments

Simulation is an important means of studying driver behavior, but the driving task cannot be rendered completely and realistically in a driving simulator. Therefore, a crucial issue for simulator design is to ensure that the results obtained in a driving simulator study are similar to a real traffic environment [19]. In previous studies, efforts of measures and comparisons of lateral displacement both on the road and in a driving simulator have been made [20], and simulator validity and fidelity were investigated [21]. However, no proper explanation of the differences between the results from the simulator and real-world experiments, such as lateral offset and time to lane crossing (TLC) [22] were obtained. Our work intends to fill this gap and to calibrate the simulation results with real-world data.

Comparison of Driving Behavior

Twelve drivers were invited to take both experiments for simulated and on-road driving. The data in both the simulator and the instrumented vehicle were recorded and analyzed. In this research, parameters for consideration include THW, TTCi–THW, lateral offset, and TLC.

(1) THW

THW is the parameter that can significantly reflect the driver's longitudinal characteristics. As Figure 5.13 shows, the THW distribution in simulator tests is similar to that in real-world tests. Most data concentrate on 0.5–2 s, and the peak is close to 1 s. Nevertheless, it can be observed that the THW from the simulator data is slightly smaller than that from the real-world data, which means that drivers tend to follow a little closer to the leading vehicle on a simulator than in real-world driving. This could be explained by: (1) drivers feeling safer when driving in a

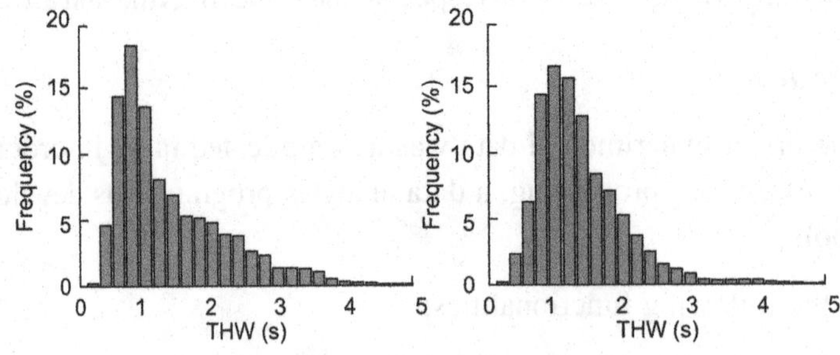

Figure 5.13: THW Distribution of the Driving Simulator Test (left) and the Real-World Test (right).

simulator than in the real world, i.e. the drivers think it unnecessary to pay similar attention to a potential collision on the simulator as in the real world; and (2) the relative distance estimation is different in the 3D virtual traffic environment from that in the real world.

(2) TTCi–THW

As Figure 5.14 shows, the TTCi distribution for the simulator test has similar tendencies as those in the real world. In spite of that, TTCi has a slightly larger and more decentralized PDD (Probability Density Distribution) in the simulator than in the real world, which reflects the driver's preference to follow closer to the leading vehicle in the simulator than in the real world. The result and explanation are consistent with those of the THW.

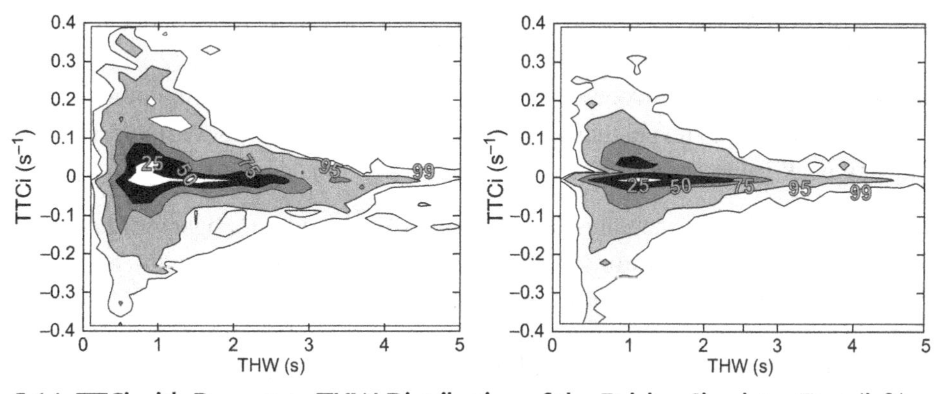

Figure 5.14: TTCi with Respect to THW Distribution of the Driving Simulator Test (left) and the Real-World Test (right).

(3) Lateral offset

Lateral offset reflects the vehicle position with respect to the lane centerline. It is the direct expression of the vehicle's lateral position (see Figure 5.15).

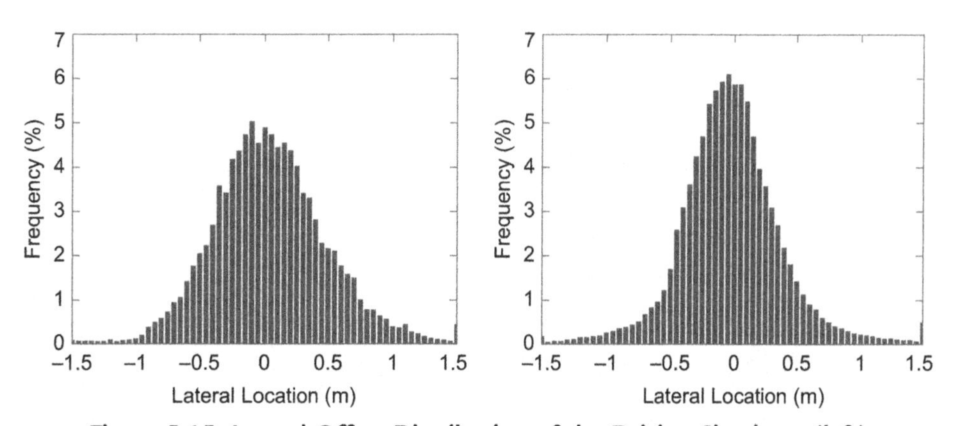

Figure 5.15: Lateral Offset Distribution of the Driving Simulator (left) and of the Real-World Test (right).

From Figure 5.15, it can be seen that, in the simulator (left panel), drivers do not have a preference for driving on the left or the right side of the lane. In addition, the PDD for simulator data is a little more dispersed. However, in real-world tests, drivers prefer driving on the left side of the lane. The PDD is much more centralized. Figure 5.15 (right) shows that the graph is high and narrow, and the peak close to −0.1 m.

(4) TLC

TLC is the most useful quantitative characteristic to express a vehicle's lane deviation. It provides a measurable criterion for a lane departure warning.

The value of TLC is significantly influenced by the speed of the driver in lane changing. TLC increases with increasing lateral offset and decreasing lateral speed.

Comparisons of the TLC value frequency distributions for the driving simulator and the real-world test (see Figure 5.16) reveal that they are similar, with peak values around TLC = 4.5 s, and frequency peak values near 1.5%.

The analysis of the test data indicates that the values of the driver's behavior parameters obtained from the driving simulator tests are different when compared to the real-world test results, but the differences are generally consistent for different data sets.

Correlation Model for Driving Simulator and Real-world Test

The aforementioned analysis shows that there exist differences in the results between the driving simulator data and the real-world data. Actually, lateral offset PDD is close to a normal distribution, while the THW and TLC PDDs are closer to a gamma distribution. Such anomalies lie in the difference between the distribution parameters.

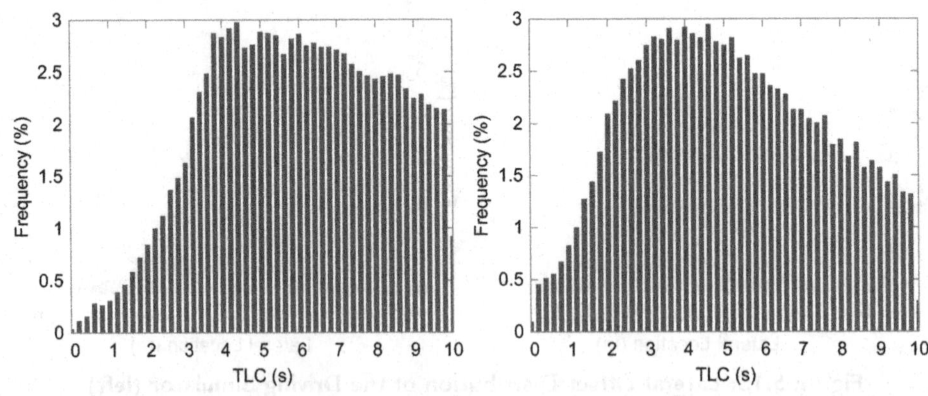

Figure 5.16: TLC Result Contrast: The Driving Simulator (left) vs. Real-World Test (right).

The normal distribution PDF is

$$\phi(x;\mu,\sigma) = \frac{1}{\sqrt{2\pi}\sigma}\exp\left[\frac{(x-\mu)^2}{2\sigma^2}\right], \quad x \in R. \tag{5.4}$$

The normal distribution is determined by μ and σ, where μ denotes the expected value and σ denotes the standard deviation.

The gamma distribution PDF is

$$f_G(x;r,\lambda) = \frac{\lambda^r}{\Gamma(r)}x^{r-1}e^{-\lambda x}, \quad x > 0. \tag{5.5}$$

The gamma distribution is determined by r and λ, where r denotes the shape parameter and λ denotes the scale parameter.

In Ref. [23], a calibration method for a correlation model between the two data sets was presented that was able to correct the driving simulator data using the real-word data with sufficient accuracy.

It can be concluded from the discussion above that an experimental tool is effective in driver behavior testing and analysis. However, the simulator test data should be calibrated with the real-world test data to improve simulator utility. The proposed model justifies the use of the simulator for the development of in-vehicle driving assistance systems as a feasible and solid approach.

5.4 Effect of Human Factors

Driver physical and mental characteristics are the critical factors affecting driver behavior, which depend on the driver's individual features such as gender, age, driving experience, education level, etc. In the first part of this section, we design an experiment to investigate the effects of human factors on driver behavior. In the second part, a comparison of relative factors is presented, e.g. male vs. female, different age groups, driving experience, education, nationality, etc.

5.4.1 Experiment Design

Driver Profile

There were 33 driver subjects involved in the driver behavior experiments. The driver's feature data including age, gender and years of driving experience were used to divide the drivers into groups.

- **Comparison of drivers' age.** The distribution of the driver age was nearly symmetrical as shown in Figure 5.17. The mean age was 44.9, and the variance and standard deviation were 116.87 and 10.81 respectively. The oldest and youngest were 69 and 30 respectively.

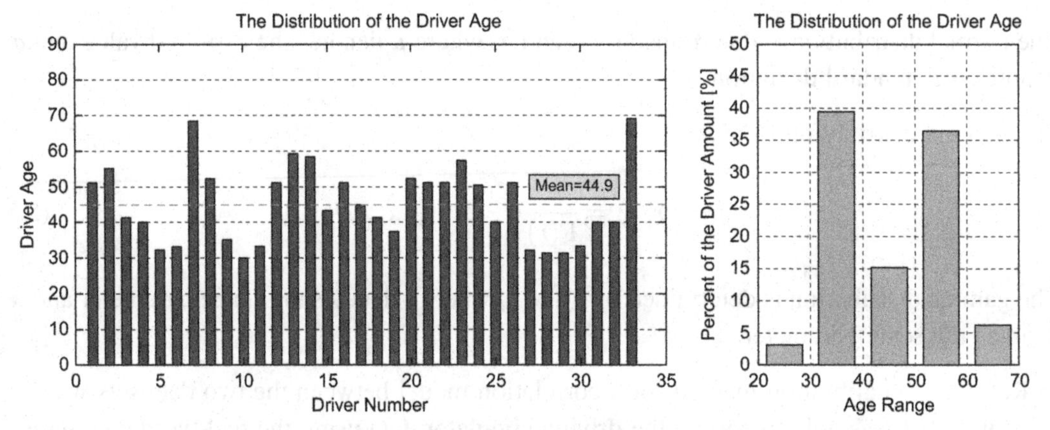

Figure 5.17: Distribution of Driver Age.

- **Drivers' gender.** The driver subjects were composed of seven female drivers and 26 male drivers. The distribution of driver gender is shown in Figure 5.18.
- **Years of driving experience.** It was difficult to define the driving experience quantitatively. For example, in China, some drivers only drive occasionally but still have good driving skills. In those cases, the number of years of driving experience was selected as the criterion. The distribution and statistics are showed in Figure 5.19. The mean value was 12.03. The variance and standard deviation were 128.28 and 11.33 respectively.

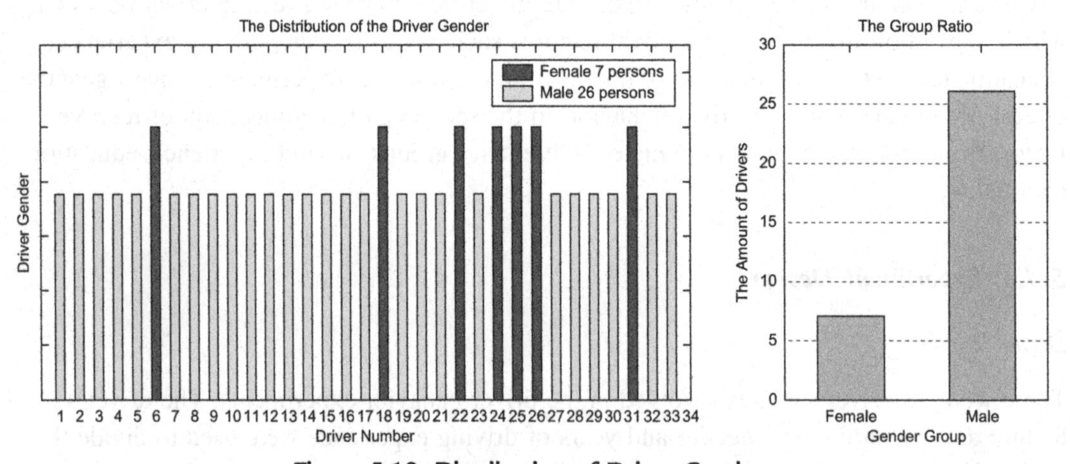

Figure 5.18: Distribution of Driver Gender.

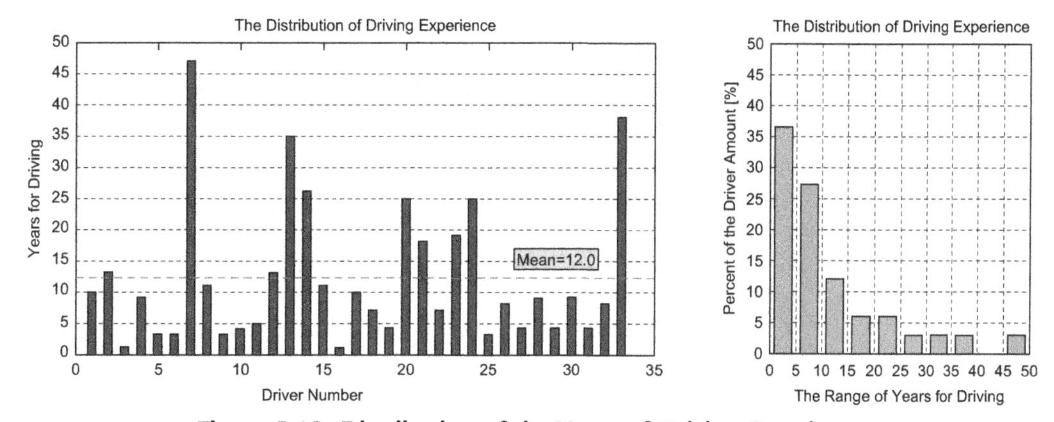

Figure 5.19: Distribution of the Years of Driving Experience.

Experimental Route Selection

Because the influence of the road conditions on driver behavior is important, different roads were selected purposely for the experiments. Based on the observation and analysis of the common roads in Beijing, three road types were chosen, including arterial urban highway ("Highway" or "Urban distributor road"), comparatively congested local roads ("City road" or "Urban access road") and intercity freeway ("Freeway" or "Through road"), and the experimental route was designed for efficiency and effectiveness. Figure 5.20 shows the route map.

During the experiment, it was found that, under most circumstances, when the instrumented vehicle crossed the intersections, the vehicle speed was so low and the distance to the leading vehicle was so small that the Lidar lost function. Therefore, the driver-following behavior at intersections was not considered in this study.

5.4.2 Comparison of Individual Driver Behavior

Figures 5.21–5.24 depict the distribution of drivers' statistic results, such as the amount of valid data sample points, mean DHW, mean THW, and mean vehicle speed for data collected on the entire highway section. Regarding the individual differences in longitudinal driving behavior, it was found that different drivers have different driving styles and characteristics.

5.4.3 Age

Age is a demographic variable frequently used in studies on driver behavior. Younger drivers have the highest rate of accidents [24]. They are significantly overrepresented among all drivers involved in traffic accidents and fatalities, and are much more likely than older drivers to be responsible for the crashes in which they are involved [25]. Golias and Karlaftis [26]

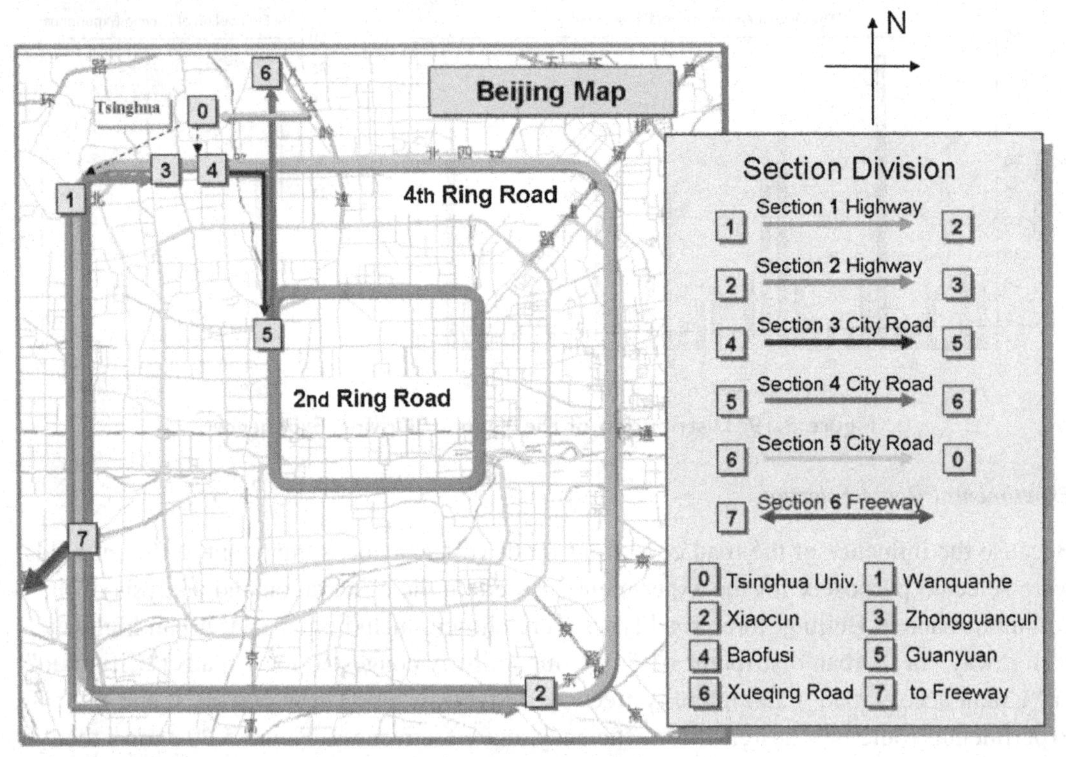

Figure 5.20: Experimental Route Map.

found that when age was considered, drivers seemed to become more law abiding and to take fewer risks as they grew older. Drivers over 55 years old seem to drive distinctly more carefully than younger drivers, while those below 25 years old seemed to exert a distinctly less law-abiding approach to driving or were more prone to regulation violations. Traffic accidents can be caused by different age-related factors, such as visual attention and risky

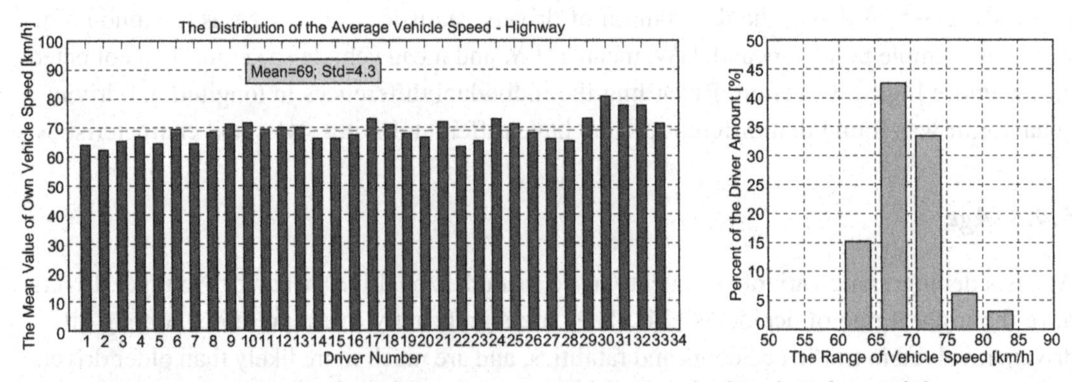

Figure 5.21: Distribution of the Average Vehicle Speed of Each Driver, Highway.

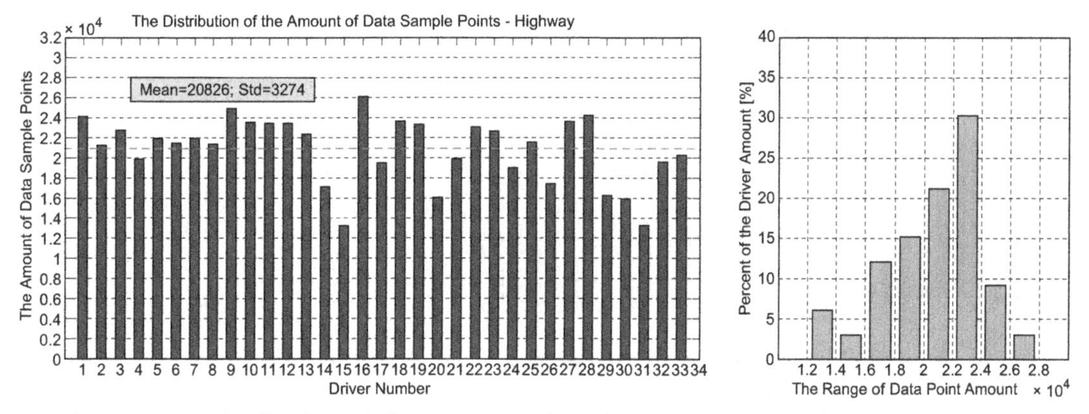

Figure 5.22: Distribution of the Data Sample Points Amount of Each Driver, Highway.

driving styles [27]. According to videotape data, Finn and Bragg [25] found that the pedestrian sequence was seen as more risky by young drivers, while the tailgating sequence was seen as more risky by older drivers. For example, Yagil [28] concluded that younger drivers and male drivers expressed a lower level of normative motivation to comply with traffic laws than do female and older drivers. The commission of traffic violations was found to be related more to the evaluation of traffic laws among men and younger drivers, compared to women and older drivers.

To compare the influence of age on driver behavior, the age groups used in this study were: 18–34 (eight subjects), 35–44 (nine subjects), 45–64 (14 subjects), and 65+ (two subjects). The results are shown in Figures 5.25–5.28. The *t*-test was used to compare the differences with the results listed in Table 5.3.

Regarding age groups, it was found that the values of THW, TTCi, and TTC were different among age groups. Also, the trend of variation was uncertain. Furthermore, the values of

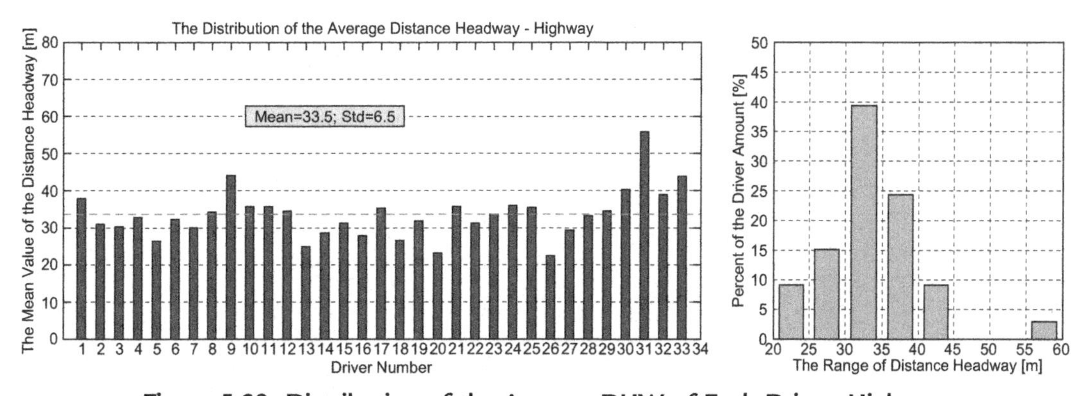

Figure 5.23: Distribution of the Average DHW of Each Driver, Highway.

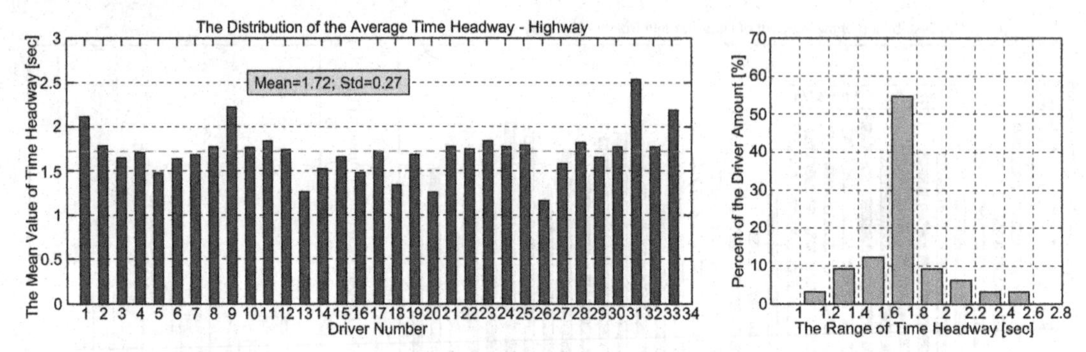

Figure 5.24: Distribution of the Average THW of Each Driver, Highway.

speed and DHW of each age group were very similar (see Figures 5.25 and 5.26), except the results for 65+ drivers on through roads. The reason for this could be that the sample size of 65+ drivers was too small, and elderly drivers might be more prudent on the through roads with high speed than young drivers. In addition, it appeared that the values of the standard deviation for each parameter were always significantly larger for the young group than those of other groups. Therefore, it could be concluded that elderly drivers had a relatively stable driving style.

5.4.4 Gender

It has been recognized that men and women exhibit different driving behaviors. A great deal of literature supports the greater driver crash rates for males when compared to females (for example, Ref. [29] and many others). The evidence of gender differences in driving behavior can be established more on a natural psychological basis than on experience and differences in capabilities and driving skills. According to two independent surveys in

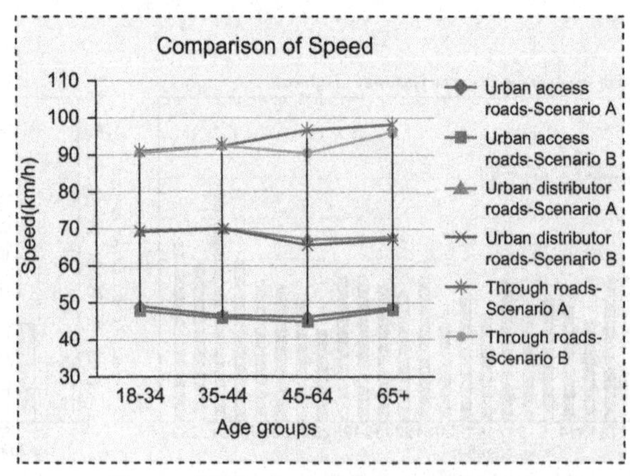

Figure 5.25: Comparison of Speed.

Table 5.3: Absolute Values of *t*-Test Variables Based on Different Age Groups

Variables			Age (18–34 and 35–44) Rejection Region: $\|t\| > 2.1315$	Age (18–34 and 45–64) Rejection Region: $\|t\| > 2.0860$	Age (18–34 and 65+) Rejection Region: $\|t\| > 2.3060$
THW	Urban access roads (Uar)	Scenario A	1.6389	0.4662	0.3855
		Scenario B	1.4890	0.2669	0.4782
	Urban distributor roads (Udr)	Scenario A	0.9127	0.5053	1.7246
		Scenario B	0.7835	0.8170	1.4520
	Through roads (Tr)	Scenario A	1.7383	0.8494	3.7164
		Scenario B	1.8861	1.3775	3.8139
TTC	Urban access roads	Accelerator release (Acc)	0.4619	0.2989	0.5938
		Brake activation (Bra)	0.4206	0.0155	0.2467
	Urban distributor roads	Accelerator release	1.3846	1.0060	0.1237
		Brake activation	0.3446	1.0293	0.9660
	Through roads	Accelerator release	0.7579	1.5552	1.1037
		Brake activation	1.0016	0.0884	0.5914

the years 1978 and 2001 in Finland, Laapotti et al. [30] found that the difference in driving behavior between males and females remained unchanged, or even increased in some aspects. The differences involved traffic accidents and offenses, although the driving times, attitudes, education, and other background factors were controlled. On average, men exhibited higher

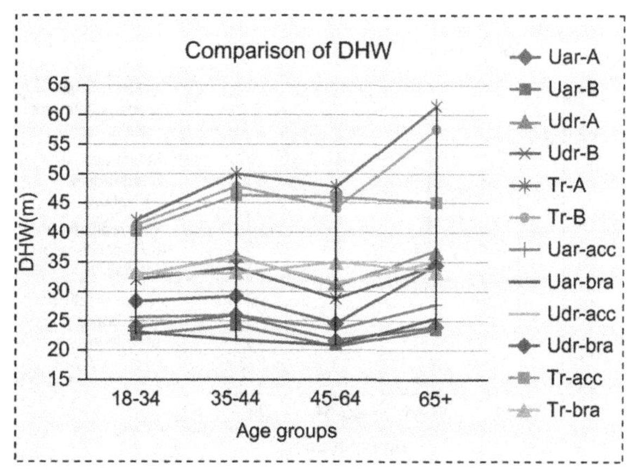

Figure 5.26: Comparison of DHW.

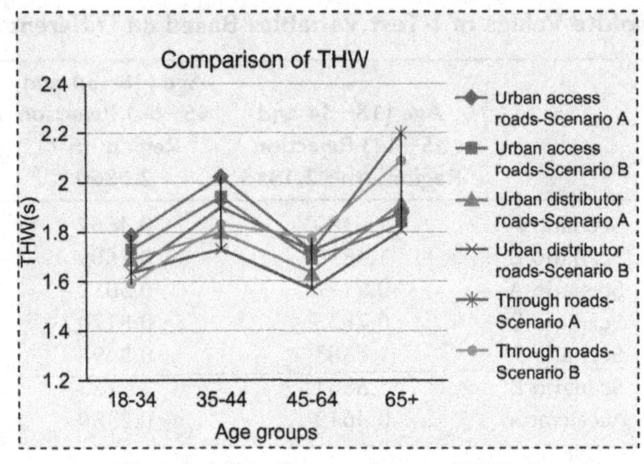

Figure 5.27: Comparison of THW.

levels of sensation-seeking, risk-taking, and deviant behavior when compared to women, which could be explained, at least in part, using a psychology perspective [31]. For frequent violators, whether the driver was male or female seemed to make no difference to the frequency of active crash involvement; they were equally at risk [32].

However, according to the literature on aggressive driving, the role of gender is a very complex issue [33]. The research presented in Ref. [31] provided a detailed review on aggressive driving. Work in Ref. [33] also contained substantial research on gender differences in drivers' aggressive behavior.

What about Chinese drivers? In this study, of the 33 drivers, 26 are male and seven are female. The comparative analysis results are shown in Figures 5.29–5.33.

t-Test results are listed in Table 5.4. According to the analysis, no significant differences between gender groups were found. The difference in THW between males and females is only about $\pm 1\%$, which indicates that gender is not the principal factor in driver behavior

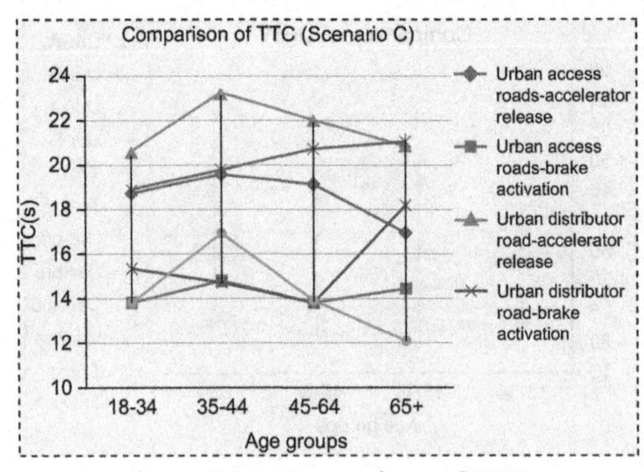

Figure 5.28: Comparison of TTC.

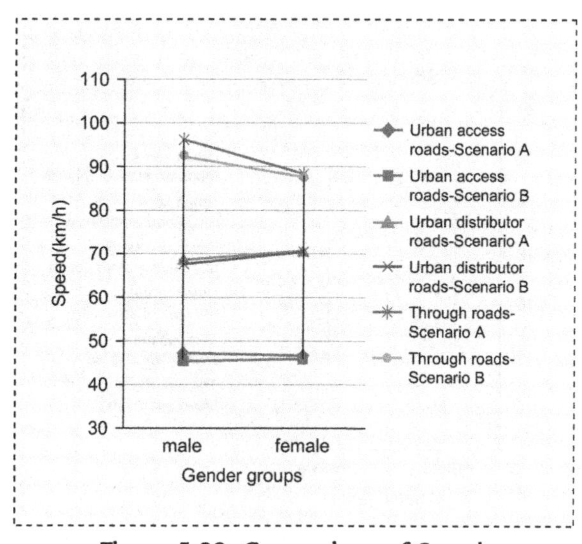

Figure 5.29: Comparison of Speed.

characteristics. It was found that the standard deviation of the THW of female groups is obviously larger than for male groups (see Figure 5.33). A plausible explanation for this could be that female Chinese drivers are generally less skillful than male Chinese drivers in very complex mixed traffic conditions (vehicles, non-motorized vehicles, and pedestrians share the same lane). Most female drivers were less experienced than male drivers before the time of the test (in 2005). Even if female drivers had driving licenses for many years, they had less chance to drive. The other reason might be that the number of female participants was less than that of male participants.

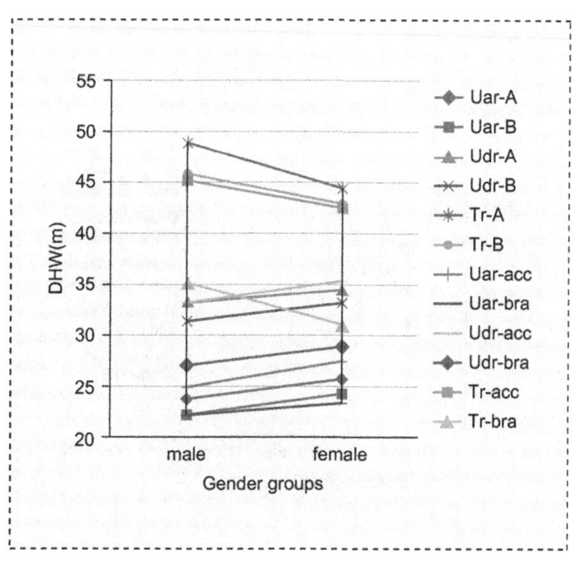

Figure 5.30: Comparison of DHW.

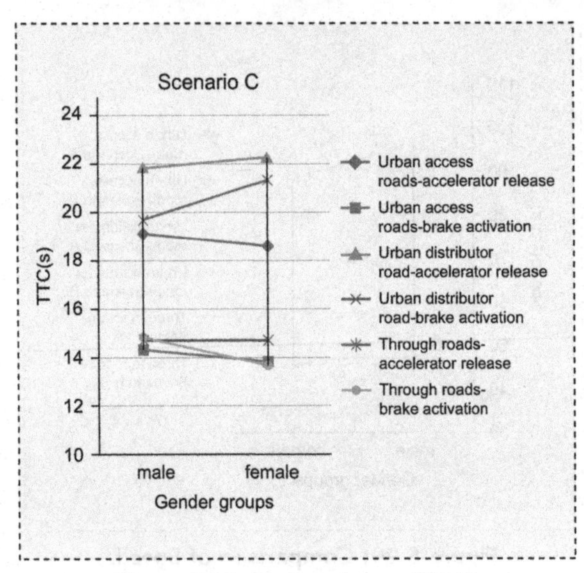

Figure 5.31: Comparison of TTC.

5.4.5 Driving Experience

Mourant and Rockwell [34] found that the visual acquisition process of the novice driver was unskilled and overloaded. Compared to experienced drivers, novice drivers lacked adequate coverage of neighborhood visual scenes, and looked closer in front of the vehicle and more to

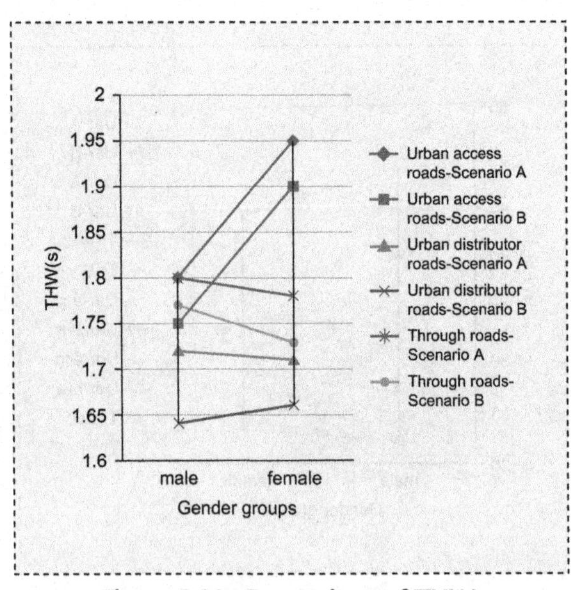

Figure 5.32: Comparison of THW.

Table 5.4: Absolute Values of *t*-Test Variables Based on Different Gender Groups

Variables Rejection Region: $\|t\| > 2.0395$			Gender (Male and Female)
THW	Urban access roads	Scenario A	1.0777
		Scenario B	1.2182
	Urban distributor roads	Scenario A	0.0838
		Scenario B	0.1938
	Through roads	Scenario A	0.1288
		Scenario B	0.2893
TTC	Urban access roads	Accelerator release	0.3896
		Brake activation	0.3171
	Urban distributor roads	Accelerator release	0.3146
		Brake activation	0.0276
	Through roads	Accelerator release	1.5323
		Brake activation	0.5281

the right of the vehicle's direction. Instead of making only eye contacts on the freeway route as experienced drivers do, novice drivers tended to pursue eye movements. Brown and Groeger's finding [35] on risk perception showed a significant role of driving experience in the development of schemata that accurately represented the spatio-temporal characteristics of vehicles and road traffic. Novice drivers initially use knowledge-based behavior to shift gears, while experienced drivers use skill based on an automatic pattern of action [36]. It was proposed that drivers at rule- or skill-based levels operated more homogeneously and predictably than those at a knowledge-based level [37]. Compared with experienced drivers, novice drivers were more likely to underestimate hazards, while experienced drivers were more likely to show anticipatory avoidance of a hazard by changing speed, direction, level of vigilance, focus of attention, and information transmitted to other road users [38].

In our study, no novice drivers were involved. In addition, the participants could not accurately evaluate driving mileages because of uncertainty in vehicle usage, but most

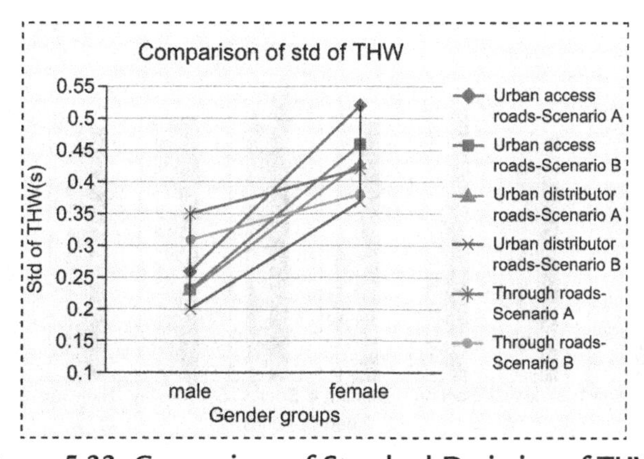

Figure 5.33: Comparison of Standard Deviation of THW.

participants (27 subjects) said they drove more than once each week. For these reasons, we simply divided the subjects into two groups: a group of drivers with driving experience over 10 years (12 subjects), and a group with 10 years or less (21 subjects).

From the results shown in Figures 5.34 and 5.35 and Table 5.5, no significant differences exist between the two groups. The differences in THW between the two driving groups are about 1%, and the differences in TTC accelerator release are about 3%. We can conclude that driving experience is not the principal factor affecting driving characteristics except for novice drivers.

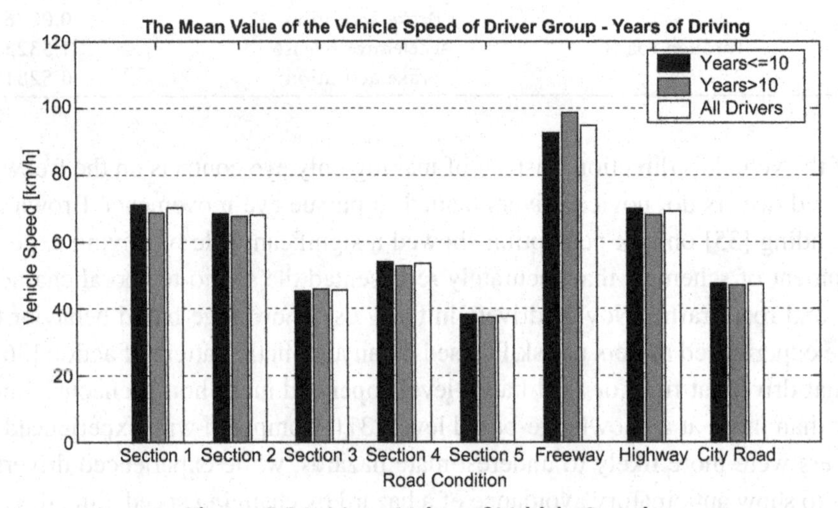

Figure 5.34: Mean Value of Vehicle Speed.

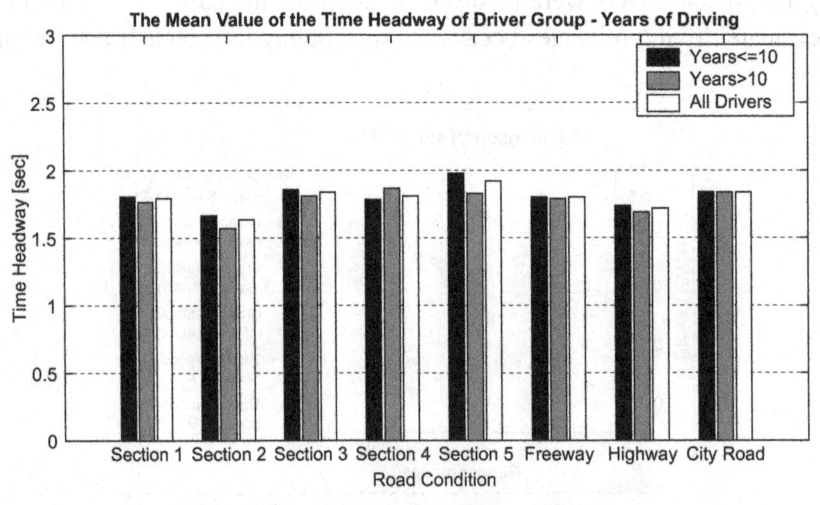

Figure 5.35: Mean Value of Time Headway.

Table 5.5: Absolute Values of *t*-Test Variables Based on Different Driving Experience

Variables Rejection Region: $\|t\| > 2.0395$			Driving Experience (≤10 years and >10 years)
THW	Urban access roads	Scenario A	0.1689
		Scenario B	0.0910
	Urban distributor roads	Scenario A	0.2022
		Scenario B	0.2337
	Through roads	Scenario A	0.1492
		Scenario B	0.7369
TTC	Urban access roads	Accelerator release	0.5345
		Brake activation	0.0146
	Urban distributor roads	Accelerator release	0.7449
		Brake activation	0.6662
	Through roads	Accelerator release	0.7949
		Brake activation	0.9940

5.4.6 Workload

In this study, workload means the degree of drowsiness. The concept of "drowsiness" is a qualitative factor that is very difficult to quantify. No "gold standard" of drowsiness exists to express drowsiness numerically and accurately. Wierwille and Ellsworth [38] examined the method of evaluating the degrees of drivers' drowsiness using their facial video. This subjective evaluation has been used in many research studies and has proved to be consistent and accurate. In this approach psychologically trained raters assess the driver's drowsiness from the signs of their facial video. These signs are indicative of drowsiness, and include rubbing the face or eyes, facial contortions, moving restlessly in the seat, slow eyelid closures, etc.

In this study, we developed this approach further and created a subjective evaluation criterion called VSCD (Video Scoring Criterion of Drowsiness), as shown in Table 5.6. The VSCD focused on four features of facial expression/movement to assess driver's level of drowsiness [39]:

1. Eye features such as pupils movement, blinking speed, and blinking frequency.
2. Breathing features such as yawns and deep breaths.

Table 5.6: Video Scoring Criterion of Drowsiness (VSCD)

Drowsiness Level	Description	Score
Not Drowsy	Eyes open as normal; Quick blinking; Active pupils movement; Upright head/body posture, etc.	1
Some Drowsiness	Eyelids become heavy; Less active pupil movement; Yawn, Deep breath, Adjusting posture, etc.	2
Very Drowsy	Eyelids become very heavy; Long eye close duration; Unintentional nodding head; Less upright posture, etc.	3

3. Movements such as scratching (eyes/face/head), adjusting posture (head/body), and nodding.
4. The rater's subjective evaluation of the whole facial expression.

Three raters were trained to look into these features of drowsiness and make a subjective, but specific, assessment of the level of drowsiness. Data from the facial video had been acquired from a CCD camera fixed on the car panel for data collection. The camera view covered both the driver's face and parts of the body, so that the rater could see the drowsy driver fidgeting, stretching, having heavy eyes, etc. After the experiment, the video record was digitized and cut into 1-minute segments by special software called Fatigue Driving Scoring System. It was then screened for the raters in random sequence. Each segment was scored by three raters independently using the VSCD. To reduce subjectivity, if the difference of the scores given by the two raters was no larger than 1, the average score was used, otherwise a third rater would be introduced and all three raters were asked to rate again after a discussion. The average score of the three was taken as the final value. After the video scoring, the test data of each 1-minute segment were classified into groups of (1) not drowsy, (2) some drowsiness, and (3) very drowsy based on their scores.

According to the method introduced above, we find the state of each driver in each 1-minute segment based on the scores of their facial video. It should be noted that the experiment for workload is different from the aforementioned experiments for other features. Figure 5.36 shows the experiment route. Figure 5.37 shows the final result for one driver. The relation between road and time is shown in Table 5.7.

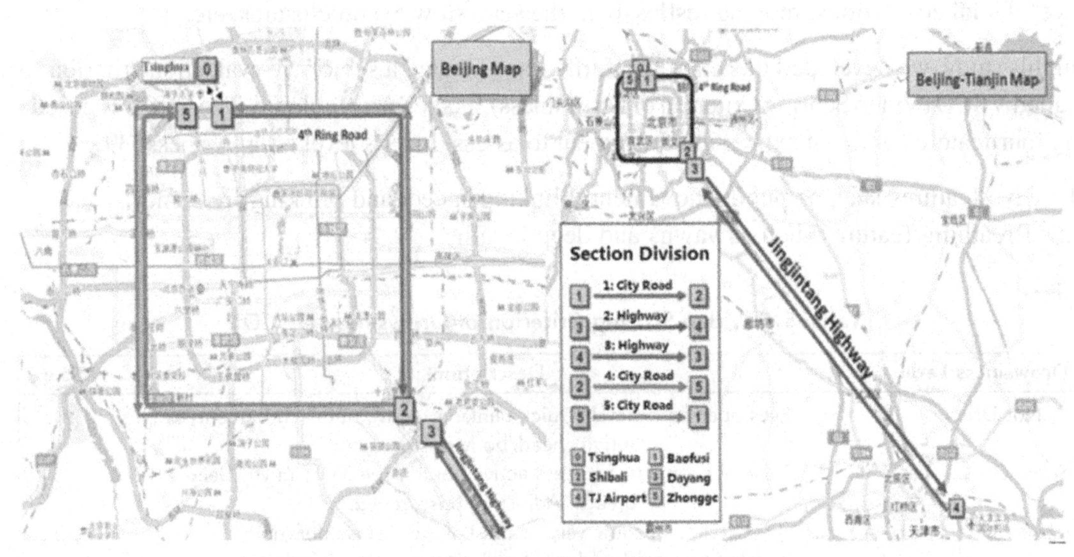

Figure 5.36: Experiment Route for Workload.

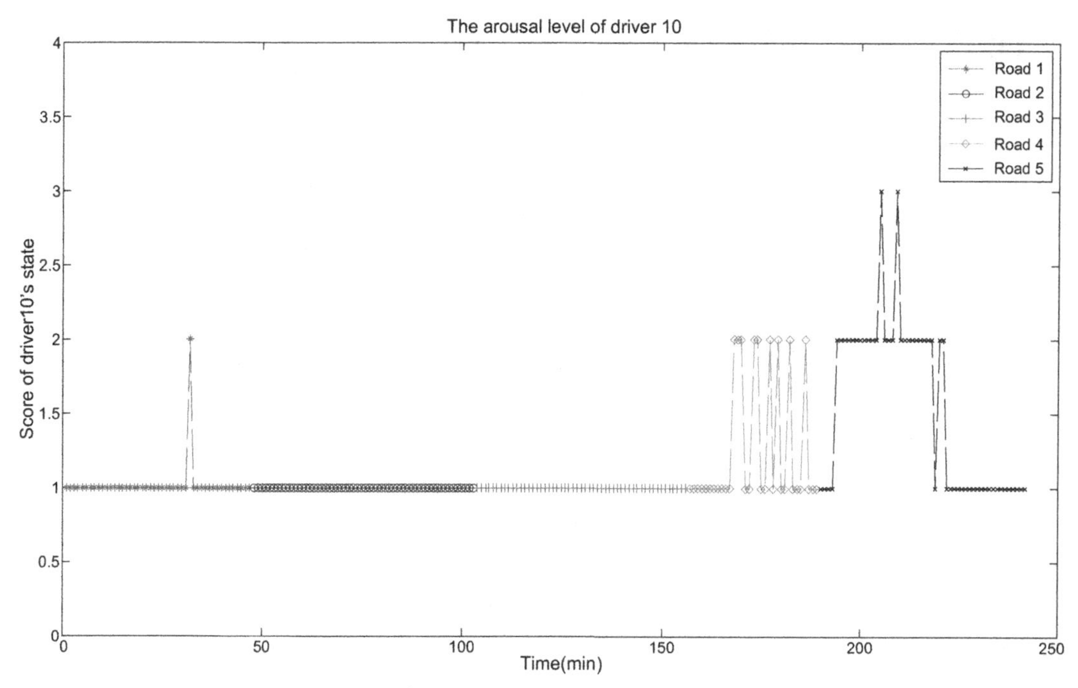

Figure 5.37: Arousal Level.

Figure 5.37 shows the arousal level on each road section in Figure 5.36 (Roads 1, 4, and 5 belong to the 4th Ring Road, and Roads 2 and 3 belong to the Jingjintang Highway-freeway). It can be observed that the driver is seldom tired on Roads 1, 2, and 3. The reason might be that drivers usually do not feel tired in the morning, and the experiment time is not long enough. In addition, drivers show some drowsiness on Roads 4 and 5 because of the longer experiment time and postmeridian drowsiness after lunch.

The values of THW for non-restricted following and TTC for approaching on braking were calculated and compared under arousal levels. The effect of arousal level on driving characteristics has been analyzed according to the statistical results.

The average values of THW at levels 1, 2, and 3 were calculated on the highway shown in Figure 5.38 and Table 5.8. The trend is that the higher the level, the longer the THW, and the more risk-compensated behavior drivers intend to take.

Table 5.7: Relation of the Road and Time

Road	Road 1	Road 2	Road 3	Road 4	Road 5
Time (min)	1–47	48–103	104–156	157–189	190–242

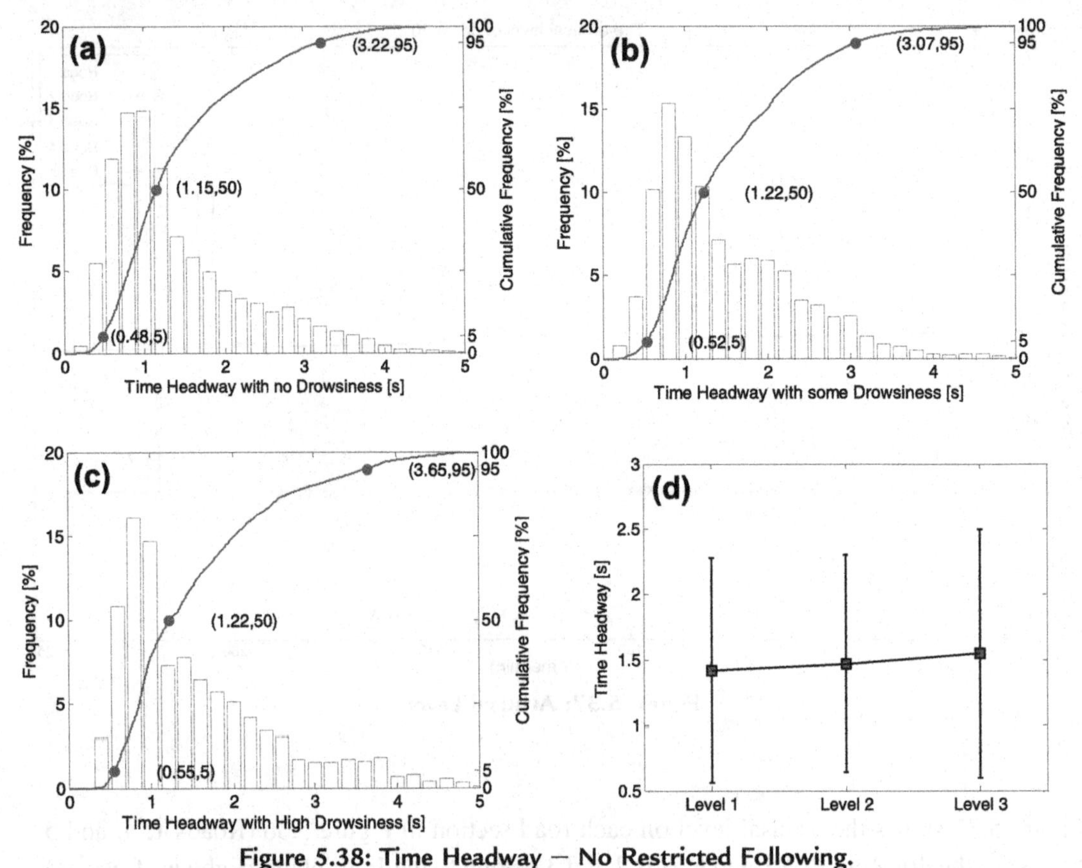

Figure 5.38: Time Headway — No Restricted Following.
(a) Level 1. (b) Level 2. (c) Level 3. (d) Comparison of THW in different drowsiness levels.

5.4.7 Level of Education

Several studies have been conducted on the relationship between driver behavior and their levels of education. Hemenway and Solnick [40] found that drivers who had had a higher

Table 5.8: Data Information for Time Headway — Non-Restricted Following

	4th Ring Road			Freeway		
Parameter	**Level 1**	**Level 2**	**Level 3**	**Level 1**	**Level 2**	**Level 3**
Number of data	119,304	56,813	12,631	92,009	31,017	8724
Average (s)	1.57	1.69	1.61	1.42	1.47	1.55
Standard deviation (s)	0.9	0.87	0.82	0.86	0.83	0.95
Median (s)	1.33	1.5	1.43	1.15	1.22	1.22
Mode (s)	1.8	1.8	1.2	0.9	0.9	0.76
5% tile value (s)	0.55	0.64	0.53	0.48	0.52	0.55
50% tile value (s)	1.33	1.5	1.43	1.15	1.22	1.22
95% tile value (s)	3.46	3.49	3.14	3.22	3.07	3.65

level of education were more likely to be involved in an accident. However, different conclusions were also drawn by other researchers. It was also found that the level of education was not directly related to accident involvement [41]. In the work of Turner and McClure [42], the level of education did not show any significant association with the likelihood of a crash. In terms of safe driving behavior, it was found that drivers who received a higher level of education tended to speed more often and were not less likely to commit to drinking and driving. It was reported that there was an increase in the use of safety belts for drivers with a higher education level for both men and women. However, complete avoidance of drinking and driving hardly varied across groups with different education levels, and people with higher education levels were even less likely to abide by the speed limit all the time [43]. Also, Laapotti et al. [44] showed that a low level of education increased the odds for committing traffic offenses.

5.4.8 Nationality

Marsden et al. [45] studied the differences in motorway driving behavior between three sites in the UK, France, and Germany. Data used in their study was collected using an instrumented vehicle, equipped with an optical speedometer, radar, and a video-audio monitoring system that measured the driver's behavior in a 10-Hz range.

The analysis of THW showed that what was observed in Lille were lower than those at Hamburg, which were, in turn, lower than those observed on the M3 in the UK at free flow speeds. In terms of the analysis of relative speed tolerances, a comparison of TTC for different nationalities was conducted. Statistical differences between the number of low occurrences to collision events were found at each site. The study revealed significant differences between driver behavior at the three sites. Such differences could impact on the effectiveness of roadside telematics systems as well as the design of advanced vehicle control and safety systems.

Golias and Karlaftis [46] used a combination of factor analysis and tree-based regression to identify driver groups with homogeneous self-reported behavior and to determine whether regional differences in driving behaviors exist. The study was based on a large database of more than 20,000 questionnaires from 19 European countries through a SARTRE survey. Important differences and similarities among drivers in different regions in Europe were found. It was concluded that Northern European drivers had a much higher compliance rate with drinking and driving laws and seat-belt use regulations than do Southern and Eastern European drivers.

Work in Ref. [47] conducted a comparitive study of the car-following behavior between Southampton (UK) and Tsukuba (Japan). Data were collected on public roads at the two sites. A fuzzy logic car-following model was used to evaluate the dynamics. It was found that

Southampton drivers had short DHW on motorways, while Tsukuba drivers had long DHW on rural roads. In addition, Southampton drivers tended to increase their speeds when DHW was greater with a higher acceleration rate than drivers from Tsukuba.

5.5 Objective and Subjective Evaluation

For the design and the evaluation of in-vehicle driving assistance systems, classification of driver behavior is necessary. With different driver groups, proper control algorithms can be designed to adapt their characteristics for better performance. In behavioral studies with this purpose, drivers were classified by gender, age, and driving experience (number of years in driving), and the behavior analysis was conducted for all driver groups [48]. In earlier studies, analysis was mainly based on the objective experimental data from real-world driving experiments or driving simulators. An assignment procedure was investigated by Canale and Malan [49] to classify driver behavior with respect to a stop-and-go task. Othman et al. [50] introduced a method for detection and classification of abnormal driver behaviors with estimated jerk from a driving simulator. Although the relationship between driving behavior and the internal state of the driver was verified, the questionnaire for the survey was very simple: it only requested feedback by the subjects regarding their driving condition from level one to five. Work in Refs. [51,52] attempted a driver style classification by using the ratio of the standard deviation and the average acceleration extracted from the acceleration profile within a specified time window. The authors then incorporated the predicted driver style into their power management strategy as a practical application. Similarly, Ref. [53] developed an algorithm for classifying driving style with statistical information from the jerk profile, road type, and traffic congestion level prediction.

Though many research efforts in the past focused on driver classification, they almost always were based on the objective data, and the parameters used in these studies were not, in general, comprehensive. There was little research focusing on both objective experimental data and subjective evaluation simultaneously. In using some in-vehicle driver assistance systems (e.g. adaptive cruise control), drivers usually set the parameters (e.g. time headway in steady car following) according to their own driving characteristics. The subjective evaluation could affect driving safety and stability. Thus, it is essential to objectively verify the validity of the subjective evaluation; the comparison of classifications based on the objective experimental data and subjective evaluations can achieve this purpose [54].

The answers to the following questions are sought in our research:

- How do we determine the dissimilarity of the driving behavior regarding driving style and driving skill comprehensively taking into account longitudinal and lateral characteristics?
- How do we classify the dissimilarity of the driving behavior based on the subjective evaluation and the objective experimental data respectively?

- What is the consistency of classification based on the subjective evaluation and the objective experimental data?

5.5.1 Participants

A sample of 52 participants with full Chinese driving licenses were recruited from Haidian District, Beijing. All of the participants took the self-reported survey and then real-world driving experiments. There were 48 males (92.3%) and four females (7.7 %). The average age of the participants was 44.60 years (standard deviation, S.D. = 9.18), and the age range was from 27 to 62 years. On average, the participants had held their driving licenses for 14.19 years (S.D. = 7.94, range 3–38 years).

5.5.2 Driver Classification Based on DBQ

Because of the complication and variability of the human driver, it is challenging to convert descriptive concepts of driver behavior into quantitative mathematic variables. A self-reported survey was conducted as the method for studying the driver behavior of the sample group. The answers to the questionnaires were then used to analyze driver behavior, especially the relationship between abnormal behavior and driver characteristics.

Design of DBQ

A self-reported survey was designed based on the DBQ [55] with some modifications made to adapt to the traffic situation and driving conditions in China. For example, the traffic density in China is much greater than that in many other countries. The number of overtaking and undertaking maneuvers was one factor in the survey. The survey consisted of two parts:

1. Individual information, including gender, age, and driving experience.
2. The main body of the survey, including 30 items describing possible abnormal behaviors that could occur in daily driving, which were extracted and transferred from the original questionnaire, as shown in Table 5.9. In order to cover the most important driver characteristics, driving style and driving skill, abnormal longitudinal and lateral behaviors respectively were listed in the survey.

Statistical Data Analysis

In the statistical data analysis, reliability and validity were the two main indices that could determine whether the results of such self-reported questionnaires are reliable and accurate [56]. Cronbach's alpha reliability coefficient method was the most commonly used method that reflects the internal consistency of scale [57]. If it was greater than 0.6, the scale was considered persuasive. The number in our study was 0.884, which meant that it was persuasive with good consistency.

Table 5.9: Statistical Analysis Results and Factor Score Coefficients of DBQ

No.	Item	Mean	S.D.	r with Total Score	F_{style}	F_{skill}
1	Drive so close to the car in front that it would be difficult to stop in an emergency	0.731	0.770	0.418**	0.149	−0.073
2	Disregard the speed limit on a residential road	0.865	0.864	0.598**	0.150	−0.061
3	Disregard the speed limit on an intercity highway	0.942	0.938	0.474**	0.160	−0.075
4	Become angered by another driver and show anger through aggressive driving	1.404	1.015	0.545**	0.063	0.030
5	Overtake a slow driver on the inside	1.308	0.853	0.422**	0.096	−0.022
6	Become angered by another driver and give chase	0.788	0.750	0.443**	0.145	−0.065
7	Sound your horn to indicate your annoyance to another road user	1.269	0.819	0.445**	0.153	−0.069
8	Stay in a closing lane and force your way into another	0.577	0.667	0.282*	0.047	0.001
9	Underestimate the speed of an oncoming vehicle when overtaking	0.923	0.813	0.597**	−0.025	0.140
10	Get into the wrong lane when approaching a roundabout or a junction	1.096	0.774	0.623**	0.017	0.093
11	Misread the signs and exit from a roundabout on the wrong road	1.096	0.891	0.664**	−0.011	0.145
12	Forget where you left your car in a car park	0.962	1.047	0.512**	0.016	0.062
13	Hit something when reversing that you had not previously seen	0.885	0.732	0.547**	0.062	0.045
14	Have no clear recollection of the road along which you have just been traveling	0.808	0.864	0.458**	−0.099	0.193
15	Cut off the vehicle in the next lane when car following on a residential road	2.192	0.841	0.389**	−0.058	0.140
16	Repeatedly try to pass, but fail	1.231	0.899	0.571**	−0.051	0.155
17	Accelerate through the intersection when the lights change	1.615	1.013	0.553**	0.126	−0.025
18	Overtake always	1.692	0.981	0.609**	0.128	−0.014

Table 5.9: Statistical Analysis Results and Factor Score Coefficients of DBQ—cont'd

No.	Item	Mean	S.D.	*r* with Total Score	F_{style}	F_{skill}
19	Change lanes frequently	1.308	0.829	0.519**	0.093	0.001
20	Forget to check the instrument panel and trouble lights when start	0.942	1.018	0.573**	0.002	0.110
21	Go to the wrong lane when turning left	0.712	0.750	0.567**	−0.071	0.184
22	Forget to wear a seatbelt or to release the handbrake when starting	1.038	0.885	0.679**	0.039	0.087
23	Change lanes because of a slow driver on the inside	2.135	1.121	0.561**	0.081	0.023
24	Change lanes from the outside lane to the inside lane because of a fast driver that is following	1.538	1.038	0.083	−0.100	0.109
25	Force your way into another lane when coming across some barrier	1.635	0.950	0.380**	−0.018	0.072
26	Underestimate the speed of an oncoming vehicle when changing lanes	1.212	0.848	0.520**	0.040	0.059
27	Drive near the left side of the lane	1.250	1.266	0.521**	0.060	0.020
28	Change lanes slowly with a large distance from the lead vehicle	1.577	1.538	0.273	−0.017	0.051
29	Change lanes quickly with a large distance from the lead vehicle	1.654	1.136	0.563**	0.074	0.022
30	Accelerate until a small distance away from the lead vehicle, then change lanes quickly	0.846	0.849	0.344*	0.049	0.008
	Total score	1.208	1.015	1.000		

*$P < 0.05$;
**$P < 0.01$.

The validity of the scale can be determined by the correlation analysis of each item score and the total score. The closer the correlation that exists between each item score and the total score, the more the items involved reflect the same theme. The Pearson correlative coefficients (r) between each item score and the total score of the DBQ are shown in Table 5.9, and 93% of the items have a high positive interrelation at the 0.05 and 0.01 significance levels, which indicate that this DBQ has good validity and may be applied to driver behavior analysis.

The item scores of the questionnaire were analyzed, and the mean score, the standard deviation of the tested samples and that of the total scores are shown in Table 5.9.

Driver Characteristics Quantification Based on Factor Analysis

Factor analysis theory [58] has been applied to the data processing and extraction of this 30-item DBQ. A two-dimensional DBQ factor structure has been established. Principal components analysis with oblique rotation has been implemented as the factor analysis method to investigate the factor structure.

The factor analysis method distinguishes different types of behaviors distinctively. The first factor accounts for 18.47% of the total variance and contains 16 items (i.e. items 1–8, 13, 17–19, 23, 27, 29, 30), which reflect the intentional abnormal behaviors. The second factor contains the other 14 items, which mainly describe unintentional abnormal behaviors. It can be assumed that the driver's intention is attributed to driving style. For instance, considering the 23rd item "change lanes because of a slow driver on the inside", this behavior occurs because the driver is overconfident and aggressive whilst driving. Thus, the first factor is defined as F_{style} to describe the intentional abnormal characteristics. On the other hand, the driver's unintentional passive characteristics are caused by a lack of driving skill. Therefore, the second factor can be defined as F_{skill} to describe the unintentional abnormal characteristics of driving skill.

Based on the scale point data of DBQ, the factor score coefficients of each driver can be estimated using regression analysis. The quantified factors are normalized so that the mean is 0 and the standard deviation is 1. The drivers with larger F_{style} have higher frequencies of intentional abnormal behaviors and drive in more aggressive styles, and those with a larger value of F_{skill} show more unintentional abnormal behaviors and are less skillful. The factor scores of the 52 participants are distributed over all four quadrants, which are shown in Figure 5.39.

Driver Classification and Analysis

Considering the cluster regularity of this population, there were some consistencies in the trend of characteristics among the same type cluster or pattern. Thus, based on the cluster analysis, different clusters of drivers were matched with different parameters in the algorithm design so that the results would be more representative and applicable, which could eliminate dissimilarities between individuals. Based on this idea, the two factors were taken as the basis of cluster analysis, and then the distribution of space of the pilot model was established.

In this section, the K-means clustering algorithm [59] is used to classify the drivers. This algorithm treats each observation in the data as an object with a location in space. Each cluster

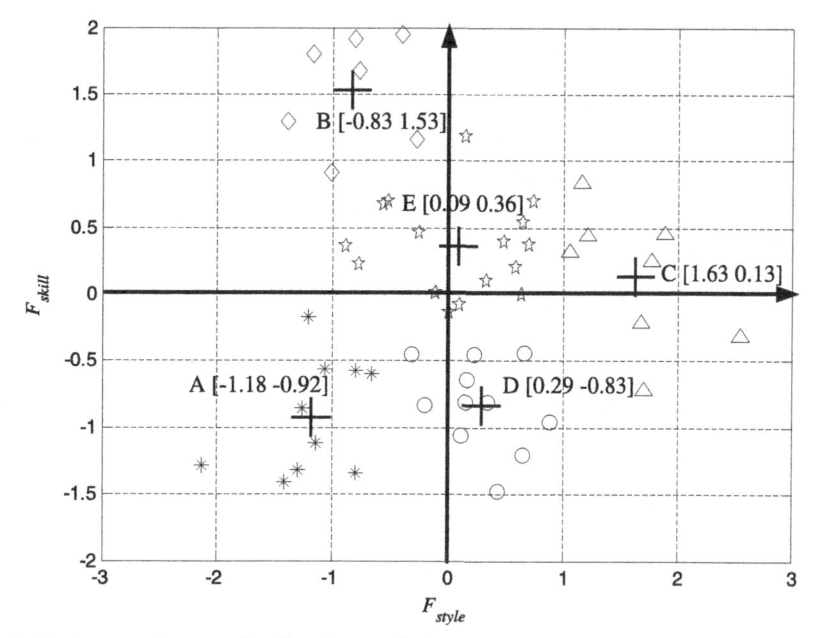

Figure 5.39: Factor Score Distribution and the Result of K-Means Clustering Analysis.

in the partition is defined by its member objects and by its centroid. K-means uses an iterative algorithm to minimize the sum of distances from each object to its cluster centroid over all clusters.

The cluster result can be seen in Figure 5.39, and it is clear that it is not definitive. Firstly, the definition of the cluster is indistinctive. Take the drivers in cluster C, for example; the score of F_{style} is the largest relative to the other four clusters so that they can be regarded as the aggressive group. However, regarding driving skill, the group is biased toward normal because the score of F_{style} is in the middle state. Secondly, the drivers on the borderline of two opposite clusters are not clearly grouped. Taking the driver in cluster C with the lowest score of F_{skill} as an example, it can be observed that the distances from this driver to clusters C and D are comparable, but the two clusters are opposite for driving skill. Therefore, there is no sufficient reason to classify this driver to cluster C or D. Finally, the classification result is directly affected by the selection of the clustering method. The K-means clustering algorithm is a learning method without surveillance. The general drawback of standard clustering methods is that they ignore measurement errors or uncertainty associated with the data, and the outlier points in the set of experimental data can lead to local optima and be misleading in the outcome.

Considering all the shortcomings of the clustering approach, a novel clustering that accounts for these conventions is proposed and applied to the classification of dissimilarities. Common sense tells us that there are a great majority of drivers in the normal group that gather around the origin of coordinates, and the others are scattered across the four quadrants. Based on this idea, a circle with its center at the origin that contains 50% of all the participants is defined as the threshold of the normal group. The reason for the 50% figure is as follows. In verifying the consistency of two classifications, if the boundary is selected larger than 50%, e.g. 60%, the worst case result is that 40% of all the drivers based on the subjective evaluation are different from that based on the objective analysis. They will then be placed in the normal group of the objective analysis. Similarly, 40% of all the drivers based on the objective data could be in the normal group of the subjective evaluation. Then the lowest consistency of the two classifications is 20%, which has no statistical significance. Also, if the boundary is selected smaller than 50%, it is statistically insignificant. When 50% is selected, the lowest consistency is zero, which is acceptable. Once the normal group has been determined, the other four groups can be naturally determined as falling into the four quadrants.

The results of the classification can be seen in Figure 5.40. The center of the circle is verified as the peak of each factor from the distribution of the two factors' scores, whose position reflects the general characteristics of Chinese drivers. It can be seen that Chinese drivers are inclined to be aggressive in driving style and normal in driving skill from a subjective evaluation. There were five participants (9.6%) in group A, six participants (11.5%) in group B, eight participants (15.4%) in group C, seven participants (13.5%) in group D, and 26 participants in group E (which contains 50% of all the participants). Considering the mean of the two factors, group A is defined as the aggressive and non-skillful group, group B the prudent and non-skillful group, group C the prudent and skillful group, group D the aggressive and skillful group, and group E is the normal group.

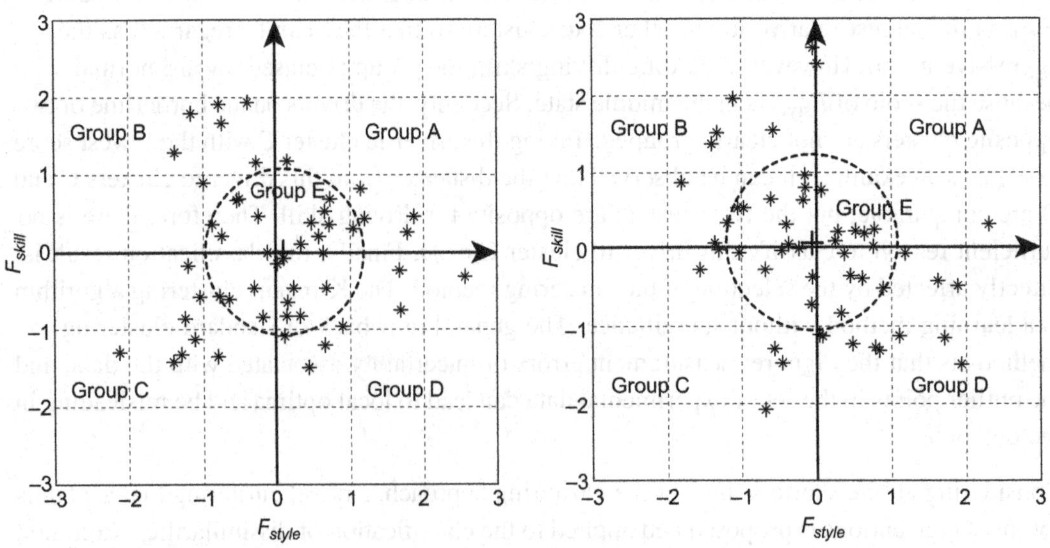

Figure 5.40: Classification Results Based on DBQ (left) and Factor Analysis (right).

5.5.3 Driver Classification Based on Real-World Driving Data

Experiment Design and Data Collection

The experimental platform is shown in Figure 5.10. The experimental conditions are as follows: intercity highway (freeway), daytime, and good weather. Experimental procedures are: (1) Introduction of the experiment; (2) taking a trial drive in the new instrumented vehicle on an ordinary road for about half an hour; and (3) driving on an intercity highway for 1 hour. There are no other restrictions imposed on the driver.

Measurable Parameters Selection

Considering that the two factors extracted from the DBQ are gained from aspects of abnormal driving behavior, the parameters that are selected in experimental data should reflect the same meaning. Based on the correlation analysis, six parameters were selected for factor analysis: the mean values of DHWi in steady car following, TTCi_1stBra in approaching, a_{dmax} in braking (factors related to driving style), the standard deviation of R_{ap} (the rate of the acceleration pedal position for every acceleration scenario in accelerating), R_{sa} (the steering angle rate), and a_{lat} (the lateral acceleration). It can be assumed that drivers with larger mean values have a more aggressive driving style, and those with larger values of standard deviation are less skillful. The six selected parameters reflect the driver characteristics reasonably well, as these parameters are related to driver operations (e.g. R_{ap}, R_{sa}), host vehicle state (e.g. a_{dmax}, a_{lat}), and the relative state of the host vehicle and the lead vehicle (e.g. DHWi, TTCi_1stBra). Likewise, the longitudinal and lateral behaviors are considered simultaneously.

Classification Based on Factor Analysis

In accordance with the classification based on the subjective evaluation, the same classification method is proposed and analyzed. The same five groups are also divided according to their factor scores. The definitions of the groups are the same as before. There are two participants (3.8%) in group A, nine participants (17.3%) in group B, four participants (7.7%) in group C, 11 participants (21.2%) in group D, and 26 participants in group E.

5.5.4 Comparison of Subjective Evaluation and Objective Experiment

In the two sections above, the 52 participants were classified into five groups from subjective questionnaire and objective data analysis. The difference is that the drivers involved in each group are not the same. It is necessary to verify the consistency of the subjective evaluation and objective data analysis.

Firstly, a comparison of the centers of the two circles is necessary. The center position can reflect the general characteristics of Chinese drivers, who subjectively are inclined to be aggressive in driving style and normal in driving skill. However, from Figure 5.41 it can be seen that the participants are objectively inclined to be a little prudent in driving style and slightly skillful at driving. The difference between the overall trends exists, but is not significant.

Secondly, the statistical results of the numbers of drivers involved in each group are analyzed. The distributions of drivers involved in the five groups are shown in Figure 5.41(a) and (b). The number of drivers involved in each group is shown. Meanwhile, the numbers of drivers in each of the five groups that coincide on both classifications are shown in Figure 5.41(c). There are 15 participants (about 28.8%) whose subjective belief coincides fully with reality. There are some differences between self-evaluation and the real driving characteristics of the other 37 participants. Since the classification based on real-world experimental data is more objective and more likely represents the truth, it can be assumed that if a prudent driver overrates his or her driving style, it is unacceptable and could affect the safety of normal driving. The reason is that a prudent driver always leaves a longer distance for reaction to danger, but if he/she overrates his/her driving style, there is not enough time to react once a collision becomes imminent. On the contrary, if an aggressive driver underestimates his/her driving style, it is acceptable and it would not affect safety though driving comfort may be sacrificed. Similarly, from the aspect of driving skill, if a driver overrates his/her driving skill, it may affect safe driving, which is unacceptable, while underestimating the driving skill is acceptable. With those unacceptable conditions, 22 participants (42.3%) are considered not to be driving unsafely, which means that the subjective evaluations of those drivers are unreliable. In other words, there are 30 participants (57.7%) who did the subjective evaluation effectively, from the consideration of safe driving.

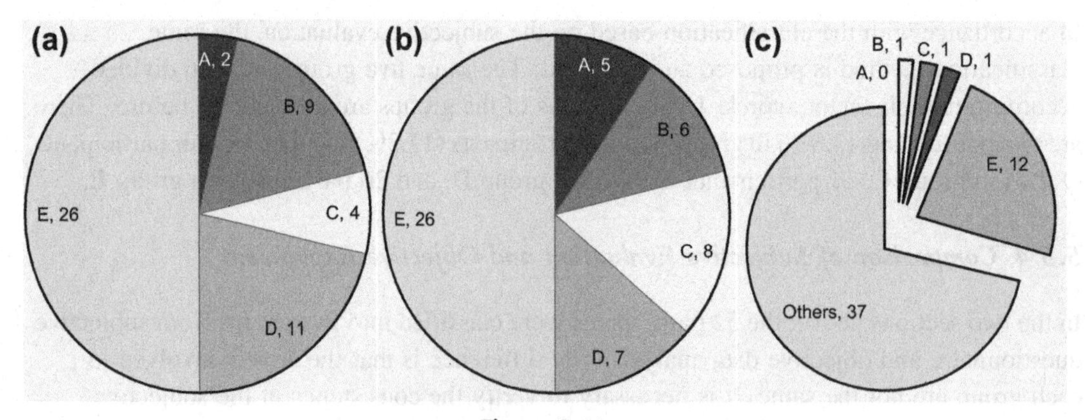

Figure 5.41
Driver distributions of the five groups classified by the objective data (a), driver distribution of the five groups classified by the subjective evaluations (b), and numbers of drivers in each group that coincide fully on the two classifications (c).

Finally, since THW and TTC are the two important parameters that can reflect the longitudinal behavior of the driver, they are taken as parameters for comparison. Considering the relative speed is zero sometimes, TTC is replaced by TTCi. The mean values and standard deviation of THW in steady car following and TTCi under non-restricted situation were calculated for the five groups classified with subjective and objective evaluations respectively. The results are shown in Figure 5.42. The mean values of THW and TTCi reflect driving style. The comparison of driving style ranges by the degree of aggressiveness from prudent to aggressive. A more aggressive driver leaves a much smaller THW and larger TTCi compared with a normal driver, especially a prudent driver. The standard deviation of THW and TTCi reflects driving skill. The comparison of driving style ranges by the degree of skill from non-skillful to skillful. A non-skillful driver controls the vehicle unstably compared with a normal driver, especially a skillful driver, and therefore the standard deviation is much

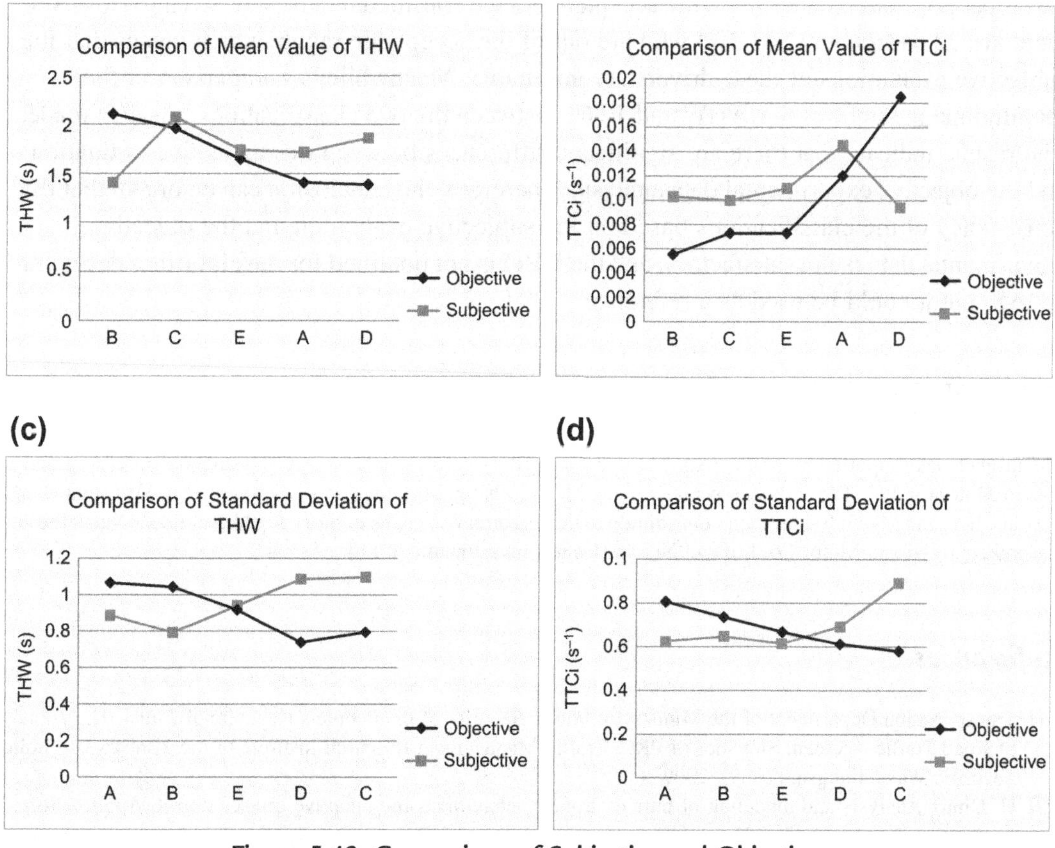

Figure 5.42: Comparison of Subjective and Objective.
(a) Comparisons of mean value of THW. (b) Comparisons of mean value of TTCi. (c) Comparisons of standard deviation of THW. (d) Comparisons of standard deviation of TTCi.

larger. It can be observed from the figure that the trend of values of these aspects is incongruent between subjective and objective tests. For driving skill in particular, the trend is almost opposite.

Therefore, it can be concluded that there is a disagreement between the subjective evaluation and the objective experimental data analysis. The reasons could be as follows. The parameters of the DBQ are inadequate to reflect driver characteristics comprehensively. The meaning of borderline scores for each item is ambiguous, which could confuse the participants. The understanding of the items also varies from driver to driver. All these causes may lead to the result that the scale is not reliable enough.

5.5.5 Conclusions

According to the statistical analysis, only 15 participants (about 28.8%) performed evaluations that coincide with the reality of their own driving in the real world, and another 15 participants (about 28.8%) are in the acceptable range considered to be safe driving. However, there are 22 participants (42.3%) who are out of the acceptable range, which means that the subjective evaluations of these drivers are unreliable. Meanwhile, a comparison of the longitudinal parameters (i.e. THW and TTCi) between the two classifications was also made. The results indicate that there are significant differences between the subjective evaluation and the objective experimental data analysis. Therefore, the conclusion can be drawn that the consistency of the classifications based on the subjective evaluation and the objective experimental data is not satisfactory, and the DBQ is not qualified for the algorithm design of ADAS, but it could be used as a reference.

Acknowledgments

The project was supported by NSFC (No. 51175290) and the joint research project of Tsinghua University and Nissan Motor Co. Ltd. The authors especially thank Mr Lei Zhang, Mr Qing Xiao, Ms Xiaojia Lu, Ms Ruina Dang, Mr Lai Chen, and Ms Yu Bai for their contribution to the research work. The authors would also like to thank the co-researchers of Nissan Motor Co. Ltd and Nissan (China) Investment Co. Ltd.

References

[1] Transportation Department of the Ministry of Public Security of the People's Republic of China. The Annals of Road Traffic Accident Statistics of PRC, Traffic Management Research Institute of the Ministry of Public Security, Wuxi, Jiangsu, (2009). 2009.

[2] H. Ohno, Analysis and modeling of human driving behaviors using adaptive cruise control, Applied Soft Computing 1 (2001) 237–243.

[3] A.M. Vadeby, Modeling of relative collision safety including driver characteristics, Accident Analysis and Prevention 36 (2004) 909–917.

[4] M. Canale, S. Malan, Analysis and classification of human driving behaviour in an urban environment, Cognition, Technology and Work 4 (2002) 197–206.
[5] H. Yoo, P. Green, Driver Behavior while Following Cars, Trucks and Buses, Technical Report UMTRI-99–14.
[6] J.C. McCall, D. Wipf, M.M. Trivedi, Lane change intent analysis using robust operators and sparse Bayesian learning. Proceedings of the 2005 IEEE Computer Society Conference on Computer Vision and Pattern Recognition, 2005.
[7] B. Cheng, M. Hashimoto, T. Suetomi, Analysis of driver response to collision warning during car following, JSAE Review 23 (2002) 231–237.
[8] D. Salvucci, E. Boer, A. Liu, Toward an integrated model of driver behavior in a cognitive architecture, Transportation Research Record vol. 1779 (2001) 9–16.
[9] A. Pentland, A. Liu, Modeling and prediction of human behavior, Neural Computation 11 (1999) 229–242.
[10] L. Zhang, A Vehicle Longitudinal Driving Assistance System Based on Self-Learning Method of Driver Characteristics. Doctoral dissertation, Tsinghua University, Beijing, China, 2009.
[11] G. Welch, G. Bishop, An introduction to the Kalman Filter, SIGGRAPH, 2001 Course.
[12] Z. Sheng, S. Xie, C. Pan, Probability Theory and Mathematical Statistics, vol. 3, Higher Education Press, 2001, pp. 225–229 (in Chinese).
[13] C. Liu, J. Wan, Probability and Statistics, Higher Education Press, Beijing, 2005 (in Chinese).
[14] Q. Yan, J.M. Williams, J. Li, Chassis control system development using simulation: Software in the loop, rapid prototyping, and hardware in the loop, Society of Automotive Engineers (SAE), 2002. Paper No. 2002-01-1565.
[15] O. Gietelink, J. Ploeg, B. De Schutter, M. Verhaegen, Development of advanced driver assistance systems with vehicle hardware-in-the-loop simulations, Vehicle System Dynamics 44 (7) (2006) 569–590.
[16] J. Wang, S. Li, X. Huang, K. Li, Driving simulation platform applied to develop driving assistance systems, Journal of IET Intelligent Transportation System 4 (2) (2010) 121–127.
[17] S. Li, J. Wang, K. Li, xPC technique based hardware-in the-loop simulator for driver assistance systems, China Mechanical Engineering 18 (16) (2007) 2012–2015 (in Chinese).
[18] L. Zhang, J. Wang, K. Li, T. Yamamura, N. Kuge, T. Nakagawa, An instrumented vehicle test bed and analysis methodology for investigating driver behavior, 14th World Congress on Intelligent Transport Systems, Beijing, China, 9–13 October 2007.
[19] J.D. Lee, D.V. McGehee, T.L. Brown, M.L. Reyes, Collision warning timing, driver distraction, and driver response to imminent rear end collisions in a high-fidelity driving simulator, Human Factors 44 (2) (2002) 314–334.
[20] E. Blana, J. Golias, Differences between vehicle lateral displacement on the road and in a fixed-base simulator, Human Factors 44 (2) (2002) 303–313.
[21] A. Kemeny, F. Panerai, Evaluating perception in driving simulation experiments, Trends in Cognitive Sciences 7 (1) (2003) 31–37.
[22] D.V. McGehee, T.L. Brown, J.D. Lee, T.B. Wilson, The effect of warning timing on collision avoidance behavior in a stationary lead-vehicle scenario, Transportation Research Record 1803 (2002) 1–7.
[23] L. Jia, M. Lu, J. Wang, Using real-world data to calibrate a driving simulator measuring lateral driving behaviour, IET Intelligent Transport Systems 5 (1) (2011) 21–31.
[24] P.F. Lourens, et al., Annual mileage, driving violations, and accident involvement in relation to drivers' sex, age, and level of education, Accident Analysis and Prevention 31 (1999) 593–597.
[25] P. Finn, B. Bragg, Perception of the risk of an accident by young and older drivers, Accident Analysis and Prevention 18 (1986) 289–298.
[26] I. Golias, M.G. Karlaftis, An international comparative study of self-reported driver behavior, Transportation Research Part F 4 (2002) 243–256.
[27] J. Elander, R. West, D. French, Behavioral correlates of individual differences in road-traffic crash risk: An examination of methods and findings, Psychological Bulletin 113 (1993) 279–294.

[28] D. Yagil, Gender and age-related differences in attitudes toward traffic laws and traffic violations, Transportation Research Part F 1 (1998) 123–135.
[29] L. Evans, Traffic Safety and the Driver, van Nostrand Reinhold, New York, 1991.
[30] S. Laapotti, et al., Comparison of young male and female drivers' attitude and self-reported traffic behavior in Finland in 1978 and 2001, Journal of Safety Research 24 (2003) 579–587.
[31] The Social Issues Research Centre, Sex Differences in Driving and Insurance Risk: An Analysis of the Social and Psychological Differences Between Men and Women that are Relevant to their Briving Behavior, 2004.
[32] M. Meadows, S. Stradling, Are women better drivers than men? in: J. Hartley, A. Branthwaite (Eds.), The Applied Psychologist, second ed., Open University Press, Buckingham, 1999.
[33] T. Lajunen, D. Parker, Are aggressive people aggressive drivers? A study of the relationship between self-reported general aggressiveness, driver anger and aggressive driving, Accident Analysis and Prevention 33 (2001) 243–255.
[34] R.R. Mourant, T.H. Rockwell, Strategies of visual search by novice and experienced drivers, Human Factors 14 (1972) 325–335.
[35] D. Brown, J.A. Groeger, Risk perception and decision taking during the transition between novice and experienced driver status, Ergonomics 31 (4) (1988) 585–597.
[36] T.A. Ranney, Models of driving behavior: A review of their evolution, Accident Analysis and Prevention 26 (1994) 733–750.
[37] R. Hale, J. Stoop, J. Hommels, Human error models as predictors of accident scenarios for designers in road transport systems, Ergonomics 33 (10–11) (1990) 1377–1387.
[38] R. Fuller, Towards a general theory of driver behavior, Accident Analysis and Prevention 37 (2005) 461–472.
[39] W.W. Wierwille, L.A. Ellsworth, Evaluation of driver drowsiness by trained raters, Accident Analysis and Prevention 26 (5) (1994) 571–581.
[40] T. Ma, B. Cheng, Detection of driver's drowsiness using facial expression features, Journal of Automotive Safety and Energy 1 (3) (2010) 200–204.
[41] D. Hemenway, S.J. Solnick, Fuzzy dice, dream cars, and indecent gestures: Correlates of driver behavior? Accident Analysis and Prevention 25 (1993) 161–170.
[42] P.F. Lourens, J.A.M.M. Vissers, M. Jessurun, Annual mileage, driving violations and accident involvement in relation to drivers' sex, age and level of education, Accident Analysis and Prevention 31 (1) (1999) 593–597.
[43] C. Turner, R. McClure, Age and gender differences in risk-taking behavior as an explanation for high incidence of motor vehicle crashes as a driver in young males, Injury Control and Safety Promotion 10 (3) (2003) 123–130.
[44] D. Shinar, E. Schechtman, R. Compton, Self-reports of safe driving behaviors in relationship to sex, age, education and income in the US adult driving population, Accident Analysis and Prevention 33 (2001) 111–116.
[45] S. Laapotti, E. Keskinen, S. Rajalin, Comparison of young male and female drivers' attitude and self-reported traffic behaviour in Finland in 1978 and 2001, Journal of Safety Research 34 (2003) 579–587.
[46] G.R. Marsden, M. Mcdonald, M. Brackstone, A comparative assessment of driving behaviours at three sites, European Journal of Transport and Infrastructure Research 3 (1) (2003) 5–20.
[47] I. Golias, M.G. Karlaftis, An international comparative study of self-reported driver behavior, Transportation Research Part F 4 (2002) 243–256.
[48] T. Sato, M. Akamatsu, P. Zheng, M. McDonald, Comparison of car following behavior between UK and Japan. ICROS–SICE International Joint Conference 2009, Fukuoka International Congress Center, Japan, 18–21 August 2009.
[49] M. Hoedemaeker, Driving with Intelligent Vehicles: Driving Behaviour with Adaptive Cruise Control and the Acceptance by Individual Drivers, Doctoral thesis, Delft University of Technology, Delft, 1999.
[50] M. Canale, S. Malan, Analysis and classification of human driving behaviour in an urban environment, Cognition, Technology and Work 4 (3) (2002) 197–206.
[51] M.R. Othman, Z. Zhang, T. Imamura, T. Miyake, A study of analysis method for driver features extraction, Proceedings: IEEE International Conference on Systems, Man and Cybernetics (2008) 1501–1505.

[52] R. Langari, Jong-Seob Won, Intelligent energy management agent for a parallel hybrid vehicle – part I: System architecture and design of the driving situation identification process, IEEE Transactions on Vehicular Technology 54 (3) (2005) 925–934.

[53] R. Jong-Seob Won, Langari, Intelligent energy management agent for a parallel hybrid vehicle – part II: Torque distribution, charge sustenance strategies, and performance results, IEEE Transactions on Vehicular Technology 54 (3) (2005) 935–953.

[54] Y.L. Murphey, R. Milton, L. Kiliaris, Driver's style classification using jerk analysis, Proceedings: IEEE Symposium Series on Computational Intelligence (2009).

[55] Jianqiang Wang, Xiaojia Lu, Qing Xiao, Meng Lu, Comparison of driver classification based on subjective evaluation and objective experiment, in: Proc. 90th TRB Annual Meeting, Paper 11-3170, 23–27 January 2011. Washington, DC.

[56] J. Davey, D. Wishart, J. Freeman, B. Watson, An application of the driver behaviour questionnaire in an Australian organisational fleet setting, Transportation Research Part F 10 (2007) 11–21.

[57] Huixin Ke, Hao Shen, The Statistical Analysis of Research, second ed., China Media University Press, Beijing, 2005.

[58] L.J. Cronbach, Coefficient alpha and the internal structure of tests, Psychometrika 16 (3) (1951) 297–334.

[59] R.L. Gorsuch, Factor Analysis, Lawrence Erlbaum Associates, 1983.

[60] J.B. MacQueen, Some methods for classification and analysis of multivariate observations, in: Proceedings: 5th Berkeley Symposium on Mathematical Statistics and Probability, vol. 1, University of California Press, 1967, pp. 281–297.

CHAPTER 6

Comparative Analysis and Modeling of Driver Behavior Characteristics

Jianqiang Wang*, Keqiang Li*, Xiao-Yun Lu†
*State Key Laboratory of Automotive Safety and Energy, Tsinghua University, China
†California PATH, ITS, University of California, Berkeley, USA

Chapter Outline

Advances in Intelligent Vehicles. http://dx.doi.org/10.1016/B978-0-12-397199-9.00006-9

6.1 Introduction

Since the pioneering theoretical study in human driver behaviors by Gibson and Crooks in 1938 [1], many researchers have contributed to driver behavior characteristics [2–5]. In order to reveal and describe driver behavior and characteristics, many researchers have contributed to driver behavior modeling, including the Gazis–Herman–Rothery (GHR) model [6], the Gipps model [7], the linear (Helly) model [8], and neural network and fuzzy logic models [9,10].

In driving behavior characteristic studies, though many valuable results have been obtained, different perspectives, goals, approaches, and assumptions may lead to quite different results. In addition, the accuracy and representativeness of the research results will heavily depend on how well relevant factors are distinguished, how appropriately the experiment is designed, and how rigorously various parameters are determined from experimental data.

This research will focus on the comparison of driver behavior characteristics influenced by the factors of vehicle, road, driving status (maneuvers), and so on.

6.2 Study Approach

In order to compare how driver behavior is influenced by different factors, the terms and defined scenarios presented are consistent with those in Chapter 5. The test vehicle shown in Chapter 5 can also be used in this experiment. Beside the parameters and scenarios listed in Chapter 5, other scenarios are defined in the following section.

6.2.1 Definitions of Vehicle Driving Scenarios

In Chapter 5, some scenarios are defined to reflect longitudinal driving behavior. Other scenarios reflecting lateral driving behavior are defined in Table 6.1.

Table 6.1: Symbols and Definitions of Driving Scenarios

No.	Symbol	Definition
1	Lane deviation	The host vehicle drives in a certain lane and reaches the lane mark some time without changing to the adjacent lane
2	Lane change	The host vehicle changes its driving route between two lanes
3	Activating turn signal	The moment when the driver activates the turn signal before lane change
4	Starting lane change	The moment when the lateral position of a vehicle in its original lane begins to get closer to the destination lane mark, which is between the original lane and destination lane
5	Aborting lane change	The moment when the left or right front wheel of the host vehicle crosses the lane mark from the original lane to the destination lane
6	Finishing lane change	The moment when the host vehicle adjusts its longitudinal direction parallel to the road centerline after reaching the center of the destination lane
7	Steering reversion point	The moment when the inflection point is from the left (right) to the right (left) lane and lateral velocity has left (right) deviation direction

6.2.2 Comparative Analysis Method

In this study, the statistical u-test is used for result validation. When the population variance is unknown and the number of samples is relatively large (which is the case in this study, such as comparison on different road categories), especially with more than 30 samples, the U-test is an appropriate test method according to the study in Ref. [11]. A hypothesis is made here that the expectation values of the two samples (such as A and B) is equal at a specified confidence level. From this, a rejection region for the hypothesis is derived. The absolute value of the testing variable u is calculated with the formula of the u-test. Finally, it is determined whether the calculated value of u is located within the rejection region or not. If it is within the region, the conclusion is drawn that the hypothesis is inappropriate, and that sample A is considered to be different from sample B. Otherwise, the conclusion is drawn that there is little difference between the two samples.

The testing variable u is defined as follows:

$$u = \frac{\overline{X} - \overline{Y}}{\sqrt{\frac{S_1^2}{n_1} + \frac{S_2^2}{n_2}}}, \tag{6.1}$$

where $\overline{X}$ and $\overline{Y}$ are the expectation values of samples A and B of the statistical variable THW for urban access roads and for urban distributor roads respectively, S_1 and S_2 are the standard deviations of samples A and B, and n_1 and n_2 are the numbers of the two samples. Assuming that the sample A is little different from sample B, we make a hypothesis H_0 and an alternative hypothesis H_1:

$$\begin{aligned} H_0 &: E_X - E_Y = 0 \\ H_1 &: E_X - E_Y \neq 0, \end{aligned} \tag{6.2}$$

where E_X and E_Y are the expectation values of the two samples.

6.3 Vehicle Factors

6.3.1 Vehicle Type

Vehicle type is associated with a proportion of accidents. Figure 6.1 shows the proportion of accidents caused by different vehicle types [12].

Wenzel and Ross studied the dependence of risk on vehicle type and especially on vehicle model [13]. The risk was measured by the number of driver fatalities per year per million vehicles registered. Following the work in Ref. [14], two risks were considered, i.e. the "risk to drivers" of the model of the subject vehicle (or vehicle type) and the "risk to drivers of other vehicles" (risk to others) that crashed with the subject vehicle. The differences in risk between types are less than 10% and are not statistically significant.

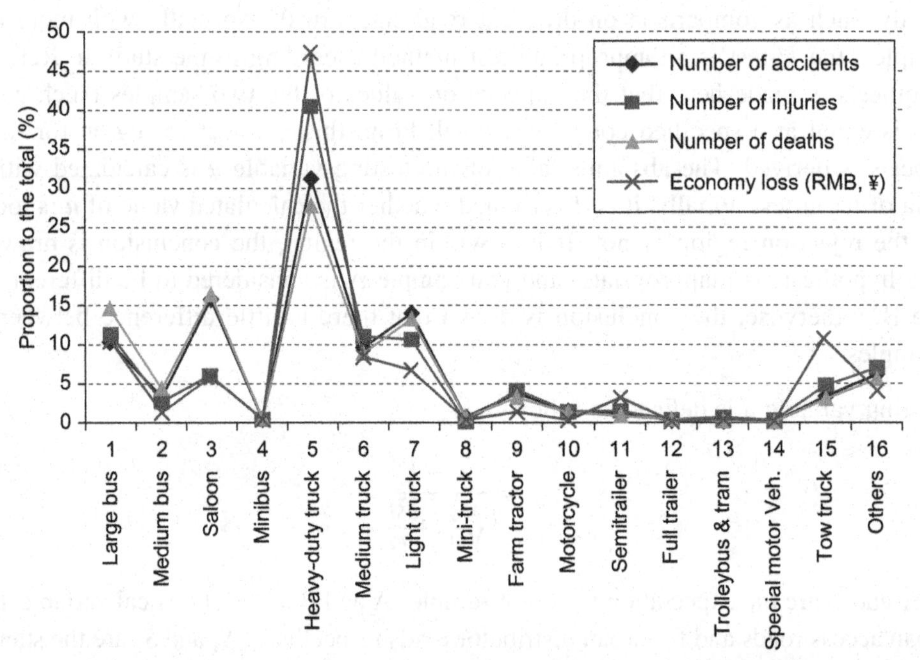

Figure 6.1: Proportion to Total Caused by Different Vehicle Types.

Vehicle type can affect driver behavior. The effect of car size on headways in high-flow traffic has been investigated [15], in which observed cars were classified into three size categories based on the observer's judgment. The relation of vehicle mass and average headway was studied by providing a more precise measure of vehicle size [16]. To determine the statistical significance of the mass effect, the observations have been divided into three categories: <1600 kg, 1600–1900 kg, and >1900 kg. The resulting F-statistic is $F = 6.64$ with two degrees of freedom, which is significant at the $p = 0.001$ level. Results significant at $p = 0.03$ were obtained using the mass categories <1500 kg, 1500–1900 kg, and >1900 kg, which was similar to the classification used by Wasielewski [15]. To estimate the effect of car size on freeway capacity, consideration must also be given to the effect of lead car size on headways.

In this study, we have conducted car-following experiments on the highway with three drivers using a tractor as model DFL4181A2. Figures 6.2–6.4 show the statistical results of driver behavior.

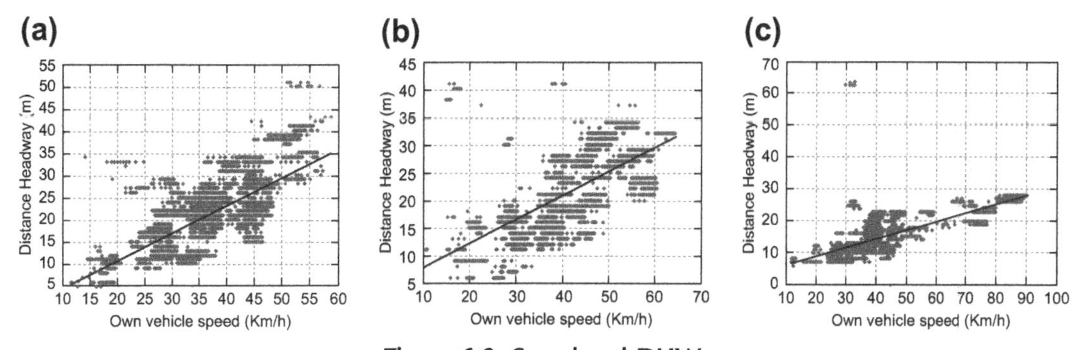

Figure 6.2: Speed and DHW.
(a) Driver 1. (b) Driver 2. (c) Driver 3.

From Figure 6.2, in steady car-following scenario, the closer is distance headway, the higher is the host vehicle speed, which can fit well a straight line.

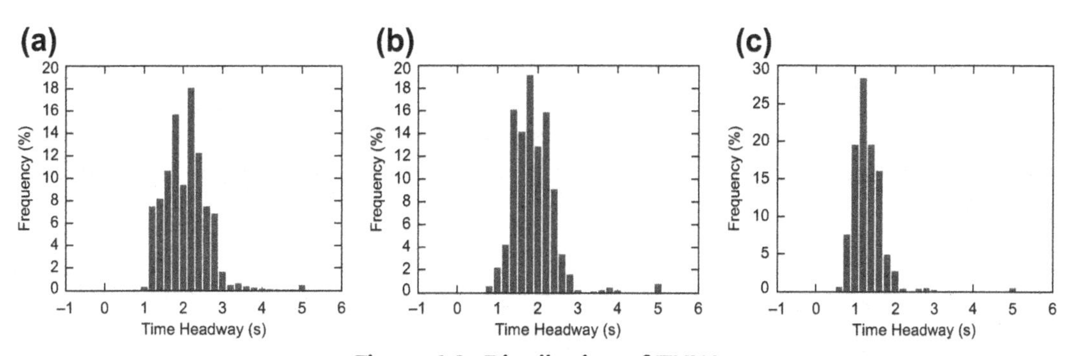

Figure 6.3: Distribution of THW.
(a) Driver 1. (b) Driver 2. (c) Driver 3.

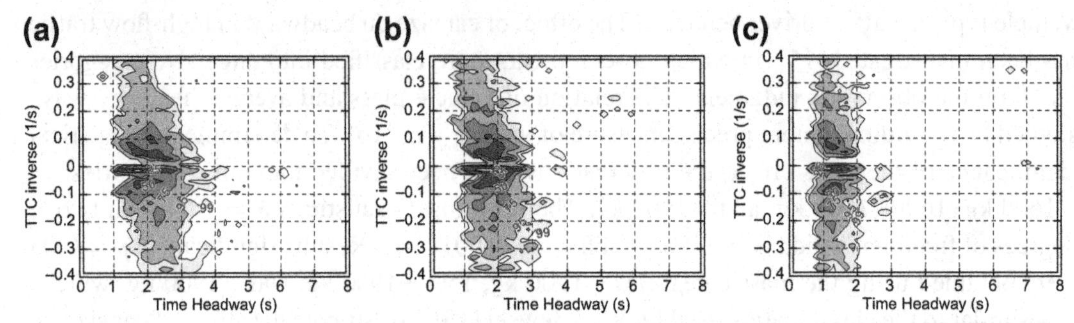

Figure 6.4: Distribution of THW and TTCi.
(a) Driver 1. (b) Driver 2. (c) Driver 3.

From Figure 6.3, the THWs used by the drivers in car-following were 1–2.5 s.

From Figure 6.4, in a non-strict car-following scenario, the 25% percentile was the area that the drivers used in a car-following scenario.

Though the experiments were conducted on truck platforms by only three drivers, it could be preliminarily observed that the driving behaviors were different from driver to driver. Compared with the results shown in Chapter 5, they were also different from the driving behavior using a saloon.

6.3.2 Vehicle Safety Technology

In this research, the experimental platform was a passenger car equipped with ITS systems. There are two ITS systems with different functions on this vehicle: Distance Control Assistance System (DCA) and Lane Departure Prevention System (LDP). The DCA system supports the driver to maintain an appropriate distance by raising the gas pedal to push the driver to shift to braking. The LDP system detects unintentional lane departure, and assists the driver to return to the center of the lane.

Driver Profile

There were 40 drivers involved in the driver behavior experiments. All the participants were male, and most were professional drivers such as taxi drivers. Driving time per day is listed in Table 6.2.

Table 6.2: Distribution of the Driving Time Per Day

Driving time per day (hours)	<2	2–4	4–8	>8
Number	7	20	8	5

- **Age of drivers.** The mean age was 44.7. The variance and standard deviation of age were 90.03 and 9.49 respectively. The oldest was 63 and the youngest 27.
- **Years of driving.** The mean number of years of driving was 17.25. The variance and standard deviation were 58.86 and 7.67 respectively.

Experiment for Comparison

Three comparative experiments were carried out for 4 hours in daytime with good weather, and with both DCA and LDP off, DCA on, and LDP on respectively.

Comparison of All Systems On and DCA On

The experimental route was the 4th Ring Road in Beijing. The experimental results are listed in Table 6.3. Figure 6.5 shows the THW distribution, THW–TTCi distribution and longitudinal acceleration–TTCi distribution in the condition of all systems off. Figure 6.6 shows the experimental results in the condition of DCA on.

Table 6.3: Statistical Results with Both Systems Off and DCA On

Parameters	THW_{mean} (s)	σ_{THW}	$TTC\_1stAccR_{mean}$ (s)	$\sigma_{TTC\_1stAccR}$	TTC_1stBra_{mean} (s)	σ_{TTC_1stBra}
System off	1.61	0.92	22.87	11.54	14.08	9.21
DCA on	1.75	1.03	22.76	11.26	14.07	8.87

From Figure 6.7(b) and (c), it can be observed that, with the DCA system on, the values of TTC with the driver in action are similar to the results obtained with the systems turned off.

In order to evaluate the effect of the DCA system (including positive and negative), we defined three indicators to reflect the characteristics:

$$P_{DCA} = \frac{N\_manual}{N\_alloff} \tag{6.3}$$

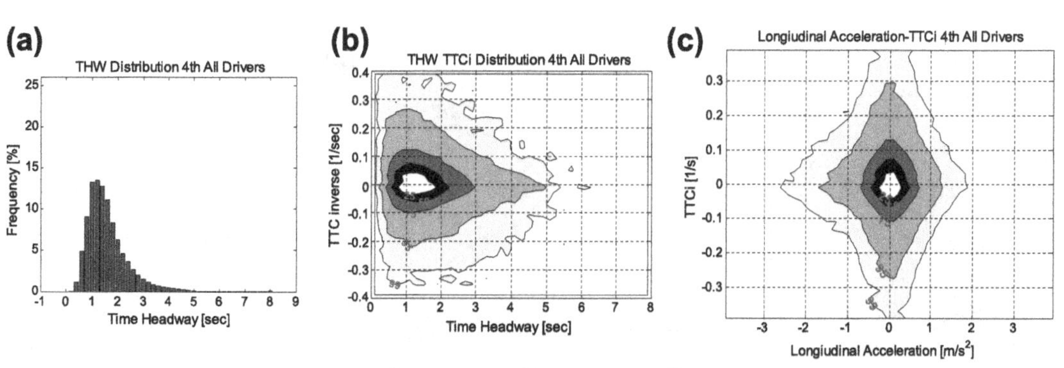

Figure 6.5: All Systems Off.
(a) THW distribution. (b) THW–TTCi distribution. (c) Longitudinal acceleration–TTCi distribution.

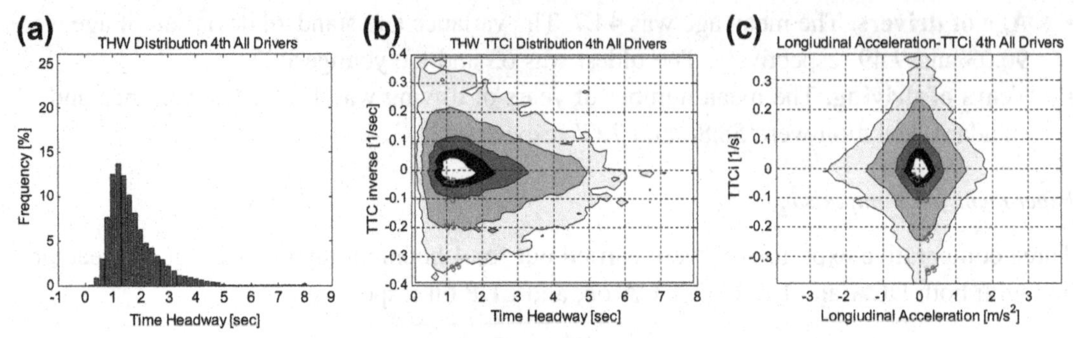

Figure 6.6: DCA On.
(a) THW distribution. (b) THW–TTCi distribution. (c) Longitudinal acceleration–TTCi distribution.

$$N_{\mathrm{DCA}} = \frac{N\_manual + N\_system - N\_alloff}{N\_alloff} \tag{6.4}$$

$$Q_{\mathrm{DCA}} = \frac{N\_manual}{N\_manual + N\_system}. \tag{6.5}$$

It is expected that the smaller the value, the better the effect. If P_{DCA} is less than 1 this means that the DCA system has some positive effect – with less braking. It therefore has reduced the labor intensity to some extent. If N_{DCA} is greater than 0 this means that the amount of braking with DCA on is more than with the system off. It indicates that the DCA system does assist in normal driving by braking. Q_{DCA} indicates the proportion of manual braking when the DCA system is on. If Q_{DCA} is less than 0.5 then the DCA system can reduce the braking times effectively.

Test results are shown in Table 6.4. The effect on the former 20 drivers is not apparent. The reason could be that they did get used to the system very well in the first phase experiment

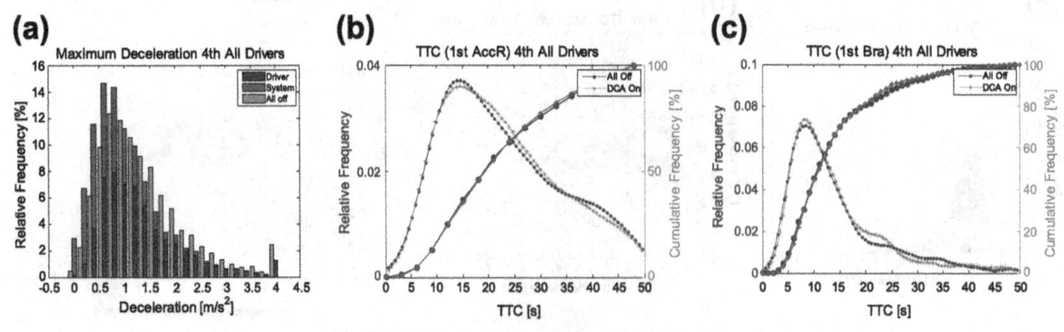

Figure 6.7: System Off and DCA On.
(a) Comparison of maximum deceleration at braking. (b) TTC distribution (A-pedal release). (c) TTC distribution (brake pedal action).

Table 6.4: Values for Different Drivers

Driver Number	P_{DCA} Values	N_{DCA} Values	Q_{DCA}
1	0.20	−0.52	0.40
2	0.42	−0.22	0.53
3	0.27	−0.45	0.49
4	0.15	−0.75	0.62
5	1.13	0.35	0.83
6	0.52	−0.01	0.52
7	0.84	−0.16	— —
8	0.51	−0.41	0.88
9	0.58	−0.25	0.78
10	1.44	0.89	0.76
11	0.97	−0.03	— —
12	1.61	0.61	— —
13	0.44	−0.53	0.93
14	—	—	—
15	0.16	−0.74	0.63
16	1.12	0.80	0.62
17	0.97	0.24	0.79
18	0.73	−0.23	0.94
19	1.77	0.77	— —
20	0.89	−0.02	0.91
21	0.51	0.02	0.50
22	0.41	−0.11	0.46
23	0.63	0.27	0.50
24	0.24	−0.39	0.39
25	0.19	−0.50	0.39
26	0.55	0.60	0.34
27	0.24	−0.43	0.43
28	0.33	−0.17	0.40
29	0.25	−0.45	0.44
30	0.28	−0.39	0.46
31	0.16	−0.52	0.33
32	0.78	0.65	0.47
33	0.44	−0.08	0.47
34	0.22	−0.47	0.42
35	0.77	0.57	0.49
36	0.56	−0.15	0.66
37	0.15	−0.67	0.44
38	1.51	2.02	0.50
39	0.41	0.12	0.36
40	3.05	9.01	0.30

Remarks: '—' means the data is incomplete; '— —' indicates that the system does not work.

without a preliminary training experiment. The reason why P_{DCA} is greater than 1 could be that the drivers did not adapt to the system, although the drivers were prudent when driving. Also, 82.1% of the values of P_{DCA} are less than 1. This indicates that the DCA system played a better role on the experimental roads.

Comparison of all Systems Off and LDP On

The experimental route was intercity highway (freeway) between Beijing and Tianjin. From Figures 6.8 and 6.9(a), it can be seen that almost all the drivers were used to driving a little left of the center line (the mean value of Position is negative). This could be a driving characteristic of Chinese drivers.

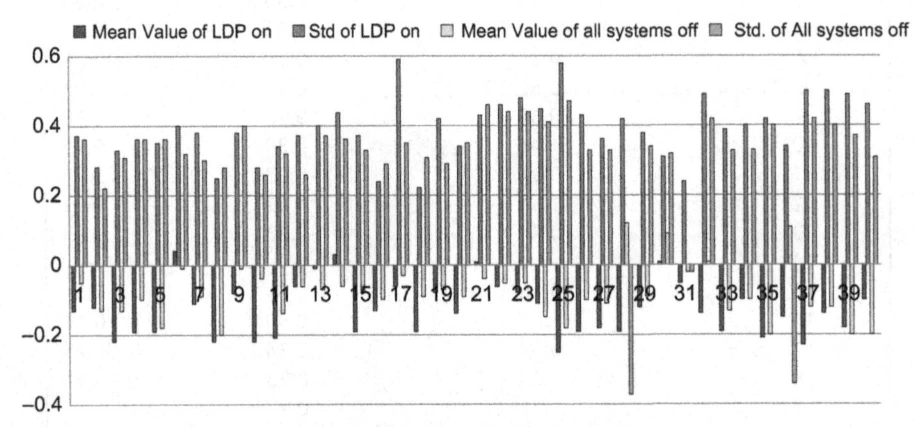

Figure 6.8: Vehicle Lateral Position for Different Drivers.

From Figure 6.9(b), we can see that the original maximum lateral acceleration in lane recovery is concentrated around 0.06 m/s^2. The difference between All-Off and LDP-On is not obvious, which means that the LDP system did not affect the lateral acceleration.

From Figure 6.9(c), it can be observed that the difference between All-Off and LDP-On is not obvious and that the value of TLCi is concentrated around 0 s^{-1}, which means that the LDP system did not affect the TLCi.

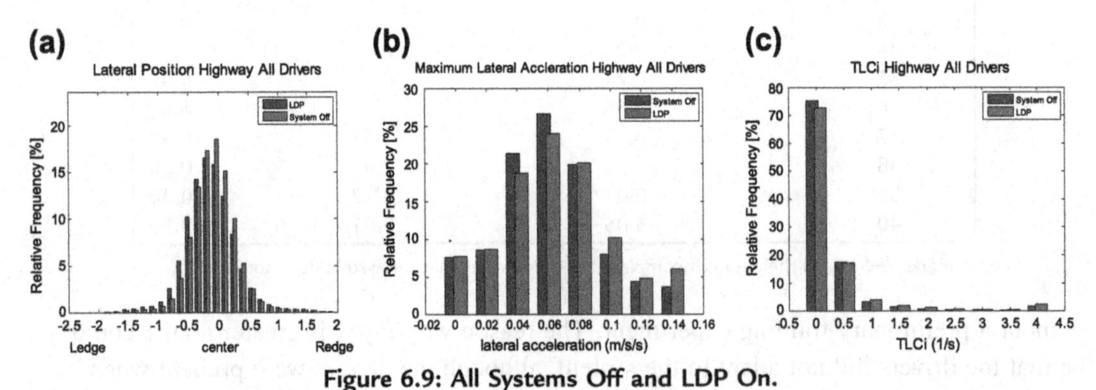

Figure 6.9: All Systems Off and LDP On.
(a) Comparison of lateral position. (b) Maximum lateral acceleration at lane recovery. (c) Comparison of TLCi.

Two indicators were used to evaluate the effectiveness of the LDP system (including positive and negative): Position (Vehicle Position) and

$$N_{\mathrm{LDP}} = \frac{N\_lanedeviation\_LDP}{N\_lanedeviation\_alloff}. \tag{6.6}$$

The smaller the value, the better the effect. The results are listed in Figure 6.10. When N_{LDP} is less than 1 or equal to 0, this means that the LDP system played a positive role. The reason why N_{LDP} is greater than 1 could be that the driver did not adapt to the system.

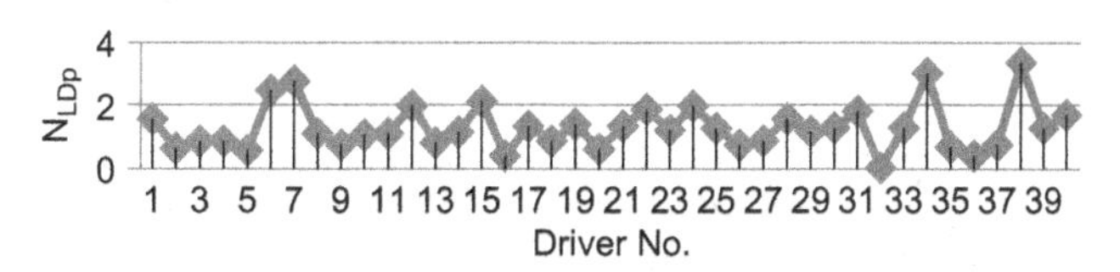

Figure 6.10: N_{LDP} Values for Different Drivers.

6.4 Road Factors

6.4.1 Road Categories

This research focuses on the microscopic analysis of longitudinal driving behavior on three types of roads in Beijing (PRC): freeways, distributor roads, and (local) access roads. The distinction is meaningful as different road categories involve different types and levels of traffic risk. In general, a specific countermeasure often relates to and is designed for a specific road category. Therefore, road categorization is useful for research on selection of optimal routes from a safety perspective.

Experimental Results

The longitudinal driving scenarios have been analyzed using the developed experimental platform. The objects of analysis were the assembled data sample points from data sets collected on different road categories. Statistical results concerning longitudinal driving behavior for each scenario (i.e. A: steady-state car following; B: non-restricted car following; or C: approaching) on different road categories are as follows.

In Scenario A, the distributions of DHW diverge with different vehicle speeds for each road category, i.e. DHW $= 0.5628V - 2.0599$ (urban access roads), DHW $= 0.9742V - 33.6788$ (urban distributor roads), and DHW $= 0.6847V - 17.5224$ (freeways) respectively. This indicates that with the increase of the host vehicle speed, the values of DHW increase accordingly. The minimum restricted vehicle speed on different roads is not the same. For example, the vehicle speed of the equation when on urban access roads is greater than 20 km/h. Therefore, the resulting DHW values will all be positive.

The linear regression model for Scenario B for correlations between vehicle speed and DHW are as follows: DHW $= 1.5262V - 1.3879$ (urban access roads), DHW $= 0.9037V - 30.139$ (urban distributor roads), and DHW $= 0.6881V - 19.015$ (freeways). The results for Scenario B contrast with those of Scenario A.

The linear regression results for Scenario C for correlations between DHW and vehicle speed are: (1) accelerator release, DHW $= 0.5847V - 2.0307$ (urban access roads), DHW $= 0.6858V - 15.015$ (urban distributor roads), and DHW $= 0.5600V - 7.1271$ (freeways); (2) brake activation, DHW $= 0.5088V + 0.2451$ (urban access roads), DHW $= 0.4586V - 2.4098$ (urban distributor roads), and DHW $= 0.4299V + 0.0027$ (freeways).

Comparative Analysis

The above results show road category dependence for variations in longitudinal driving behavior. The u-test method is used for statistical analysis in validation. Considering the restriction of Scenarios A and C that are included in Scenario B, the differences between road categories can be counteracted and Scenario B is chosen as the compared sample. THW and TTC were considered before as important parameters to measure longitudinal driver behavior in Ref. [17]. Since the relative speed can sometimes be zero, TTC is replaced by TTCi here.

$\alpha = 0.05$ was the typical level of significance used in the u-test. The results are shown in Tables 6.5 and 6.6. The rejection region is $|u| > 1.65$. For the u-test of variables based on different road categories, compared with the rejection region, the absolute values of almost all are larger (only those italicized are less), i.e. they are within the rejection region except for the values marked with bold italics. Therefore, the hypothesis H_0 is rejected and the conclusion is drawn that longitudinal driving behavior is affected by road categories.

By further analyzing the results presented in Tables 6.5 and 6.6, some significant phenomena are found. For different road categories, the mean values of DHW simultaneously increase with vehicle speed from urban access roads to urban distributor roads and to freeways (see Figure 6.11(a)). Meanwhile, the values of the standard deviations of these parameters also simultaneously increase with different roads. It seems that the driving stability decreases when the vehicle speed and DHW increase, which is shown in Figure 6.11(b). Meanwhile, longitudinal driving behavior (measured by THW, TTCi, and TTC) varies for different road categories (see Figure 6.11(c) and (d)).

In this research, the longitudinal driving behavior is measured by host vehicle speed, DHW, THW, TTC, relative speed, accelerator pedal release, brake pedal activation, and gap closing. The results show that the longitudinal driving behavior varies with road categories. In addition, road categories influence the correlation between host vehicle speed and DHW in the car-following scenario, but do not influence the correlation between relative speed and DHW.

It has been found that longitudinal driving behavior is different for different road categories, and these differences can be quantified using a data-processing program. In addition, it has

Table 6.5: Absolute Values of *u*-Test Variables Based on Different Road Categories (Parameter: THW)

Driver Number	Urban Access Roads and Urban Distributor Roads	Urban Access Roads and Freeways	Urban Distributor Roads and Freeways
1	6.9561	1.3958	13.4147
2	36.4094	41.3795	15.9786
3	44.4871	1.7786	99.2302
4	10.5652	14.2633	12.4475
5	11.2940	8.5239	4.1502
6	44.7998	67.3504	50.1572
7	32.2364	8.4551	103.6063
8	8.3393	27.8315	62.1012
9	14.0531	19.2531	13.6999
10	10.6438	14.4555	10.1004
11	11.4161	39.8083	97.0732
12	13.0054	23.1922	23.6187
13	5.2334	50.7283	58.7001
14	5.7070	14.4148	34.1414
15	14.2043	15.5245	1.851
16	25.6831	10.9557	40.5881
17	12.2499	5.3268	34.3462
18	31.1119	7.6987	41.78
19	32.3644	20.7317	24.6619
20	16.7769	8.023	13.7928
21	27.9054	10.3097	50.9016
22	3.3306	11.2747	21.9681
23	14.5222	14.7068	0.9013
24	11.4765	34.4025	56.5802
25	39.2769	48.6192	55.1755
26	39.491	16.9235	32.1954
27	16.7955	1.5894	19.3263
28	28.4096	30.7226	8.2991
29	4.4246	4.1819	16.3476
30	7.2276	14.5314	23.6253
31	10.2368	5.4859	8.7816
32	0.9347	9.3137	26.3953
33	25.5432	19.3219	11.8391

been found that the lower the host vehicle speed, the shorter the distance headway (DHW). However, the change of host vehicle speed is not always in the same direction as the change of the time headway (THW), which can be explained from the fact that drivers tend to be inconsistent in their choice of headway [18]. Due to different perspectives, goals, research approaches, and parameter settings, the results obtained here may be different from other studies.

Table 6.6: Absolute Values of *u*-Test Variables Based on Different Road Categories (Parameter: TTCi)

Driver Number	Urban Access Roads and Urban Distributor Roads	Urban Access Roads and Freeways	Urban Distributor Roads and Freeways
1	7.9507	7.9600	0.3186
2	9.7178	4.8134	11.6548
3	14.0946	9.1887	13.3268
4	11.3276	19.4584	15.747
5	4.7760	9.4072	14.5693
6	3.6912	1.1318	6.6802
7	18.3322	10.1289	21.5564
8	5.1799	0.3604	15.4514
9	11.3713	3.0731	37.5491
10	3.1255	9.3010	23.1081
11	7.2110	5.3371	5.5613
12	2.2928	6.7915	11.5126
13	7.7454	11.1861	11.0772
14	3.5249	7.4738	10.6219
15	34.544	10.4058	53.0379
16	5.5331	0.8049	18.7247
17	14.8603	13.1136	3.6537
18	1.6867	8.5749	20.2412
19	16.8363	15.5033	0.4244
20	14.2633	10.0043	11.7663
21	5.8154	9.3964	38.9002
22	5.3174	11.2824	44.772
23	6.2247	5.6265	2.032
24	17.8596	15.3124	5.8275
25	12.6612	10.7626	8.5319
26	31.5498	32.3261	3.3505
27	11.3669	33.4324	39.7057
28	3.3338	3.5822	24.4785
29	1.1243	1.5782	7.3341
30	9.7426	18.5558	39.0048
31	7.3964	0.1126	21.7154
32	5.5081	0.1503	22.7224
33	1.9863	4.7303	15.5774

6.4.2 Road Geometry

To study driver behavior on curved roads and to compare it with driver behavior on straight roads, real-vehicle experiments on both curved and straight roads were designed and carried out. A large number of experimental data was obtained through these experiments. The driver behavior in approaching the two road geometries was analyzed based on the data.

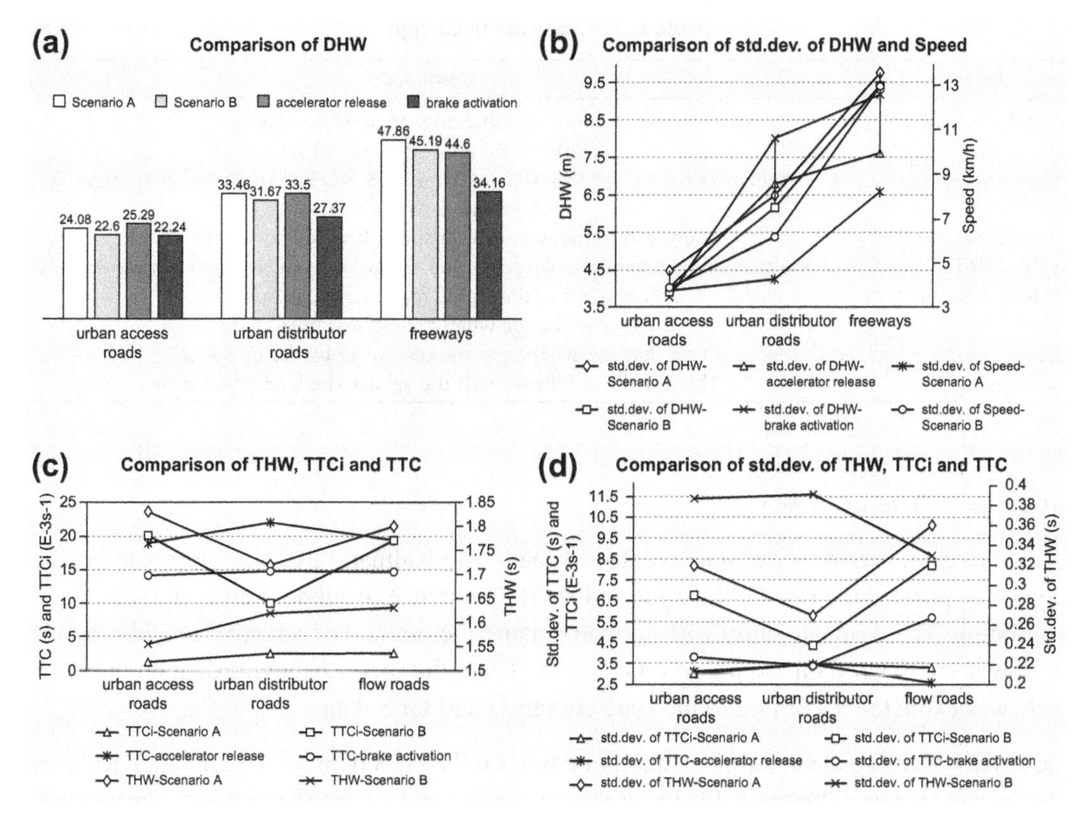

Figure 6.11: Comparison on Different Road Categories.
(a) Comparison of DHW. (b) Comparison of standard deviation of DHW and speed. (c) Comparison of THW, TTCi, and TTC. (d) Comparison of standard deviation of THW, TTCi, and TTC.

Experiment Design

The experimental approaches of driver behavior data collection on both curved and straight roads are shown in Table 6.7.

Eighteen male drivers were involved in the tests. Each test driver had a warm-up driving period for 10 kilometers to become familiar with the vehicle. At the same time, the equipment operator explained the objectives, requirements, and test procedures. In order to ensure the comparability of driver behavior data, the arrangements of all test drivers were in full agreement.

The experiment route is shown in Figure 6.12. The total mileage for each test driver was about 80 kilometers. The data obtained mainly included time, relative distance, speed, relative speed, brake signal, GPS latitude, GPS longitude, longitudinal acceleration, lateral acceleration, steering angle, steering speed, mileage, and so on.

Table 6.7: Experiment Design

Items	Contents
Weather condition	Sunny weather and dry road surface
Time	8:00–11:00 a.m., 2:00–5:00 p.m.
Experiment route	City highway in Beijing (the road section from Nankou to Badaling in Jingzang highway, a round trip)
Traffic condition	Smooth traffic with vehicle speed from 60 to 120 km/h
Staff	One test driver, one lead driver, and one equipment operator in each experiment
Requirement	1. Strictly obey the traffic regulations, no speeding 2. No lane change when driving on curved road 3. No disturbance of the test driver by the operator
Approach	The test driver follows with the vehicle the lead driver drives

Curved Road Data Extraction

To extract the driver behavior data on curved roads, the Kalman filter algorithm was firstly adopted to process the raw GPS data to reduce the influence of random error in GPS positioning. The road curvature was estimated using the method of adaptive neighborhood window growth based on the filtering GPS data. Then, the driver behavior data on curved roads was extracted according to the road curvature and GPS data.

The results of the road curvature estimation based on the GPS data before and after Kalman filtering are shown in Figure 6.13. From this figure, it can be seen that the road curvature estimation result based on the filtering GPS data was more accurate than that based on the raw GPS data. Therefore, the accuracy of road curvature estimation was improved through GPS data filtering.

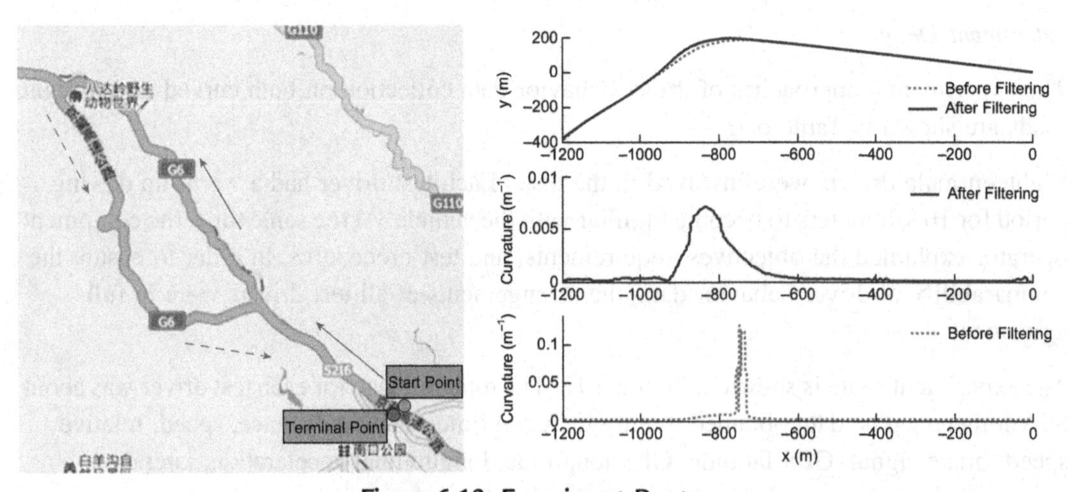

Figure 6.12: Experiment Route.

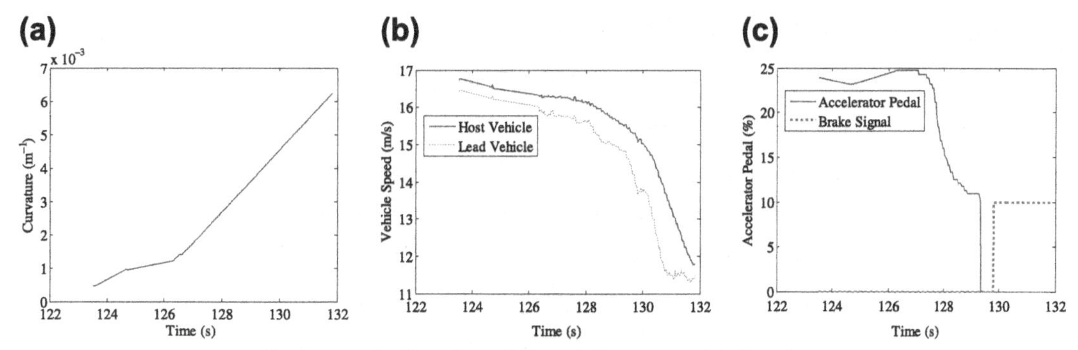

Figure 6.13: Results of Road Curvature Estimation.

GPS coordinates of road curves were determined according to the GPS data and road curvature estimation result, and based on this the curve data was extracted. According to Ref. [19], the operating speed will not be influenced by the road curvature if the road radius is greater than 2000 m. Due to the fact that all the radius of road curvature for the experiment were bigger than 100 m, the radius range is selected from 100 to 2000 m. The driver behavior data including 17 curved road sections were extracted, as listed in Table 6.8.

The road section from Nankou to Juyongguan is straight, so the driver behavior data on this road section was considered as the straight road data.

Table 6.8: Curved Road Characteristics

Number	Minimum Curvature (m)	Curve Length (m)*
1	142	284
2	160	304
3	158	252
4	157	363
5	260	339
6	210	229
7	194	215
8	217	288
9	179	406
10	249	156
11	186	347
12	207	407
13	134	247
14	213	303
15	157	365
16	444	370
17	550	489

*Criterion of curve length calculation: the radius of road is less than 2000 m.

Driver Behavior Analysis for Vehicle Approaching

A Driver Data Analysis Tool (DDAT) in MATLAB with GUI has been developed to study driver behavior in vehicle approaching.

The driver behavior data for this scenario was extracted according to the definition of vehicle approaching. Figure 6.14 shows such data on a curved road section including: host vehicle speed, lead vehicle speed, accelerator pedal, brake signal, THW, and TTCi with respect to time.

It can be observed that the host vehicle speed was larger than the lead vehicle speed before entering the curved section, and the driver released the accelerator pedal gradually to control the vehicle speed as it got closer to the lead vehicle and as the road curvature was increasing. When DHW was decreased to a certain value, the driver started to release the accelerator pedal and braked as necessary to ensure safety.

Statistical analyses about continuous car-following data when approaching on a curved road and a straight road were conducted. The comparisons of THW and TTC distribution are shown in Figure 6.15.

It can be seen that THW distribution is much wider in approaching on a curved road than on a straight road. In addition, TTC has a dispersive distribution and the difference is not significant.

Figure 6.16 shows the comparison of mean and standard deviation for THW and TTC. The mean of THW is bigger and TTC is smaller on a curved road than those on a straight road. The standard deviations of THW and TTC on a curved road are slightly greater than those on a straight road.

To verify whether there is a significant difference between THW and TTC when approaching on a curved road as compared to those on a straight road, the Wilcoxon rank-sum test method was used to compare the two independent samples.

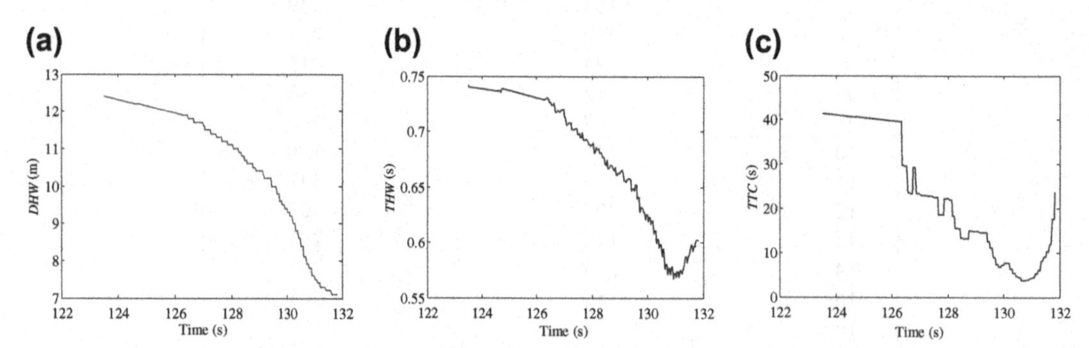

Figure 6.14: Driver Behavior Data in Approaching on Curved Road.
(a) Curvature. (b) Vehicle speed. (c) Acceleration pedal.

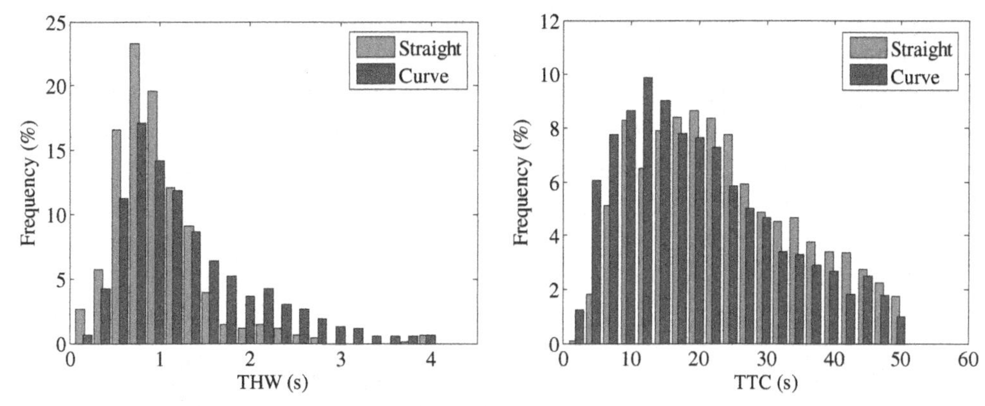

Figure 6.15: Comparison of THW and TTC Distribution.

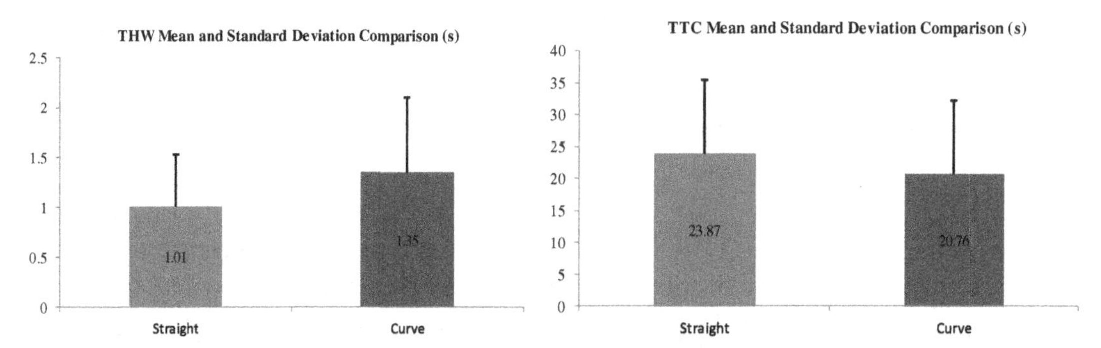

Figure 6.16: Comparison of Mean and Standard Deviation for THW and TTC.

In Table 6.9, α is the significance level, which is set to 0.05. P is the probability of equal means for the two samples. The test results indicate that there is significant difference between THW and TTC when approaching on a curved road as compared to those on a straight road. In other words, drivers have quite different driving behaviors when approaching a curved road section compared to that on a straight road.

Table 6.9: Test Results of THW and TTC

Test Object			Significance Level α	P
Approaching scenario	Straight road	THW	0.05	<0.01
	Curved road	THW	0.05	<0.01
	Straight road	TTC	0.05	<0.01
	Curved road	TTC	0.05	<0.01

The dependence of THW and TTC on road curvature was also studied. Figure 6.17 shows the contour distributions of THW and TTC with respect to road curvature. It can be observed that the THW distribution is rather centralized and decreased gradually with the increase of road curvature. On the contrary, the TTC distribution is dispersive. Also, there is no obvious change with an increase of road curvature.

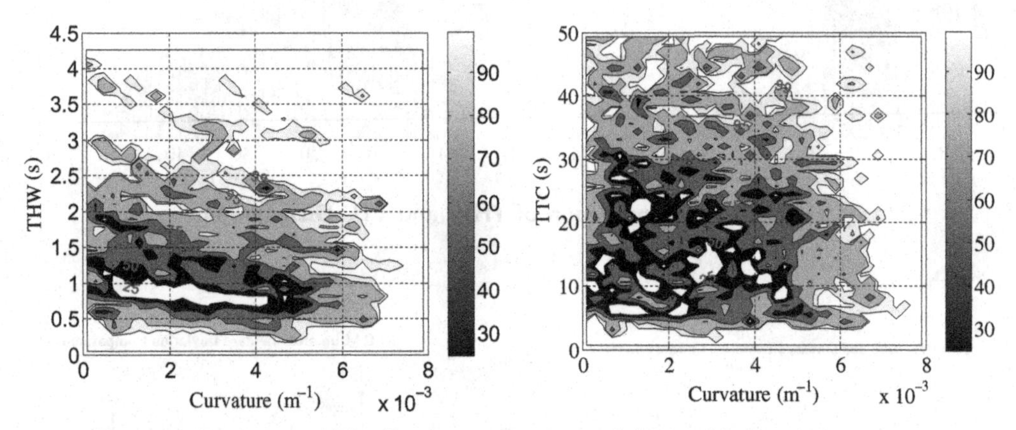

Figure 6.17: Contour Distributions of THW and TTC with Road Curvature.

Analysis of Driver Actions

Compared with a straight road, THW is increased and TTC is reduced on a curved road, and the two parameters have wider distributions. This indicates that the traffic condition on a curved road is more complicated than that on a straight road and drivers' responses vary significantly. Figures 6.18–6.21 show the differences in mean and standard deviation for THW and TTC distribution on a curved and on a straight road.

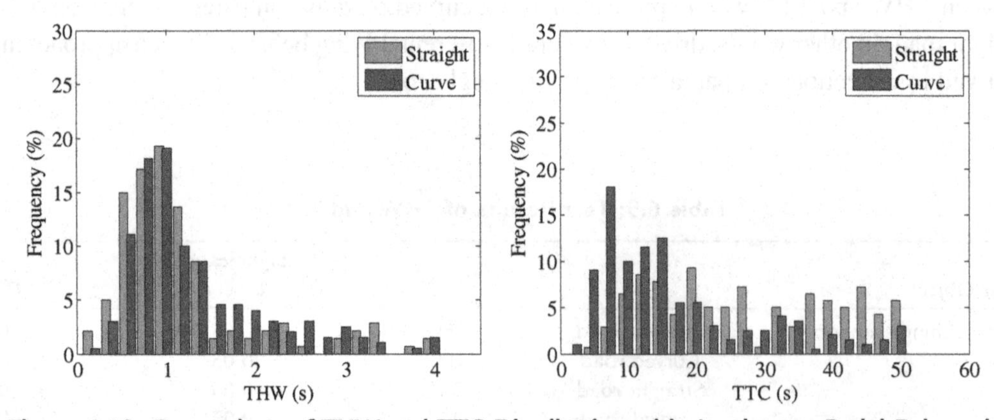

Figure 6.18: Comparison of THW and TTC Distribution with Accelerator Pedal Released.

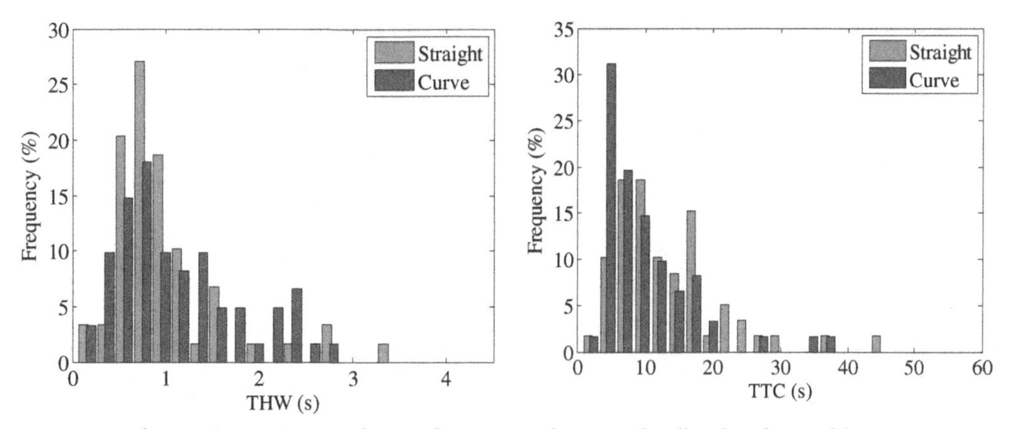

Figure 6.19: Comparison of THW and TTC Distribution in Braking.

Similarly, to verify whether there is a significant difference between THW and TTC in manual driving on a curved road and on a straight road, the Wilcoxon rank-sum test method is also used to test the significance. The test result is shown in Table 6.10. In Table 6.10, α is the significance level, which is set to 0.05. P is the probability of equal means for the two samples.

The test results indicate that there is also a significant difference between THW and TTC for manual driving on a curved road and on a straight road. Therefore, drivers have different behaviors on roads with a two-road geometry.

6.4.3 Lane Effect on Driver Behavior

To understand lane-keeping behavior, driving time distribution, longitudinal speed, lateral position, and time to lane crossing inverse have been analyzed for each lane on the 4th Ring Road in Beijing (with four lanes) and an intercity highway (with two lanes) between Beijing and Tianjin.

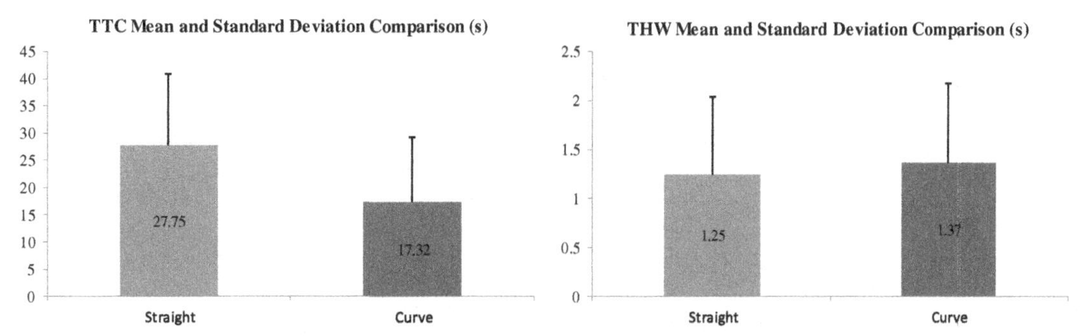

Figure 6.20: Comparison of Mean and Standard Deviation for THW and TTC as Driver Releases Accelerator Pedal.

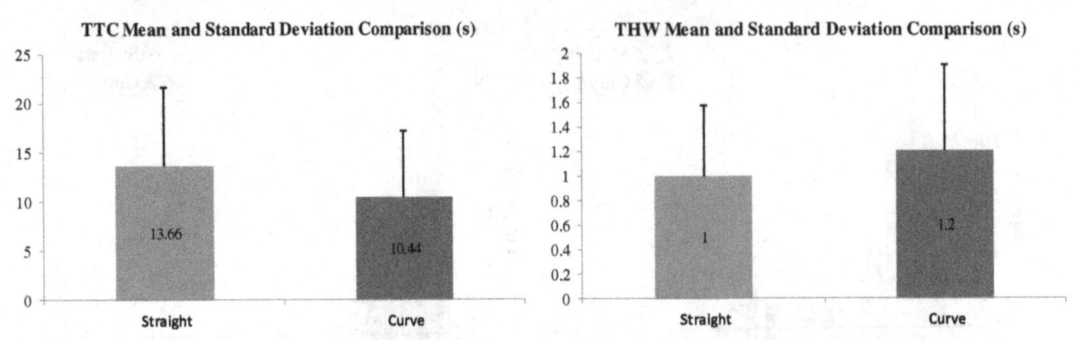

Figure 6.21: Comparison of Mean and Standard Deviation for THW and TTC in Braking.

Driving Time Distribution

In Figure 6.22, lane 1 is the far left lane. The driving time ratio on each lane is the driving time on the right lane divided by the total driving time. Here, the total driving time includes the lane change time and the driving time in each lane, so the sum of the driving time ratio in each line is smaller than 100%.

It is obvious that the driving time in the left lane is much greater than that in the right lane regardless of road types. This indicates that Chinese drivers are used to driving in the left lane instead of the right lane. The reason could be that the driving seat is on the left, which is the nearside.

Vehicle Speed Effect

In China, it is generally believed that traffic speed in the left lane is higher than that in the right lane on the highway, which means higher traffic flow in the left lane (Figure 6.23(b)). However, in Figure 6.23(a), there is no significant difference for each lane on the 4th Ring Road in practice. Although it is clear to see that, for some drivers (five of 13 drivers,

Table 6.10: Test Results of THW and TTC

Test Object			Significance Level α	P
Releasing the accelerator pedal	Straight road	THW	0.05	0.06
	Curved road	THW	0.05	
	Straight road	TTC	0.05	<0.01
	Curved road	TTC	0.05	
Braking action	Straight road	THW	0.05	0.25
	Curved road	THW	0.05	
	Straight road	TTC	0.05	<0.01
	Curved road	TTC	0.05	

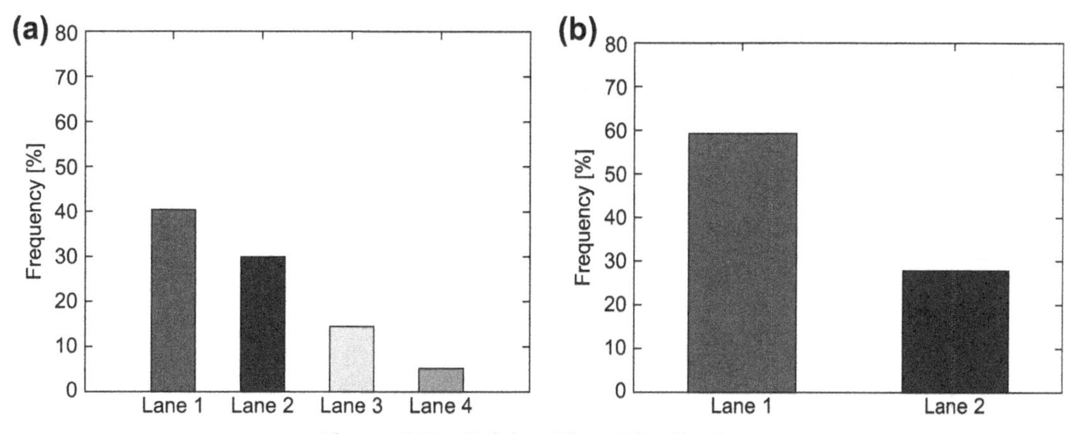

Figure 6.22: Driving Time Distribution.
(a) 4th Ring Road. (b) Highway.

i.e. about 38%), the average speed in lane 3 is the highest, but this does not infer anything because the vehicle speeds in different lanes are not comparable for different road sections and at different times.

Lateral Position Effect

The lateral position is another factor to represent driver behavior.

From Figures 6.24–6.26 and Tables 6.11 and 6.12, it can be seen that the average lateral position of the host vehicle are all negative. That is, drivers are used to driving on the left. However, things are different in each lane. The lateral position is almost in the middle in lane 1, and shifts more to the left within the lane as the lane number increases.

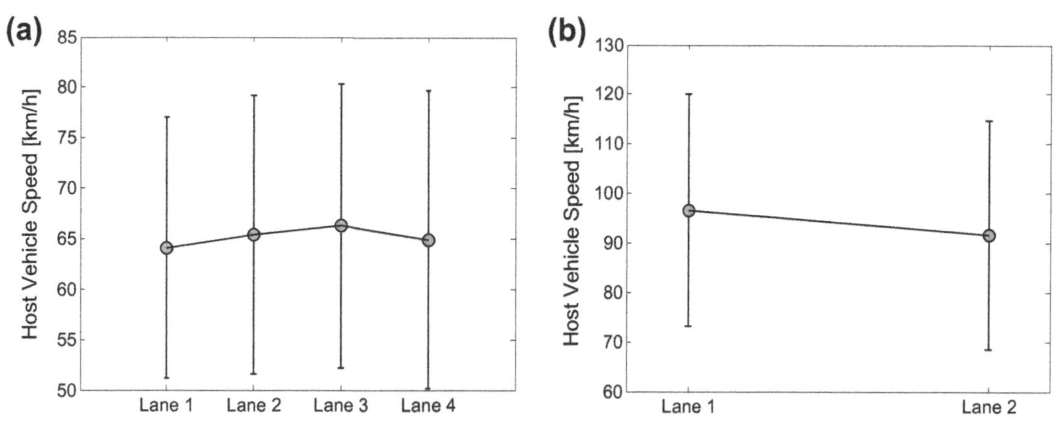

Figure 6.23: Host Vehicle Speed – Variance Analysis.
(a) 4th Ring Road. (b) Intercity Highway.

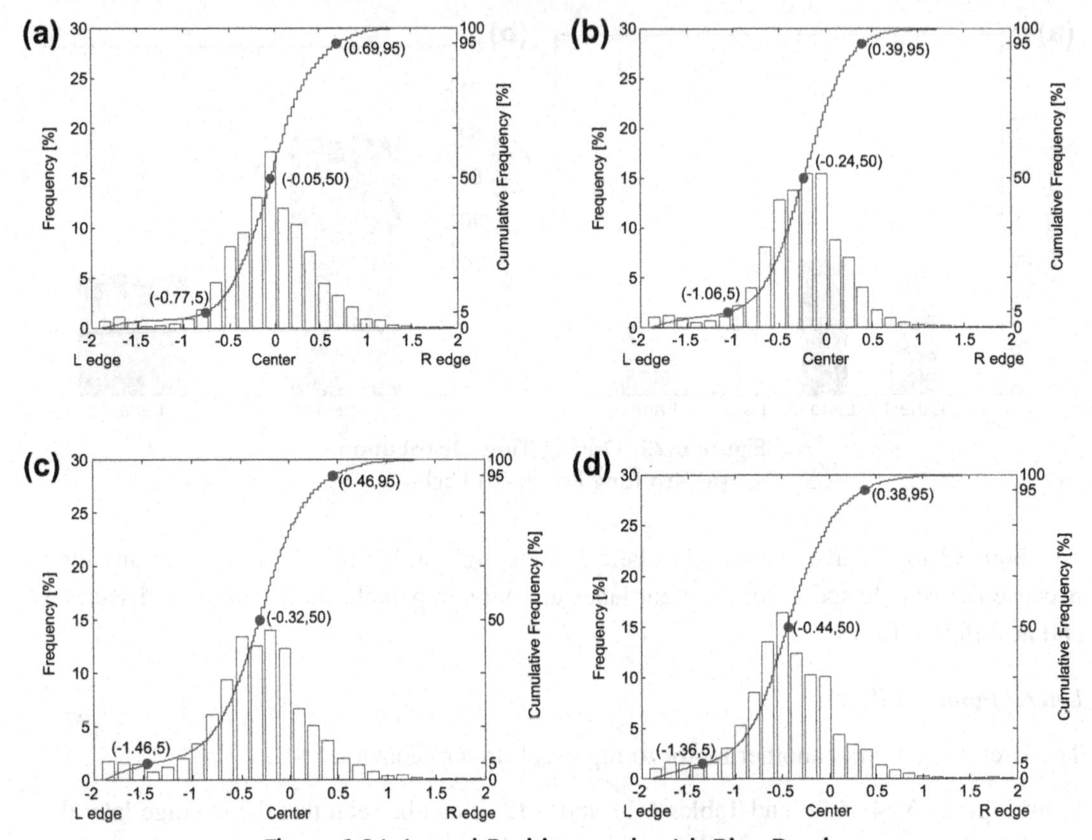

Figure 6.24: Lateral Position on the 4th Ring Road.
(a) Lane 1. (b) Lane 2. (c) Lane 3. (d) Lane 4.

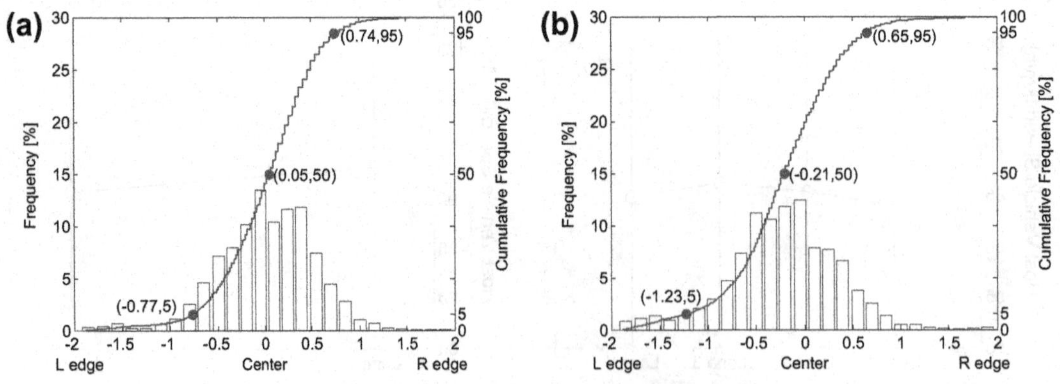

Figure 6.25: Lateral Position on the Highway.
(a) Lane 1. (b) Lane 2.

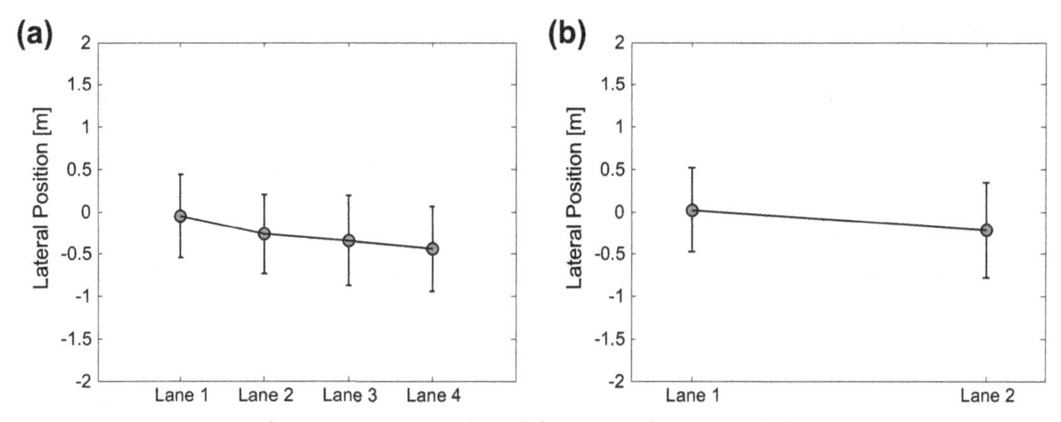

Figure 6.26: Lateral Position — Variance Analysis.
(a) 4th Ring Road. (b) Highway.

Table 6.11: Data Information for Lateral Position on the 4th Ring Road

Parameter	Lane 1	Lane 2	Lane 3	Lane 4
Number of data	149,568	107,820	51,409	18,558
Average (m)	−0.05	−0.26	−0.34	−0.44
Standard deviation (m)	0.49	0.47	0.53	0.51
Median (m)	−0.03	−0.25	−0.31	−0.45
Mode (m)	0.13	−0.08	−0.2	−0.43
5% tile value (m)	−0.77	−1.06	−1.46	−1.36
50% tile value (m)	−0.05	−0.24	−0.32	−0.44
95% tile value (m)	0.69	0.39	0.46	0.38

Table 6.12: Data Information for Lateral Position On the Highway

Parameter	Lane 1	Lane 2
Number of data	206,551	88,559
Average (m)	0.02	−0.21
Standard deviation (m)	0.5	0.57
Median (m)	0.06	−0.22
Mode (m)	0.34	0.02
5% tile value (m)	−0.77	−1.23
50% tile value (m)	0.05	−0.21
95% tile value (m)	0.74	0.65

6.5 Driving States

6.5.1 Car Following

The host vehicle state data for characterizing the longitudinal driving behavior of the test drivers under normal driving conditions and in emergency situations were recorded and comprehensively analyzed. Six different driving behavior phases are distinguished: (1) steady following, (2) approaching, (3) accelerating, (4) braking, (5) accelerator operation, (6) pedal switch.

Steady Following

Steady following is defined by four conditions: relative speed is very low, the absolute value of TTCi is less than 0.05 s^{-1}, the brake pedal is not pressed, and the host vehicle speed is below a certain threshold depending on the road category.

The experimental results in steady following are shown in Table 6.13. The average value of THW is 1.68 s on the 4th Ring Road (distributor road) and 1.72 s on the freeway, which means that drivers are more aggressive on the 4th Ring Road compared to the freeway. The reason is that the traffic flow speed on the 4th Ring Road is much lower than on the highway, and it is perceived to be safer with less nervousness on the 4th Ring Road. Figure 6.27 shows examples of the THW frequency distribution on the 4th Ring Road and freeway. In following other driving studies such as approaching, the authors present only the analysis results.

Approaching

The host vehicle approaching a preceding vehicle can be judged by two conditions: (1) the duration of gap reduction lasts for more than 5 s; and (2) TTC is less than 50 s.

- **THW on braking.** Under the approaching maneuver, the average value of THW is 0.35 s on the 4th Ring Road and 0.43 s on the freeway, which means that drivers take more

Table 6.13: Data Information for Time Headway

Parameter	4th Ring Road	Highway
Number of data	469,680	278,838
Average (s)	1.68	1.72
Standard deviation (s)	0.92	0.99
Median (s)	1.46	1.47
Mode (s)	1.8	0.9
5% tile value (s)	0.6	0.55
50% tile value (s)	1.46	1.47
95% tile value (s)	3.6	3.73

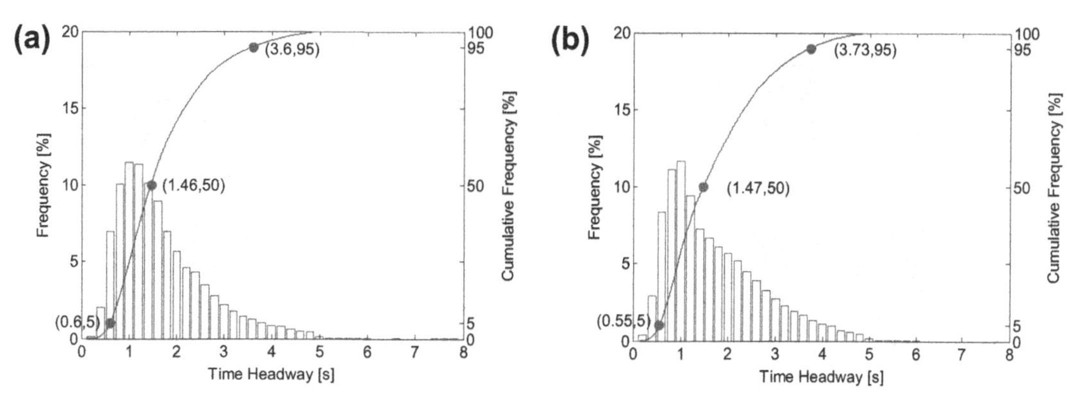

Figure 6.27: Time Headway.
(a) 4th Ring Road. (b) Freeway.

risk on the 4th Ring Road compared to the highway. One reason for this could be that the speed on the 4th Ring Road is much lower than on the freeway. Drivers feel safer on the 4th Ring Road than on the freeway in approaching.

- **TTC on braking.** In the approaching maneuver, the average value of TTC is 13.23 s on the 4th Ring Road and 10.74 s on the freeway. Therefore, drivers react later on the freeway than on the 4th Ring Road. One reason could be that the vehicle speed is very high on the freeway and the inter-vehicle distance may become very small before drivers realize the approaching condition. Another reason could be that, since the average speed on the 4th Ring Road is much slower than on the freeway and the speed range is also smaller, the relative speed will be smaller than on the highway. Another reason could be that drivers are aggressive on freeways.
- **Maximum deceleration on braking.** In the approaching maneuver, the average absolute value of maximum longitudinal deceleration is 1.17 m/s^2 on the 4th Ring Road and 1.67 m/s^2 on the freeway. This is because the vehicle speed on the freeway is greater and drivers have to brake fiercely to be safe when the inter-vehicle distance becomes smaller. This is also one reason for higher THW but lower TTC on freeways. Meanwhile, the number of braking actions on the 4th Ring Road is much greater than that on the freeway. The reason for this may be that the traffic conditions on the 4th Ring Road are more complicated and there are more traffic jams, which leads to more braking.
- **Speed–TTC.** Figure 6.28 shows the TTC distributions for different speeds with acceleration pedal release and brake pedal activation. In the approaching maneuver, TTC does not change significantly with speed. In fact, there is no linear relationship between these two parameters. The value of TTC appears discrete at low speed and almost constant at high speed, and more distributed on the 4th Ring Road and more concentrated on the freeway.
- **Relative speed–DHW.** Figure 6.29 shows the relationship between DHW and relative speed. For the approaching maneuver, DHW has a linear relationship with the relative

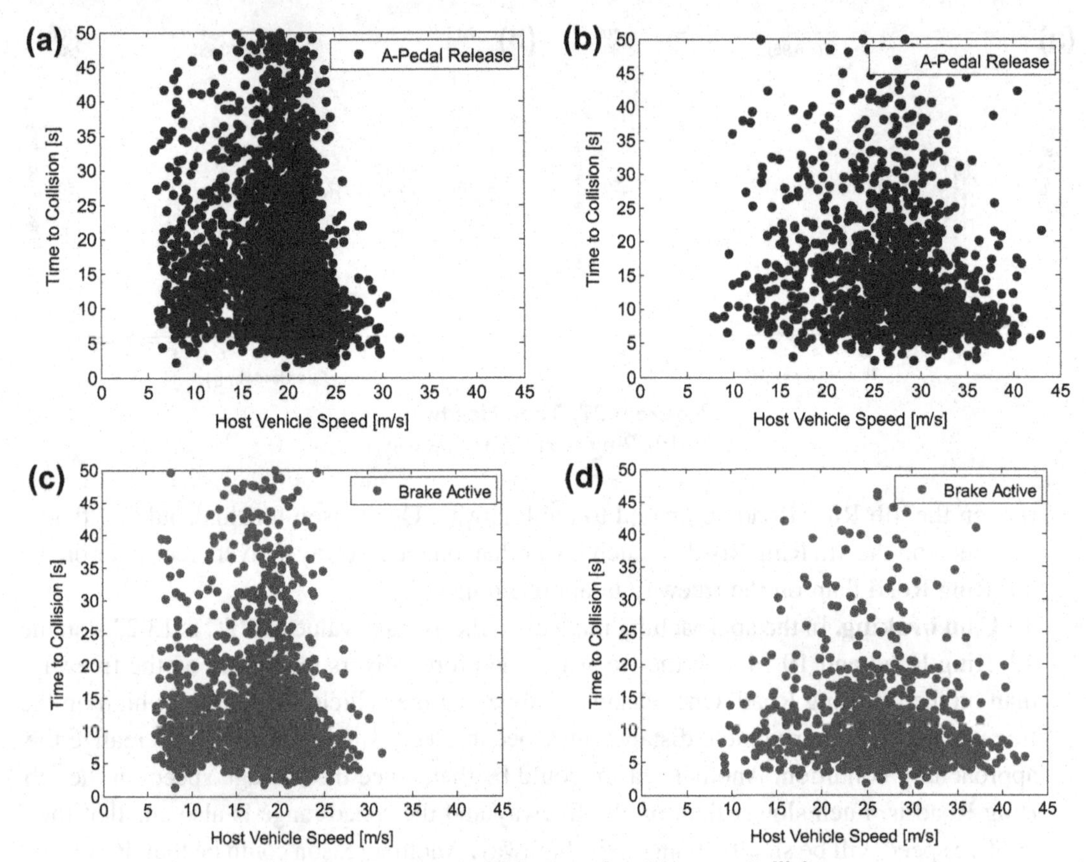

Figure 6.28: Host Vehicle Speed − Time to Collision.
(a) Accelerator pedal release, 4th Ring Road. (b) Accelerator pedal release, freeway. (c) Brake pedal activation, 4th Ring Road. (d) Brake pedal activation, freeway.

speed. For the same relative speed, the distance headway is larger on the freeway and smaller after braking.

- **Non-restricted following.** The situation in car following without any restriction on TTCi is defined as *non-restricted following.*
- **THW−TTCi.** Figure 6.30 shows the relationship between TTCi and THW. Under non-restricted following, the range of THW is 0−5 s, and the range of TTCi is −0.3−0.4 s^{-1}. 50% data of THW and TTCi on the 4th Ring Road and the freeway appear to be similar. For the rest of the data, if the TTCi value is negative, the TTCi values are almost the same on both the 4th Ring Road and the freeway, and if the TTCi value is positive, the values on the freeway are greater than those on the 4th Ring Road.
- **Acceleration−TTCi.** Figure 6.31 shows the relationship between TTCi and longitudinal acceleration. Under non-restricted following, the range of longitudinal acceleration is in the range of −2 to 1.3 m/s^2 on the 4th Ring Road and −3 to 1 m/s^2 on the freeway. Drivers

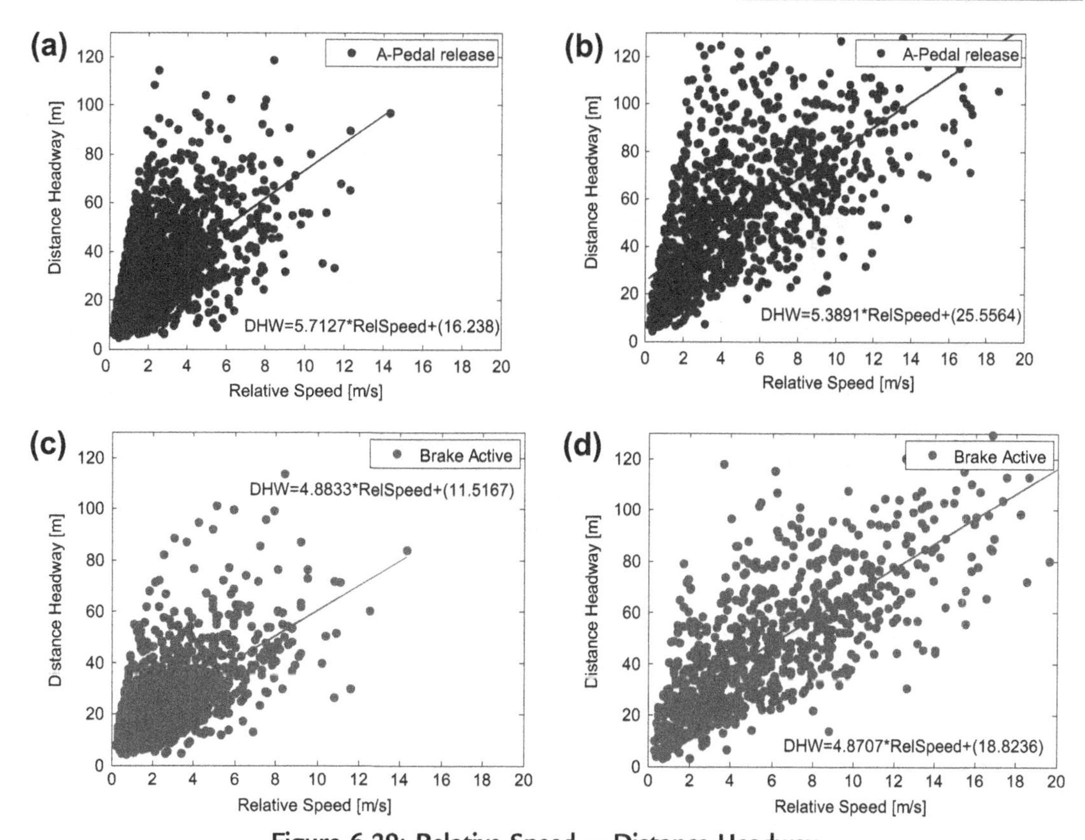

Figure 6.29: Relative Speed — Distance Headway.
(a) Accelerator pedal release, 4th Ring Road. (b) Accelerator pedal release, freeway. (c) Brake pedal activation, 4th Ring Road. (d) Brake pedal activation, freeway.

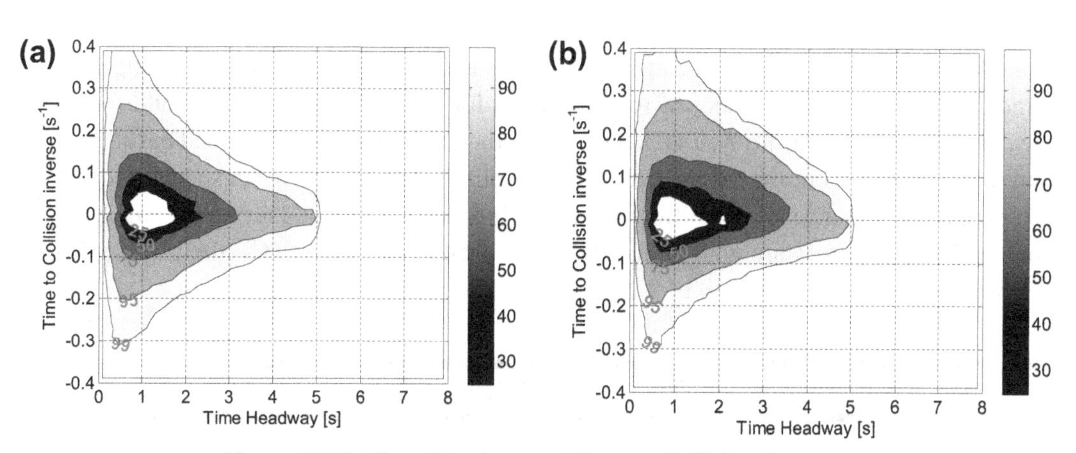

Figure 6.30: Time Headway — Time to Collision Inverse.
(a) 4th Ring Road. (b) Freeway.

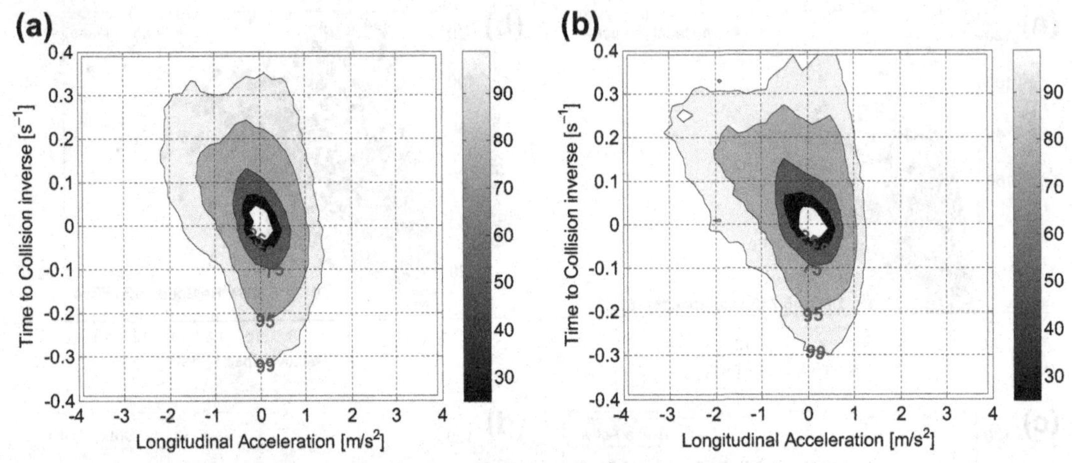

Figure 6.31: Longitudinal Acceleration − Time to Collision Inverse.
(a) 4th Ring Road. (b) Freeway.

choose smaller acceleration and larger deceleration on freeways for safety. The range of TTCi is -0.3 to $0.35\ s^{-1}$ on the 4th Ring Road and -0.3 to $0.4\ s^{-1}$ on the freeway, so the driving situation is a little more dangerous on the freeway.

6.5.2 Lane Change

In this section, the data for the freeway comes from route 3. The lane change process is divided into six time points: activating turn signal, starting lane change, aborting lane change, steering reversing, finishing lane change, and turning signal off. The six points are marked in Figure 6.32.

In the data analysis, the useful sections of the left lane change and right lane change are extracted separately. The number of route sections for left lane change is 325 and for right lane change is 390. Then, THW, TTCi, and some other parameters are analyzed based on data from the extracted sections. All the void data are eliminated based on the special request of each parameter, which leads to a different data set for each parameter. During the lane change, some of the TTC values are widely distributed or even infinite, so TTCi (the inverse of the TTC value) is calculated instead of TTC.

Figure 6.32: Lane Change Process.

Activating Turn Signal

This is the time when the driver activates the turn signal before lane change and is called the activating turn signal moment. At this time, the vehicle is still in the original lane.

- **THW to rear vehicle.** Figure 6.33 and Table 6.14 show the experimental and statistic results for THW. The average value of THW for a left lane change is greater than the value for a right lane change. This indicates that the rear vehicle drivers in adjacent lanes take less risk for the host vehicle left lane change. For following other moments such as starting lane change, the authors present only the analysis results.
- **TTCi to rear vehicle.** The average values of TTCi are negative. In other words, the host vehicle speed is larger than that of the rear vehicle when activating a turn signal, which indicates that drivers accelerate as soon as they decide to make a lane change.

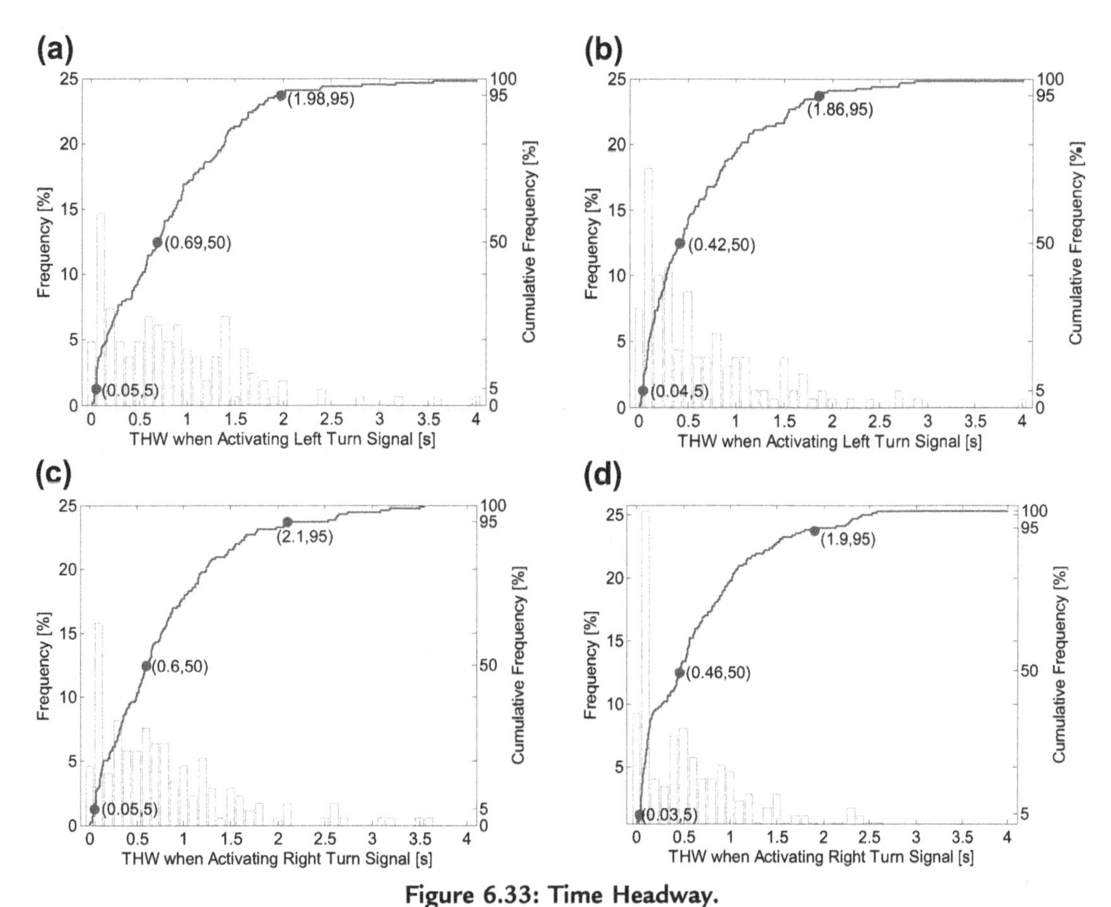

Figure 6.33: Time Headway.
(a) Left lane change, 4th Ring Road. (b) Left lane change, highway. (c) Right lane change, 4th Ring Road. (d) Right lane change, highway.

Table 6.14: Data Information for Time Headway

Parameter	4th Ring Road		Highway	
	Left Lane Change	Right Lane Change	Left Lane Change	Right Lane Change
Number of data	163	171	160	174
Average (s)	0.82	0.79	0.65	0.62
Standard deviation (s)	0.73	0.72	0.69	0.61
Median (s)	0.7	0.6	0.43	0.46
Mode (s)	0.27	0.01	0.01	0.44
5% tile value (s)	0.05	0.05	0.04	0.03
50% tile value (s)	0.69	0.6	0.42	0.46
95% tile value (s)	1.98	2.1	1.86	1.9

- **THW to lead vehicle.** The average value of THW for a left lane change is greater than the value for a right lane change. This indicates that the host vehicle driver takes less risk for a left lane change after the signal is turned on.
- **TTCi to lead vehicle.** All the average values of TTCi are positive, which indicates that the host vehicle speed is larger than the lead vehicle when activating the turn signal. In other words, drivers start to accelerate before they switch on the turning signal. The average value of TTCi for a right lane change is less than that for a left lane change. Therefore, drivers in the host vehicle tend to react later, or drivers feel there is greater danger in making left lane changes for collision avoidance.

Starting Lane Change

This moment is defined as starting the change of the lateral position. The vehicle location offset is checked after lane change start but before aborting the maneuver. This is the moment when the offset is larger than a quarter of the road width and smaller than half of the road width.

- **THW to rear vehicle.** The average value of THW for a left lane change is almost the same as that for a right lane change.
- **TTCi to rear vehicle.** All the average values of TTCi are negative, which indicates that the host vehicle speed is larger than the rear vehicle at the moment of starting a lane change. The absolute value of TTCi for a right lane change is larger than that for a left lane change, which indicates that drivers in the rear vehicle in the adjacent lane show a later reaction, or it is more dangerous to make a right lane change considering a potential collision with the following vehicle.

Aborting Lane Change

This moment is defined by the lane position offset. The time when the offset value is larger than 3 m is defined as the moment of aborting lane change.

- **THW to rear vehicle.** The average THW for a left lane change is smaller than that for a right lane change. It indicates that the driver in the rear vehicle in the adjacent lane has more risk for the host vehicle to make a left lane change.
- **TTCi to rear vehicle.** All the average values of TTCi are negative, which indicates that the host vehicle speed is still larger than the rear vehicle at the moment of aborting a lane change. The absolute value of TTCi for a right lane change is greater than that for a left lane change, which indicates that drivers in the rear vehicles show later reaction time or it leads to more dangerous situations for the rear vehicle.

Finishing Lane Change

The moment of finishing a lane change is also defined by the value of the lane location offset. This is the moment when the offset is larger than a quarter of the road width and smaller than half of the road width.

- **THW to rear vehicle.** The average THW for a left lane change is less than that for a right lane change on both roads. Drivers in the rear vehicle are in more danger as the host vehicle makes a left lane change.
- **TTCi to rear vehicle.** All the TTCi values are negative, which indicates that the host vehicle speed is still larger than the rear vehicle at the moment of finishing the lane change. The absolute value of TTCi for a right lane change is greater than that for a left lane change, which indicates that drivers in the rear vehicle have a later reaction time, or it is more dangerous for the subject vehicle to make a right lane change.

6.6 Driver Behavior Modeling

6.6.1 Neural Network Driver Model

This model is designed based on the assumption that the driver will adjust the pedal according to his/her estimation of the safety status and the pedal position. The estimation is based on the driver's perception of the movement relationship between the host vehicle and the leading vehicle. Because THW and TTCi are two key parameters for indicating driver car-following behavior [19], they are used as the model input to represent the car-following situation in the driver model. The output of the model is pedal depression. There is a feedback between the model input and output, which closes the loop. Considering the delay time, the structure of the model is shown in Figure 6.34. In this work, the delay times T_1 and T_2 are set to 0.5 s. The two-layer back-propagation neural network implements the function of an integrated driver–vehicle system as a "black box". The real traffic data are used to train the network.

Because of the difference between drivers, a particular neural network is assigned to each driver. The training data is extracted from the driver's data. The preprocessing of the data is

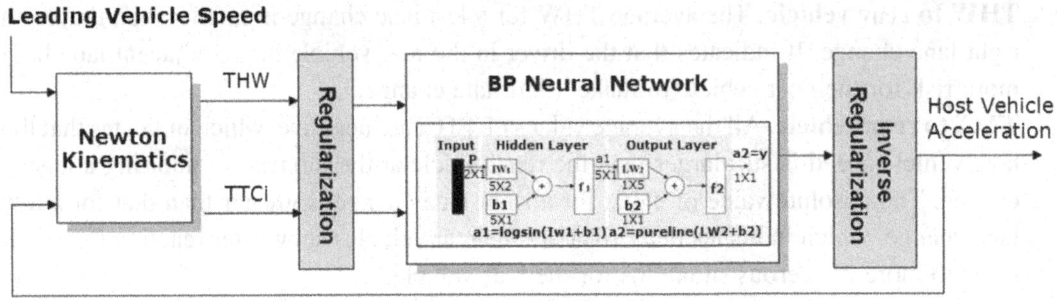

Figure 6.34: Structure of the Neural Network Model.

normalization. The distributions of time headway, time-to-collision inverse, and pedal depression are then analyzed, with the results shown in Figure 6.35. The data ranges of THW and TTCi obtained are 0–5 s and −0.065 to 0.065 s^{-1} respectively. Based on those results, the data segments are preprocessed according to Eq. (6.7) to scale the data to the range between −1 and +1:

$$\begin{aligned} \mathrm{THW_{in}}(n) &= (\mathrm{THW}(n) - 2.5)/2.5 \\ \mathrm{TTCi_{out}}(n) &= \mathrm{TTCi}(n) \times 15 \\ G_{\mathrm{in}}(n) &= (g(n) - 0.5)/0.5 \\ G_{\mathrm{out}}(n) &= (g(n+5) - 0.5)/0.5 \\ n &= 1,\ 2,\ \ldots,\ N-5. \end{aligned} \tag{6.7}$$

The transfer functions of the two layers are set as “logsig” and “pureline” respectively. The Levenberg–Marquardt (LM) algorithm is used to train the network because of its advantage of higher convergent speed compared to other methods. The performance measurement is MSE (mean square error) with the training threshold set as 0.005. Figure 6.36(a) shows the

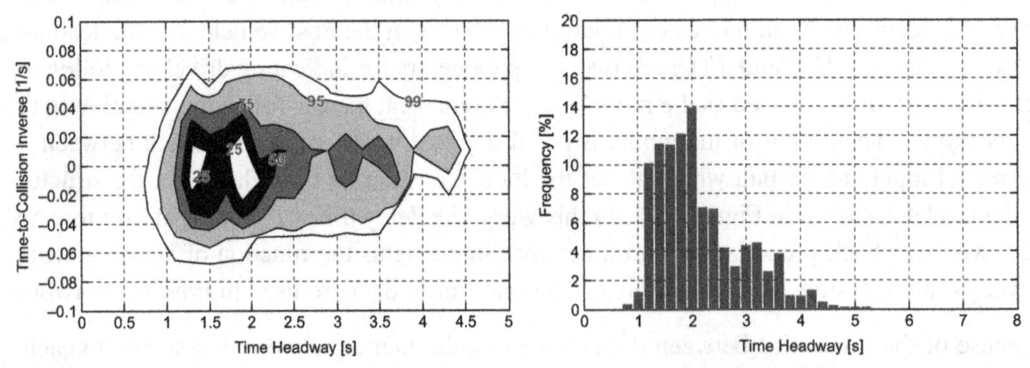

Figure 6.35: Distribution of the Model Variables.

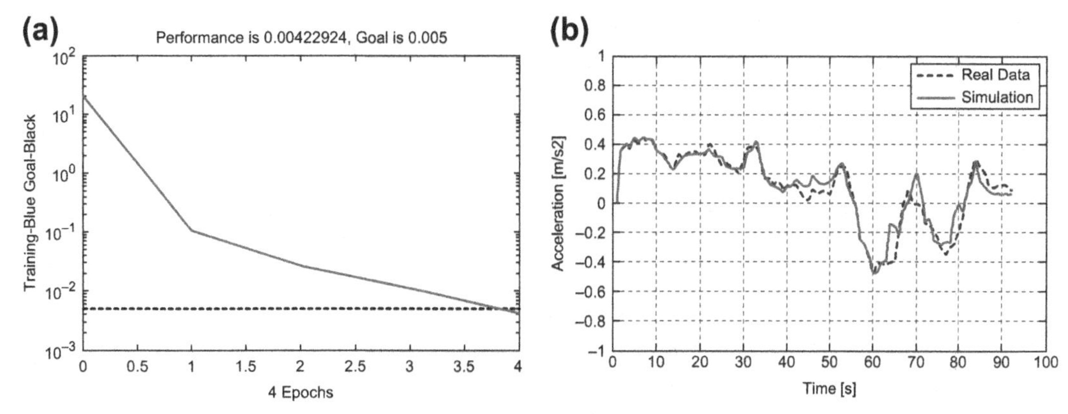

Figure 6.36: (a) Training Performance. (b) Comparison Between the Trained Network Output and Real Data.

MSE. After only five epochs, the sum-squared error falls below 0.005. Figure 6.36(b) shows a comparison between the network output and real data. The results indicate that the relationship between THW/TTCi inputs and accelerations can be modeled by this neural network and that the input selection is feasible.

6.6.2 Model Simulation

Based on the model structure shown in Figure 6.34, a Simulink model in MATLAB is established to carry out simulation with the input of real leading vehicle speed, which is shown in Figure 6.37.

The leading vehicle speed data of the training sample are imported into the Simulink model with the simulation output shown in Figure 6.38. At the beginning of this data segment, there

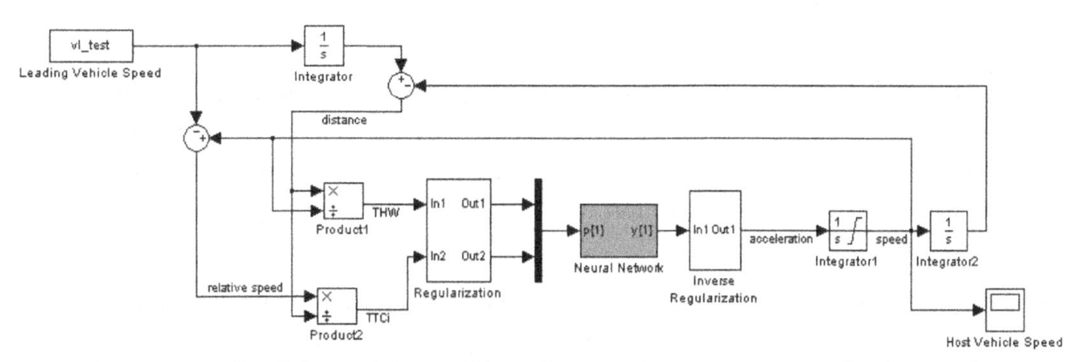

Figure 6.37: Simulink Model to Validate the Neural Network Car-Following Model.

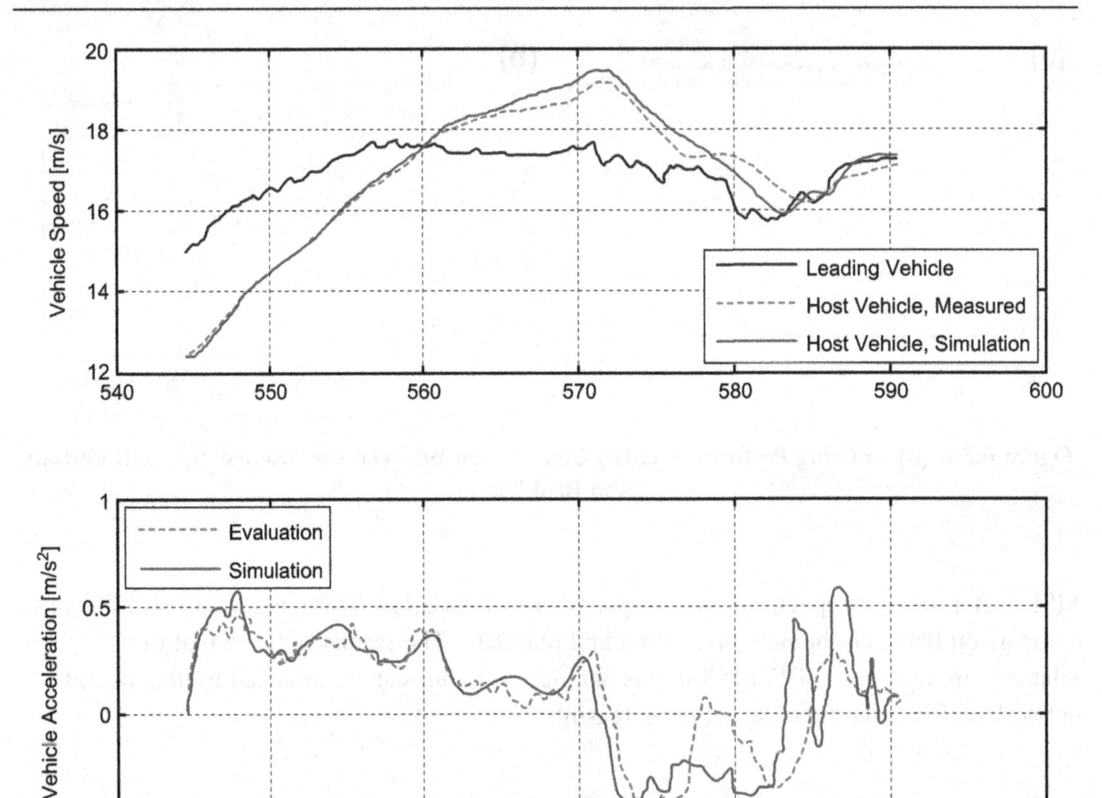

Figure 6.38: Model Simulation with the Training Sample Data.

is a large difference between the two vehicles. At the end of the segment, the host vehicle as well as its model output follows the leading vehicle more closely. In the simulation, the network uses the calculated THW and TTCi as the inputs and the result shows that the model can utilize driver behavior to produce similar output.

The step signal, sinusoid signal, and ramp signal are also imported into the model for validation, with the results shown in Figures 6.39–6.41. A short response time and a small overshoot are observed, and the leading vehicle's speed can be tracked in car following. From Figure 6.39, the initial leading vehicle speed is 15 m/s and the final speed is 20 m/s.

Another data segment is imported into the Simulink model to test the model's adaptability. This data segment includes a long-term car-following maneuver. The leading vehicle accelerates and decelerates frequently with the corresponding host vehicle following.

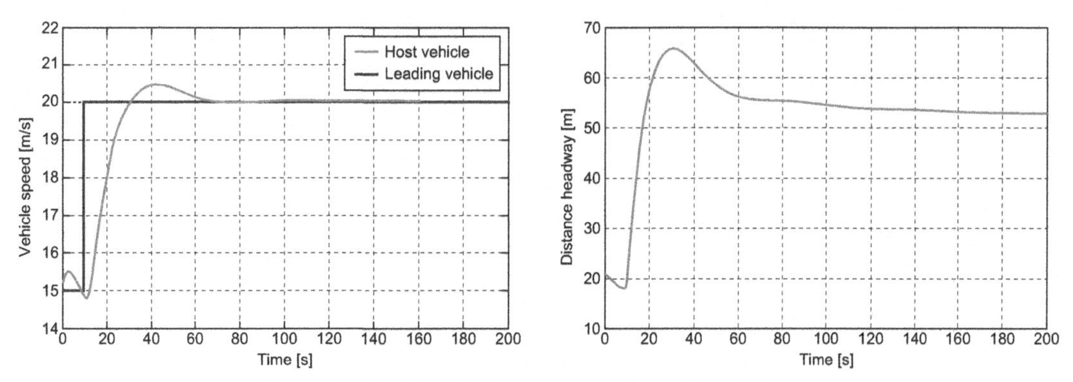

Figure 6.39: Model Response to Step Signal Input.

Figure 6.42 shows the simulation result using the neural network model. It is clear that the model can track the leading vehicle in timely fashion and reasonably well. In the training, only the information for the driver-following maneuver is used. The data is inadequate to reflect the delay in driver following to the leading vehicle. This may be the cause of the model response time being less than that of the driver.

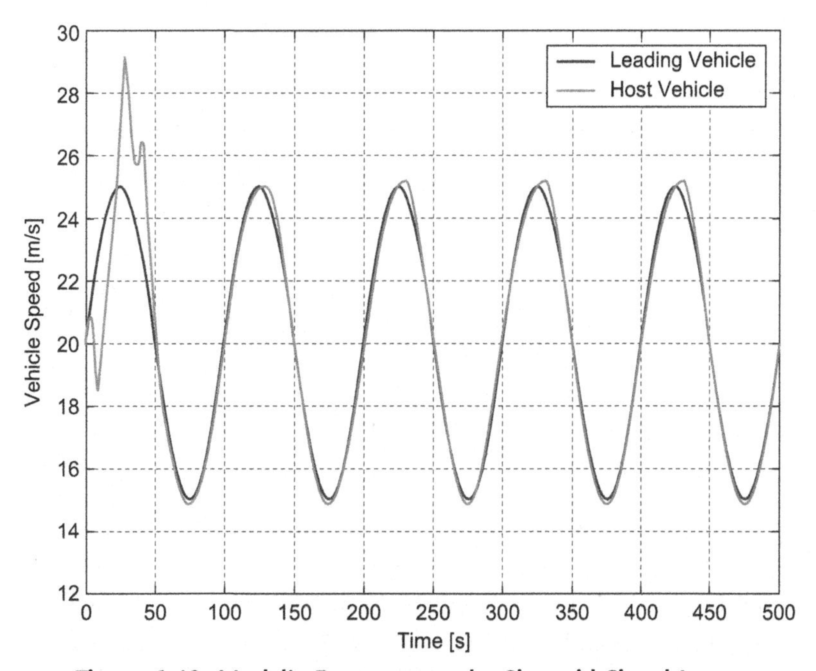

Figure 6.40: Model's Response to the Sinusoid Signal Input.

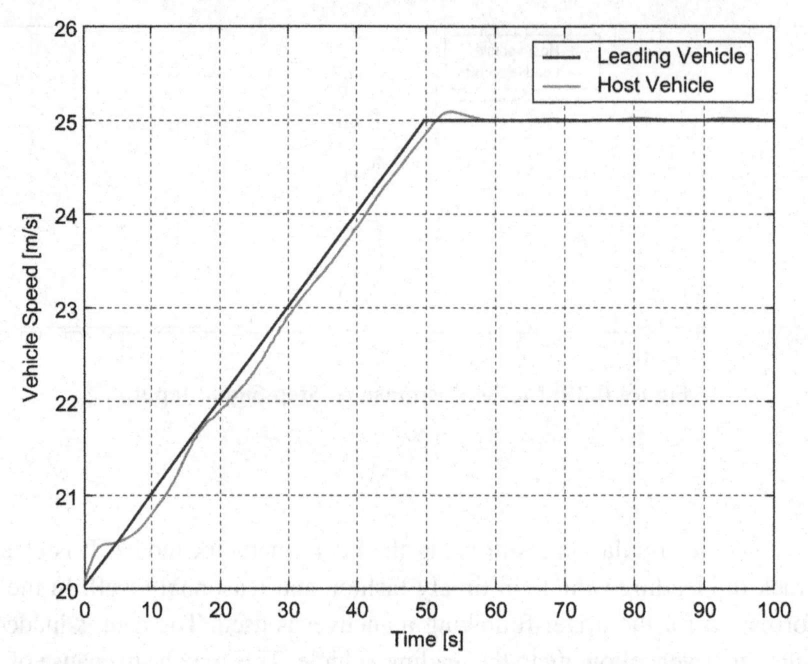

Figure 6.41: Model's Response to the Ramp Signal Input.

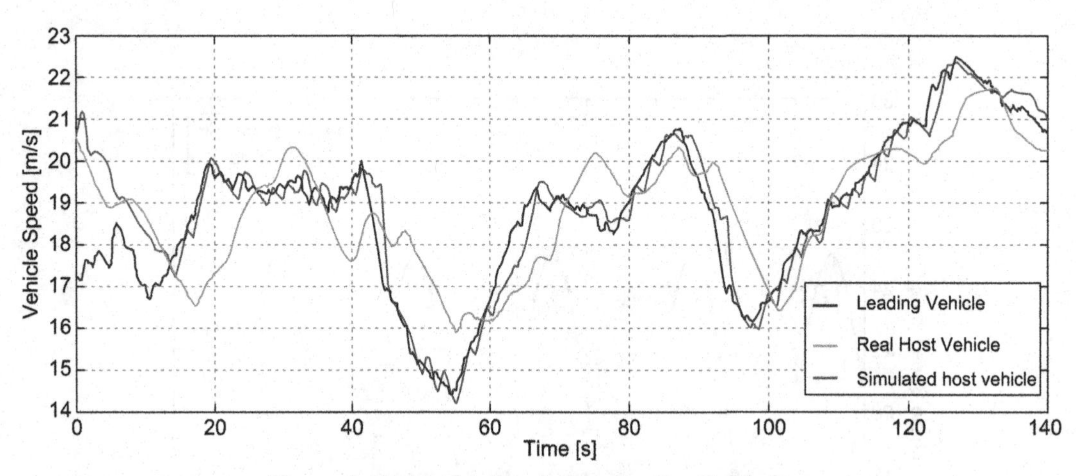

Figure 6.42: Model Simulation for Car Following.

6.6.3 Conclusions

In this chapter, a driver–vehicle model based on the analysis of driver behavior is established with a neural network. The data segments of steady-state car following are extracted for model training. Simulation results show that this model can represent a driver's car-following behavior.

Because of the complexity of human driver characteristics, modeling driver behavior is much more difficult than modeling the vehicle system. However, the driver model and characteristics are the most important parts in the design of driver assistance systems and other intelligent vehicle systems. In future work, it is necessary to investigate driver behavior more thoroughly based on more extensive analysis of experimental data. A driver model that is more understandable, generic, and closer to the driver characteristics is expected for driver behavior prediction and safety improvement.

Acknowledgments

The project was supported by NSFC (No. 51175290) and the joint research project of Tsinghua University and Nissan Motor Co. Ltd. The authors especially thank Mr Lei Zhang, Mr Qing Xiao, Ms Xiaojia Lu, Ms Ruina Dang, Mr Lai Chen, Ms Yu Bai, Ms Jieyun Ding, Ms Chenfei Yu, Ms Qianwen Yu, Ms Meng Lu, Mr Jinglei Yang, and Mr Lisha Jia for their contribution to the research work. The authors would also like to thank the co-researchers of Nissan Motor Co. Ltd and Nissan (China) Investment Co. Ltd.

References

[1] J.J. Gibson, L.E. Crooks, A theoretical field-analysis of automobile-driving, American Journal of Psychology 51 (3) (1938) 453–471.
[2] L.A. Pipes, An operational analysis of traffic dynamics, Journal of Applied Physics 24 (1953) 271–281.
[3] A. May, H. Keller, Non-integer car-following models, Highway Research Record 199 (1967) 19–32.
[4] G.A. Bekey, G.O. Burnham, Control theoretic models of human drivers in car-following, Human Factors 19 (1977) 399–413.
[5] F. Ray, A conceptualization of driving behaviour as threat avoidance, Ergonomics 27 (11) (1984) 1139–1155.
[6] D.C. Gazis, R. Herman, R.W. Rothery, Nonlinear follow the leader models of traffic flow, Operations Research 9 (1961) 545–567.
[7] P.G. Gipps, A behavioral car following model for computer simulation, Transportation Research B 15 (1981) 105–111.
[8] W. Helly, Simulation of bottlenecks in single lane traffic flow, in: Proceedings of the Symposium on Theory of Traffic Flow, Research Laboratories, General Motors (1961) 207–238.
[9] A. Ghazi Zadeh, A. Fahim, M. ElGindy, Neural network and fuzzy logic applications to vehicle systems: Literature survey, Journal of Vehicle Design 18 (2) (1997) 132–193.
[10] N. Kehtarnavaz, N. Griswold, K. Miller, P. Lescoe, A transportable neural-network approach to autonomous vehicle following, IEEE Transactions on Vehicular Technology 47 (2) (1998) 694–702.
[11] C. Zhuang, C. He, Application of Mathematical Statistics, South China University of Technology Press, Guangzhou, 2006 (in Chinese).
[12] Transportation Department of Ministry of Public Security of the People's Republic of China, The Annals of Road Traffic Accident Statistics of the PRC, Traffic Management Research Institute of the Ministry of Public Security, 2010. Wuxi, Jiangsu, 2010.
[13] T.P. Wenzel, M. Ross, The effects of vehicle model and driver behavior on risk, Accident Analysis and Prevention 37 (2005) 479–494.
[14] H. Joksch, D. Massie, R. Pichler, Vehicle Aggressivity: Fleet Characterization using Traffic Collision Data, DOT HC 808 679, University of Michigan Transportation Research Institute, Ann Arbor, MI, 1998.
[15] P. Wasielewski, The effect of car size on headways in freely flowing freeway traffic, Transportation Science 15 (4) (1981) 364–378.

[16] L. Evans, P. Wasielewski, Risky driving related to driver and vehicle characteristics, Accident Analysis and Prevention 15 (2) (1983) 121–136.
[17] L. Zhang, A Vehicle Longitudinal Driving Assistance System Based on Self-Learning Method of Driver Characteristics, Doctoral Thesis, Tsinghua University, Beijing, PRC, 2009 (in Chinese).
[18] M. Brackstone, B. Waterson, M. McDonald, Determinants of following headway in congested traffic, Transportation Research Part F: Traffic Psychology and Behaviour 12 (2) (2009) 131–142.
[19] Ke Zheng, Lisheng Jiang, Jian Rong, et al., Study on mental and physiological effects of horizontal radius of expressway on driving reaction, Journal of Highway and Transportation Research and Development 21 (2) (2004) 5–7.

CHAPTER 7

The Human Factor and Its Handling

Dana Procházková

Institute for Security Technologies and Infrastructures, Czech Technical University, Prague, Czech Republic

Chapter Outline

7.1 Introduction to Problems

From daily life experiences, from data and knowledge in professional publications mentioned below, it follows that situations in which people find themselves and which each person continually needs to handle necessitate the making of a decision. This decision may relate to matters that are vitally important (a change in the way of life, etc.) or a daily detail (whether to go in an overcrowded metro/not to go in an overcrowded metro; cross a road when the light is red/not to cross a road when the light is red, etc.). Sometimes the decision is made over a period of time (e.g. while solving work or other problems), other times it is necessary to decide immediately (in situations with a direct threat to life, with a risk in any delay). We adjudicate something either on our behalf (what I do, what I do not do) or on behalf of our subordinate workers/persons (in harmony with their interests, but also perhaps against their interests). The decision can be only the result of the arbitration of one person. It can, however, also be the output of collective intellect. The decisions may be accurate but also

Advances in Intelligent Vehicles. http://dx.doi.org/10.1016/B978-0-12-397199-9.00007-0

false. The consequences of decisions can have a different weight for both arbitrary subjects and for those around.

According to data in the professional literature and experiences from practice dealing with human behavior in different situations, human reactions to external (also internal) inputs can be very different. They may have the form of unconditioned reactions, as "automatic", inherent ways of reacting to inputs (e.g. wincing at an unpleasant input), facultative reactions (e.g. in the form of habits), or purposeful action controlled by will. In the psychological literature we mostly find difficulties with decisions connected with will and volitional processes, thinking, purposeful behavior, pertinently in connection with a fight for incentives (while solving internal conflicts). In the process of the purposeful control of human manners, not only is the decision made from a selection of the different incentives and targets, but also a selection between alternatives, whether to negotiate or not to negotiate. The person also adjudicates at the selection of the means and procedures in order to reach the aim, in a situation requiring the interruption or cessation of the activity. The capability to deal with problems correctly, prudently and in timely fashion is one of the basic conditions of practical activity and creative thinking, and it is simultaneously an important component of human personality.

According to Ref. [1], decision-making in ordinary life is mostly quick and easy. It can, however, also take a very long time, namely in situations in which all the possible alternatives are uncomfortable, and this leads to selection of the lesser evil. The tendency to choose some of the possible alternatives can be problematic and the person remains in a state of non-decision. After some time, such a state becomes so agonizing that its termination becomes a strong motive for action and it can lead to a concise solution, to an arbitrary decision. Anyone who is unable to decide for a long time becomes unable to decide rationally and reasonably and can become chronically neurotic.

After ensuring a safe decision with the accent on the protection of people and property, it is necessary to achieve the right decision or at least a decision that will not lead sooner or later to disaster, namely in the case of a decision made under stress. The decision in this concept becomes the social process. In this process the human intellect and certain inherent human knowledge and skills are put forward. Essentially, they act in the following way:

- A responsible approach to a problem and the results of its solution regarding the public or other assets.
- Moral properties as a judgment, sense of commitment, and consistency.
- The ability to analyze the problem or situation, to take an attitude in a creative approach to the solution of the problem, to use analogy, etc.
- The capability to use experiences and social skills that enable regulation of one's activity and behavior or the behavior of subordinates.

These facts form the characteristics of the human factor of a well-conditioned management worker. This means that the selection of management workers in all organizations might be chosen with the aim of respecting knowledge, capabilities, skills, and experience, and not according to political affiliation, color of a coat, or other subsidiary features.

Regarding all the above, the human factor is an aggregate of human properties, capabilities, and experiences that, in a given situation, has an influence on the safety, productivity, effectiveness, and reliability of the system in which it acts. It is evaluated from psychological, physiological, and physical viewpoints.

In this work we concentrate on the human factor's influences that are important for a safe human system[2]. If we want to reach this target, we need successfully to trade-off risks of all kinds, i.e. including those connected with the human factor. With regard to the problems connected with building a safe world, we indicate that from the viewpoint of ensuring integral safety[3], it is necessary to pay special attention to deciding situations in which there are uncertainties and vagueness[3,4], i.e. as a rule some data and results are unreliable and distinct, which means that in practice we cannot use only deterministic and probabilistic approaches, but we must apply suitable heuristics, in which there is a role for human intellect and knowledge, and experiences and intuition are interconnected, i.e. the human personality characteristics denoted as the human factor. The human, who has the capabilities just described, is very important to each organization and this is the basis of a type of management called "knowledge management".

7.2 The Human Factor

The proposition given above shows that the human factor cannot be considered only as a negative human manifestation. It is a present reality that positive manifestations of the human factor are currently used by the type of system management that is directed by knowledge – knowledge management. Humans with their own ideas are considered as an inestimable human resource. The given facts mean that humans are the most critical and simultaneously the most capable part of each system.

7.2.1 Human Error/Human Failure

Based on the results summarized in Ref. [5], human error is a deviation of human performance from that planned, demanded or given by an ideal standard. For the majority of the twentieth century the view of most organizations regarding human error consisted of a statement that blame for an incident/accident was assigned to the worker whose actions were tied up with the given incident/accident – e.g. the human who operated the system at the time at which the incident happened. At present it is possible to trace an opposite trend. The human is considered as a thinking being who is "left at the mercy" of a range of designated, organizational, and momentary factors that can lead to behavior that the external observer can comprehend (even though often in an unqualified way) as human error. It is not so simple to determine whether the error was really caused by human error, and whether the individual concerned was forced to act according to the conditions. It requires much experience in a given field to evaluate what happened, how it happened, and mainly what was the real cause of the incident/accident.

From a management viewpoint, the failure/malfunction is the result of a process consisting of the following:

- Initiation (false operation, mistakes, violation of rules, ignorance)
- Contributing effects (incorrect organization, inaccurate decisions)
- Spreading of defects leading to the accident (organizational nonfunctionality).

Because the influence of the style of management and decision-making is important, we speak of a so-called organizational accident in the form of the Reason Model[5] (Figure 7.1).

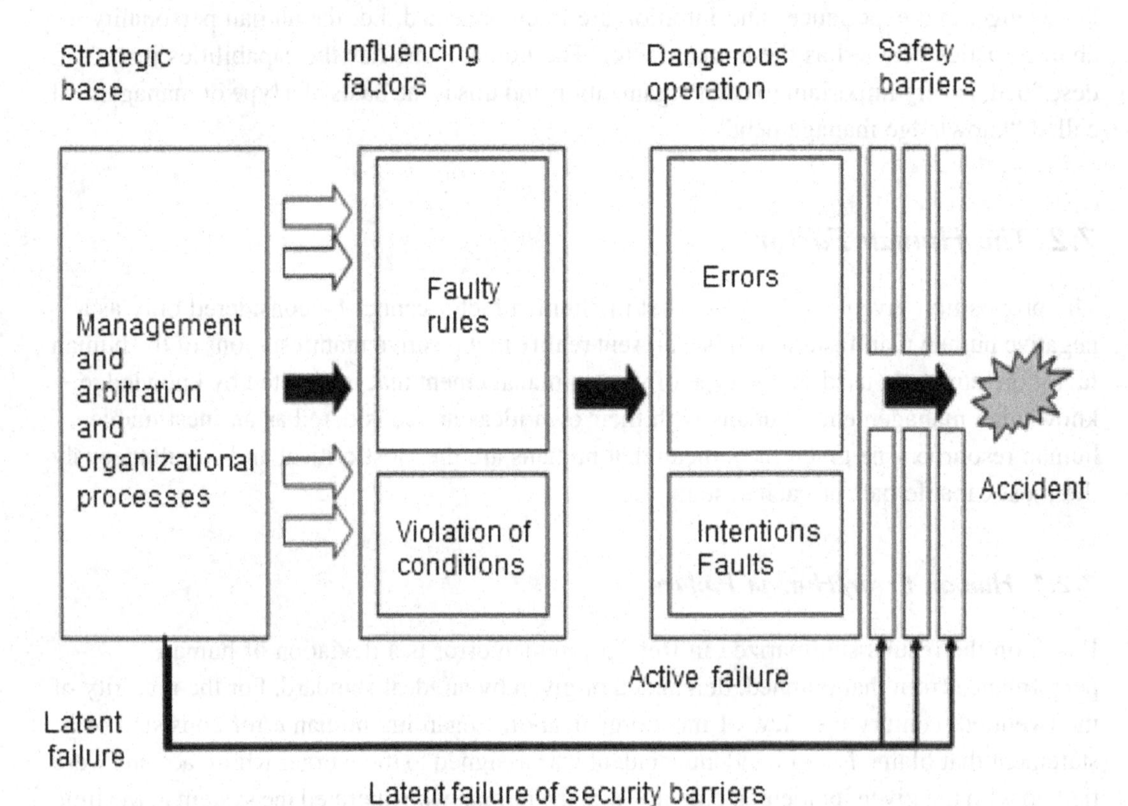

Figure 7.1: Model of Organizational Accidents According to Actions[6,7].

Safe (including dependable) system behavior arises from a condition that technical workers (operation, maintenance workers) always follow according to the requisite procedures (the procedure is formed from correct tasks/operations performed correctly). As the Reason Model shows, so-called risky operations always occur (Figure 7.2).

Therefore, in risk determination it is necessary in the context of process analysis to understand the motivation of the intended acts of both terrorists and the insiders (actual employees). Among the insider's motives are the following:

- Inconvenient security procedures (for a safe human life they must be skipped)
- Inconvenient plans (for a safe human life the modus operandi solutions must be used)
- Poor perception of security risks
- Insufficient responsibility
- Stress and management attitude or financial profit.

The insider's motivation is directly related to a safety culture.

We separate the human factor in the sense of human error (human failure) into the intentional and unintentional. Human errors originate in both:

- The performance of activities, where their sources are routine behavior, not respecting the operational and security codes, default, omission, bad state of health, bad conditions in the workplace, etc.
- The management process, where their sources are ignorance, not respecting the correct, technical, economic, and social aspects, arrogance, etc.

From the analysis of big technological accidents (e.g. Bhopal 1984, Seveso 1976, Chernobyl 1986, Mexico City 1984, Toulouse 2001, Enschede 2000, Buncefield 2005, Lvov 2007,

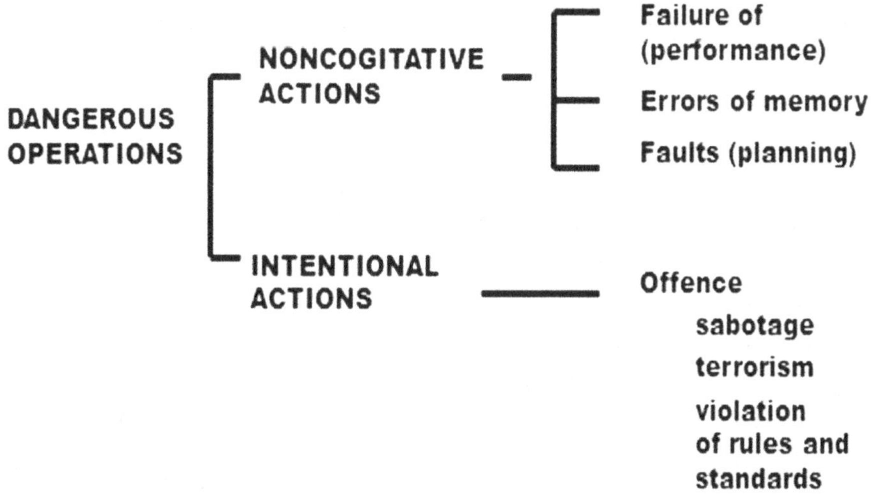

Figure 7.2: Risky Operations According to Actions[6].

Mexico Bay 2010, etc.)[8], it follows that the damage caused by human errors in management is as a rule far greater than the damage caused during the performance of activities and, therefore, in connection with the human factor, the emphasis is always on the level of safety management.

The results of research into the human factor[9–12] show the following:

1. For human error analysis, the procedure evaluating human reliability and consequences of its failure, "HRA – Human Reliability Assessment", is suitable. For analysis of the influence of the human factor, the simple model SHEL (Software, Hardware, Environment, Liveware) is appropriate:
   a. Software – poor understanding of the procedures, poorly written manuals, incorrect check lists, etc.
   b. Hardware – inconvenient equipment, insufficient maintenance, etc.
   c. Environment – poor working surroundings and conditions.
   d. Liveware – relations in the workplace, motivation, insufficient self-assertiveness.
2. The analysis of human reliability usually starts from human error assessment. For the compilation of its models, net models are often used, i.e. Bayesian, Petri, and other nets. Different scenarios of human error occurrence and consequential impacts are identified, and the occurrence of the probability of human error in real conditions is determined. To the system of systems[3], which is the human system model, is added an additional system, namely a social one. Occasionally, the fuzzy Bayesian model is used to consider the vagueness of the data.
3. In emergency situations it is important for people to be self-reliant and to get themselves and others to a safe place. This self-reliance increases if humans know the content of warning instructions and if they know how to behave. The effectiveness of warning instructions is influenced by many factors – personal characteristics, form of information processing by a responsible person, social influences, indirect information, etc.
4. In a system's vulnerability, the vulnerability that represents humans, i.e. the social system, also contributes.

The result of an examination of incidents and accidents in the context of "the error of the named real person (employee at the equipment, pilot, driver, etc.)" does not help the prevention of incidents and accidents, because it only shows *where* in the system the error happened, not *why* it happened there. The error caused by a human in the complex system might be as follows:

- Caused by a poor design/proposal
- Stimulated by inaccurate training, poorly processed operating procedures, imperfect concepts or eventually by the unfit processing of the operating procedures or manuals.

Such results then allow us to conceal the other basic causes of the incident or accident that must be considered in the prevention of further incidents and accidents.

Reduction of risks in the framework of safety management[13] covers several spheres:

- Process safety
- Protection of health and safety of an employee (work safety)
- Reduction of impacts on the environment[14].

The analysis of the impact of management on the safety of an enterprise/plant/organization/land (further only as "organization") must therefore emerge from the model of an organizational accident. An organizational accident consists of three basic elements: organizational processes, conditions that caused the origination of errors or violation of rules, and errors and/or violation of rules[15].

Organizational processes include four processes that are part of each technical or technological organization: projection and construction, building, operation, and maintenance. All processes are built as three interconnected activities: assignment of targets in the framework of the economic and social situation of the organization, setup of organization for the realization of determined long-term strategic goals, and management of operational activities.

Each of these processes and activities forms a separate, general type of failure scenario:

- Determination of targets – contradictory targets
- Organization – disproportionate arrangement (setup)
- Management – bad communication, bad planning, inappropriate inspection and monitoring
- Projection and construction – faulty projection
- Operation – bad operational procedures, bad training and education
- Maintenance – bad maintenance plan, bad maintenance procedures.

The conditions that cause the origination of errors are: insufficient training for the task; lack of time; bad separation of signal from noise; misapprehension between designer and user; irreversibility of errors; congestion of information; negative conversion between tasks (bad handover of tasks); bad perception (underestimation) of risk; bad backward link from system; lack of experience; bad instructions and procedures; insufficient check-up; unsuitable education of the person for a given task; unfriendly atmosphere; and dullness and boredom.

Conditions that cause the violation of provisions and rules are: lack of safety culture in an organization; conflicts among management workers and employees; bad morale; bad supervision and check-up; norms and standards permitting the violation of rules; bad perception of the sources of risk; perceptible lack of solicitude and interest by management workers; low pride at own work; a dab hand approach to work that encourages risks; belief that nothing bad can happen; low self-respect; recognized weakness; perceptible tolerance of the violation of rules; double-dealing, ambiguous or obviously meaningless rules; age and gender – young men execute the violation of the rules.

Dangerous pursuance can be split into errors and violation of provisions/rules:

1. Errors that are a consequence of problems in information processing may be comprehended in relation to the cognitional functions of an individual. They may be reduced by training, improvement of the workplace, interface, better information, etc.
2. The violations of provisions/rules that are based on motivation. These are social phenomena and they can only be understood in the context of a given organization. The violation can be removed by a change of approach, persuasion, norms, standards, moral and safety culture.

How to avert human failures is fundamentally important. In agreement with research results[9–13,16–18] it is possible to avert human failures of activities and management if:

1. For management of professional problems only professionals with the capability to lead the working team are authorized (they know how to explain, know how to support, to avert bullying, etc.).
2. Qualified management of processes that are the result of organization is introduced.
3. Conditions for qualified work are created.
4. Sufficient education of workers and the offer of help in the solution of complex tasks are provided.
5. The motivation and stimulation of workers as regards the adherence of operations and security provisions are ensured.
6. There is in-depth supervision of processes and their interconnections to projects and furthermore also programs that professionally and directly avert intentional and unpremeditated errors.

As was shown above, the biggest errors originate from human errors in management. Figure 7.3, adapted from Ref. [19], shows the well-known reality that cogitative management workers, while deciding the solution to a problem that is uncertain, adjudicate in their own interests if the probability of success is 0.6 (i.e. failure probability is 0.4), only in the case of losses connected with a decision (i.e. rights of recovery for a decision in their own interests, if the expected result does not materialize) are low, and with the increase in losses they are more prudent.

Among management workers two extreme categories are found:

- Gamblers who adjudicate in their own interests, even though possible losses are great
- Prudent humans who adjudicate in their own interests even though possible losses are low.

The gamblers occasionally make a huge profit.

The moral principles of human society conforming to UN principles[20] state that gambling with human lives and health is inadmissible, i.e. risk here is not permitted (tolerated). This finding might be the main principle in the selection of personnel for activities on which

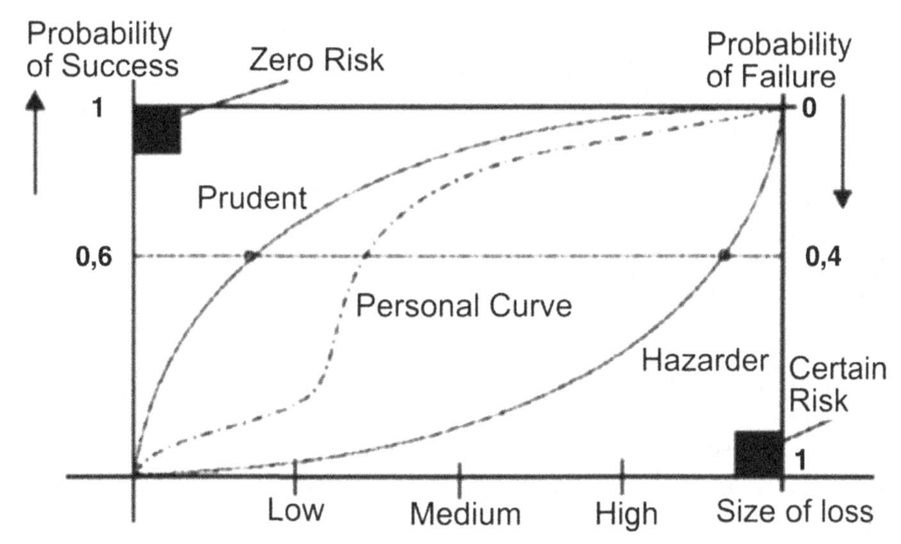

Figure 7.3: Graph Delimiting the Types of Management Workers in Risk Management.

human lives depend, either directly (drivers, pilots, and the like) and indirectly (managerial workers in enterprises and other organizations who adjudicate the activities and measures that are directly connected with unacceptable risk for persons, property, and other basic protected assets).

7.2.2 The Human Factor as the Source of Progress, Development, and Innovation

In the analysis of complex arbitrary situations, each responsible and educated person comprehends very quickly that the human intellect is irreplaceable, because of the lack of data, uncertainties, and vagueness in the data. Therefore, the discipline based on human intellect called "knowledge management" was created. The human intellect (understanding) is the capability of a human mind to generalize experiences, to work with abstract terms, and to make conclusions from assumptions.

The basic relationship that we start from is the relationship between individuals, data, information, and knowledge[20,21]. Primary items are details of the objective being followed and are collected or recorded. The data originate from their evaluation and classification. The information originates via qualified data processing (the word qualified means that data mining that consists of groundless interleaving of clusters of points by curves that allows PC software and the selection of curves with the smallest dispersion is not allowed; the emphasis must be on matter-of-fact justification). It is through qualified interpretation of obtained information in certain and real conditions that knowledge originates. It is necessary to perceive that affiliation with real conditions is important. The aim is not only knowledge accumulation but its rational use in practice. Examples of knowledge are methods of market analysis in a real situation, effective care of patients and its correct use, innovative design of a product or service, system of effective care of clients, etc. It is evident that "knowledge", "methods" and "algorithms" are closely related.

In security practice tacit (inherent) knowledge is greatly appreciated. There are many examples of such knowledge having averted colossal accidents and great human loss (see Ref. [21] and references therein).

7.3 Control of the Human Factor in Management and Engineering

Because the aim is to live in a safe world with the potential for development, it is necessary to include the human factor in human systems of safety management respecting that the human system is the system of systems[3] and that it represents territory, organization, etc. We must also include the human factor in engineering, in which we realize the targets of safety management by the way in which we negotiate/trade-off the risks. Because each human is an active element of a human system, we must systematically build a safety culture, namely in workplaces and in the society/country[20]. All given approaches are important and will be explained in the following sections.

7.3.1 Safety Culture

The safety culture must be built systematically, taking into account actual knowledge and experience. The relevant tool for its establishment is called "safety management". In each system it represents strategic, proactive, and process management based on risk management and on the results of science and advanced technologies. The present work deals with what is related to the human system and what ensures the following:

- Prevention against disasters of all kinds, i.e. natural, technological, environmental, social, and those caused by interdependencies in the critical infrastructure, including terrorist attacks and existing interactions between the assets of the human system and its vicinity.
- A preparedness to put all emergency and critical situations under control, with the capability to renovate the affected part of the human system.
- A response if the emergency or critical situation affects the human system.
- Renewal after each emergency or critical situation.

The safety management that establishes the safety culture has three basic phases:

- Standard (current) management
- Emergency management
- Crisis management.

All these phases must be reasonably interconnected and must respect characteristic features and targets. Standard management is concerned with building a safe community, safe territory, safe state, etc. Its attention is mainly focused on territorial development, prevention, and preparedness. Its main tool is strategic planning based on knowledge, experience, and good engineering practice. Emergency management is focused on coping with emergencies with the help of standard sources, forces, and means. Crisis management is focused on coping with critical situations, human survival, and stabilization of the situation so that renewal and follow-up development can be started, with the help of standard sources and additional ways and means; details are given in Ref. [20].

To attain security and sustainable development, territorial (including the human society in this territory) safety must be built. In agreement with the principles in Ref. [3], this means creating human system safety. According to present knowledge, the human system has the following assets: human lives and health; property and public welfare; environment; infrastructures and technologies, mainly the critical ones. The appropriate tool of "human system safety management" is "dynamic integral territory safety management", which proactively respects dynamic behavior globally[20,21]. For practical purposes[21] the golden rules for workplaces and territories/countries are given; they facilitate daily safety management for each entity.

7.3.2 Negotiation/Trade-Off with Risks

The aim of negotiation/trade-off with risk from the human factor viewpoint is to produce precise procedures that include knowledge and practical experience that avert human factor failure in management and engineering.

The term "risk" has its origin in the Middle Ages. There are different definitions of risk for each of several applications. The widely inconsistent and ambiguous use of the word is one of several current criticisms of management risk methods. Risk is the potential that a chosen action or activity (including the choice of inaction) will lead to a loss (an undesirable outcome). The notion implies that the choice of having an influence on the outcome exists (or existed). Potential losses themselves may also be called "risks". Almost any human behavior or endeavor carries some risk, but some are much more risky than others[22,23]. The present concept was first developed in the 1950s. In present practice we use three important terms: disaster, hazard, and risk[24]. Hazard expresses the disaster potential to cause losses, detriment, and harm to assets at a given site[3,20]. Risk expresses the probable size of undesirable and unacceptable impacts (losses, harm, and detriment) of disasters with the size of normative hazards to system assets or subsystems in a given time interval (e.g. 1 year) and a given site (i.e. it is always site specific).

Risk is a measure of the violation of monitored system security, which is a subject of possible disaster occurrence monitoring. It is a measure of disaster potential to disrupt security and sustainable development of a monitored system. The most concise definition of risk is to use the expected loss, damage, and harm to assets in a certain standardized way in order to ensure comparability (e.g. converted to unit area and unit of time[3], which are used in materials for strategic management). With dependence on the specific needs[3], we determine either the risk of one disaster or of the set of all disasters, which can affect the reference to real objects.

In determining the risk either one asset is considered and partial risk is determined, or complex assets are considered and integrated or integral risk is determined. Integrated risk only represents a certain aggregation of partial risks, which is usually determined by norms or standards. The integral risk includes both the risks associated with individual assets and the cross-cutting risks that are associated with links among the assets and with the couplings among the assets realized by flows (energy, information, instructions, commands, responses to them from top to bottom and vice versa), i.e. it represents a complex risk, the qualified management of which provides integral safety.

In a Safety Management System (SMS) concept we consider two cases, either that the risk realization is still substantially the same or it is significantly different. In the first case, we consider from safety reasons either the worst case (such an approach is found in standards based on a deterministic approach to safety provision) or we admit random uncertainties resulting from momentary local and temporal conditions of assets and as a representative variable for risk management we use the mean value obtained by evaluating the possible alternatives (arithmetic mean, median, median $+ \sigma$, where σ is the standard deviation, the probable mean value). The other procedure is now commonly considered in the preparation of documents for strategic management (the alternative scenarios for risk realization and their probabilities of occurrence are determined, and the mean and its dispersion are derived from them by a clear mathematical approach); we can find it in the norms and standards based on a probabilistic approach. In cases when we take into account the existence of vagueness in data we must use a combination of analytical and heuristic approaches that offer different theories, e.g. extreme values theory, fuzzy set theory, fractal theory, dynamic chaos theory, selected expert methods, suitable heuristics[3,4], and the recent theory of evidence[25,26].

The risk partly depends on the hazard and partly on the vulnerability of assets at a given site (i.e. on the sensitivity of each individual asset in a given place against the physical manifestation of a disaster at a given site). It expresses a possibility of what might happen. From this fact it follows that for each management it is important to know the risk, in the form of a comprehensible expression. In the practice of public administration management of risk expressed by risk analysis and assessment, in one specific case it was found that:

- 5 million euros a year are necessary for the remedy of harm caused by existing risk
- Every 10 years, 10 people die as a consequence of a given disaster
- Every 5 years, the property damage caused by disaster exceeds 5 billion euros.

Methods for the determination of the size of the risk respect both the nature of phenomena, which are their sources (i.e. the characteristics and physical nature of disasters), and the parameters of the medium in which the phenomena have an impact. There are methods used based on mathematical statistics, fuzzy sets, operational analysis, etc. that inherently assume a certain model of the phenomena, i.e. they do not permit these phenomena to be extraordinary, and methods are based on scenarios that are simulated or empirically obtained; see the data in Ref. [4]. In principle, we can formulate two basic approaches:

1. Determination of hazard from disaster H and return period t (in years) is performed by methods based on the theory of large numbers, theory of extremes, theory of fuzzy sets, theory of chaos, theory of fractals, etc.[3]. According to site vulnerability in an investigated area (e.g. around a given site: a square 10 km $\times$ 10 km; circle with radius of 5 km) the whole damage of all assets is determined for H and is denoted by S, usually expressed in money. Risk R connected with the given disaster at a given site is determined by the relation:

$$R = S/\tau.$$

   The result is very clear, i.e. "the risk from a given disaster at a given site is X euros and for a town it is mx euros".
2. Determination of a disaster scenario for a disaster with a size corresponding to the maximum expected disaster is performed (it is possible with regard to the demands of norms to use the probable size of the expected disaster, or the value of the standard size of a determined disaster or at least an unfavorable disaster). According to data for a given area, the following is determined:
   a. The value of the whole damage to all assets in the affected area SS (the method for SS determination is described in Ref. [24], usually expressed in money according to the amount of assets and their vulnerability to impacts following a disaster in the affected area, usually normalized to a certain land unit S).
   b. The frequency of the maximum expected disaster normalized to 1 year according to the professional data from databases or expert opinions. Risk R is given by the relationship:

$$R = S \times f.$$

   The result is in the same form as in the previous case. This case is often used for technological and other disasters for which we have no good long-term catalog. The EU (European Union) wants to remove this shortfall by paying special attention to the compilation of the MARS database[3].

From the facts given above, it is evident that the risk value as determined is related to a certain land unit and time unit. We say that the risk is a site-specific quantity. If we can negotiate/trade-off the risk we must know the risk size and in its determination we must respect all assets and their interfaces, as shown in Figure 7.4. Because the human system is the SoS (system of systems/systems system), we must respect this characteristic and we must also consider cross-section risks, i.e. we must determine the integral risk. For such a form of risk we do not have a simple formula respecting all human system public assets because interdependences causing cross-section risks are site specific[3]. They cause the problem to be nonlinear, irregular, and very complex.

Figure 7.4: Model of Risk Management in a Territory[3,20].

The risk is for engineering practice expressed as the probable size of losses, damage, and harm to assets that are caused by a given disaster of a specified size and that are rescheduled for a certain time unit (usually 1 year) and a certain territorial unit that is in agreement with the EU standard under preparation[23,27]. For advice in practice we distinguish between whether the risk realization continues steadily in the same way or variously depending on the immediate site and temporal conditions of assets. In the first case we determine a sort of mean value, and its validity for use in practice that is connected with the condition of its determination in a much worse case (we can find this case in the norms and standards based on the deterministic approach). The second case corresponds to a variable reality – the changing scenarios of risk realization and their probabilities of occurrence are determined; from these data, using a clear mathematical approach the mean value and its dispersion are determined (we can find it in the norms and standards based on the probabilistic approach). In present practice for complex cases the precisely defined heuristic procedures[3] are used and are considered in the preparation of the groundwork for strategic management.

The principal attributes of each risk are *uncertainty* and *vagueness*. We divide their sources into three groups: variations originating in the usual system process life cycle in normal conditions in the vicinity (uncertainties); real changes in the system process life cycle in the time and space that affects occasional extreme value occurrences – we consider normal and abnormal conditions (uncertainties and vagueness); a variable system process life cycle that is caused by process changes in the time and space induced by outside causes or by critical conditions (vagueness).

The data uncertainty relates to the dispersion of observations and measurement. It may be included in assessment and predictions by mathematical statistics apparatus. The vagueness relates to both the lack of knowledge and information and the natural variability in processes and actions that caused disasters. To process the vagueness, the mathematical statistics apparatus is insufficient and therefore it is necessary to use recent mathematical apparatus that offers, for example, extreme values theory, fuzzy set theory, fractal theory, dynamic chaos theory, selected expert methods, and suitable heuristics[4].

The data vagueness follows from the reality that data are incomplete, inhomogeneous (i.e. their accuracy depends on their size or on the time of their occurrence) and non-stationary, i.e. data have massive dispersion and are encumbered by random and sometimes also by systematic errors, the distribution functions of which cannot usually be determined. Because nothing is absolutely precise, we must generally consider data uncertainties and vagueness for each quantity that we investigate. Therefore, both safety engineering and risk engineering require that the quality of the data set ought to be verified from the viewpoint of their credibility with regard to a given task.

Risk Management and Safety Management

The strategy of management for ensuring the security and sustainable development of a managed subject consists of negotiation with risks[24,28]. In its framework, according to the present possibilities of human society, we have several ways of dealing with risk:

- Part of the risk is reduced, i.e. by preventive measures by which the risk realization is averted
- Part of the risk is mitigated, i.e. by preventive measures, activities, and preparedness (warning systems and another measures of emergency and crisis management); non-acceptable impacts are reduced or averted
- Part of the risk is re-insured
- For part of the risk there are prepared resources for response and renovation
- For part of the risk there is a prepared contingency plan, i.e. it is used for a part of the risk that is non-controllable or too expensive or of low frequency.

To this is added the distribution of the defeat of a risk among all stakeholders[26]. The distribution in good governance is performed according to the rule that all stakeholders have responsibility for the defeat of a risk and that the defeat of a real risk is assigned to a subject for which the prepared resources are best.

In practice, two risk management models are usually used:

- Classical risk management (Figures 7.4 and 7.5(a–d))
- Safety management, i.e. risk governance for security and sustainable development (Figures 7.6 and 7.7).

Figure 7.6 shows that the result for a followed system is a consensus of all considered disasters because each disaster type affects, owing to its nature, the system and its protected assets differently[20]. This is because the human factor failure, especially in risk management, belongs to disasters, i.e. phenomena that damaged the human system of a certain size. With regard to typical risk properties, which are uncertainties and vagueness, as was shown above,

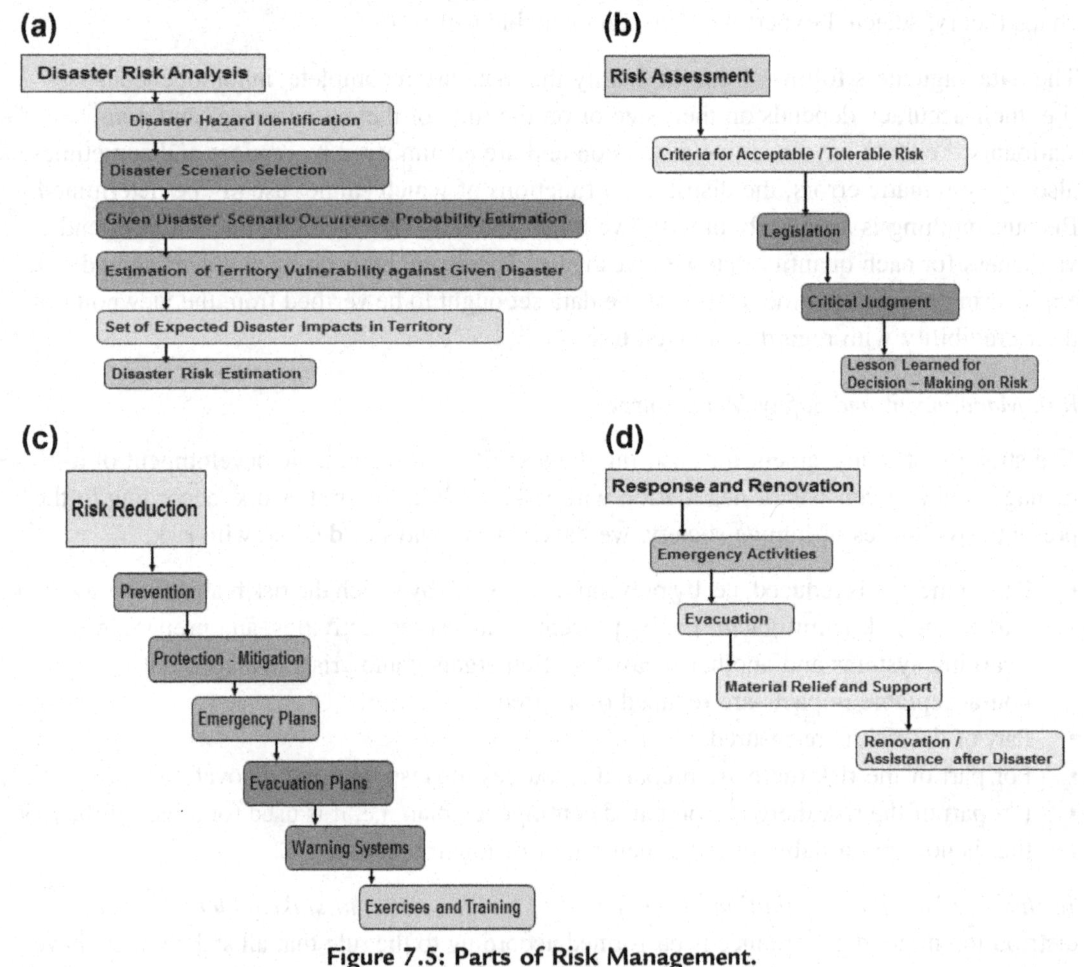

Figure 7.5: Parts of Risk Management.
(a) Disaster risk analysis. (b) Disaster risk assessment. (c) Disaster risk reduction. (d) Disaster risk reduction[3,20].

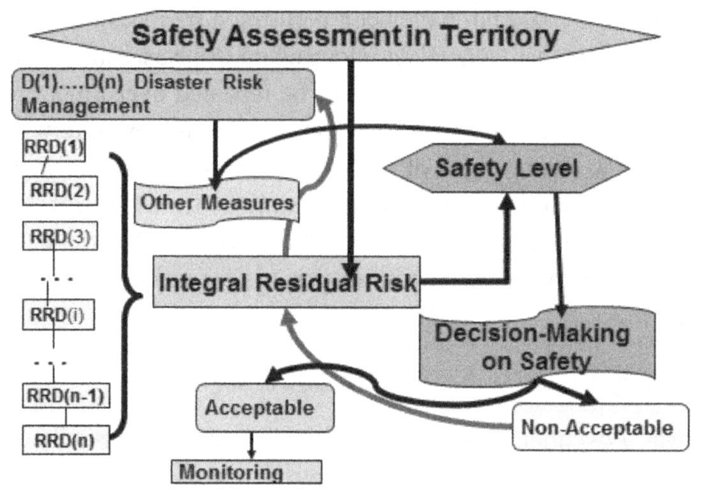

Figure 7.6: Model of Safety Assessment in Territory.
RRD(*i*), risk from the *i*th relevant disaster[2,20].

Figure 7.7: Model of Safety Assessment in Territory.
RRD(*i*), risk from the *i*th relevant disaster[2,20].

it is necessary in deciding on risk to use more possible variants of real human system behavior and multi-criteria decisions with the help of experts with verified qualifications[3,4].

From Figure 7.6 it follows that if we find that the safety level is unacceptable, the assessment process must return to the level of integral residual risk. The residual risks from individual relevant disasters (RRD(1), RRD(2), …, RRD(n)) must once again be judged and they must be revealed to be the causes of these residual risks. First and foremost, it is noted when the source of high integral risk could not have been performed by measures for reduction of risks in some of the disasters that were taken into account. Because safety changes according to the timescale, the safety assessment cycle must be repeated over time. From Figures 7.6 and 7.7 it is clear that safety management in a territory depends on safety assessment, safety monitoring and risk management of individual disasters, and on considering the links between the corrective measures for reducing the real risks from disasters in a system containing all relevant disasters in a territory. Safety management is the basis for land-use planning and for territorial development planning that is a part of strategic territorial planning[28].

Risk Engineering, Security Engineering, and Safety Engineering

Risk engineering was a twentieth century phenomenon and as its basis the groundwork for human development was set up in developed countries that are quite resistant to traditional disasters, namely natural ones, human, animal and plant diseases, technology failures, and social disasters. According to definitions used by the UN, Swiss Red Cross, The World Bank, etc., risk engineering is as follows:

- The systematic use of engineering knowledge and experiences for the optimization of the protection of human lives, environment, property, and economic assets, i.e. for the optimum reach of security and sustainable development of the human system.
- It has the main purpose to reduce all types of harm and losses by means of targeted and qualified risk management.

It is necessary to note that in current practice risk engineering has yet to be interpreted in an explicit way.

The original concept of risk engineering relies on risk management and it approaches problem solving step by step by considering individual disasters. It has to cope with all risks where the probability of occurrence is greater than or equal to 0.05. It usually includes only disasters whose sources are within the investigated system. It often solves only the technical aspects of a problem and the human factor problems have only been included in a later version from the 1980s (Figure 7.8).

The aim of the original risk engineering was to reduce the risks of technical systems connected to the internal sources of risks. As was stated above for practical purposes, the risk has been expressed as the probable size of losses, harm, and detriments of followed assets that caused a given disaster of a specified size that is calculated for a certain time unit (usually 1 year) and a certain territorial unit. For risk calculation we distinguish whether risk realization develops identically or differently, depending on the momentary local and time conditions. In the first case we determine a kind of mean risk value and its validity for use in practice. We use a condition in which the least unfavorable case is considered (the given approach uses norms and standards based on a deterministic approach). The other

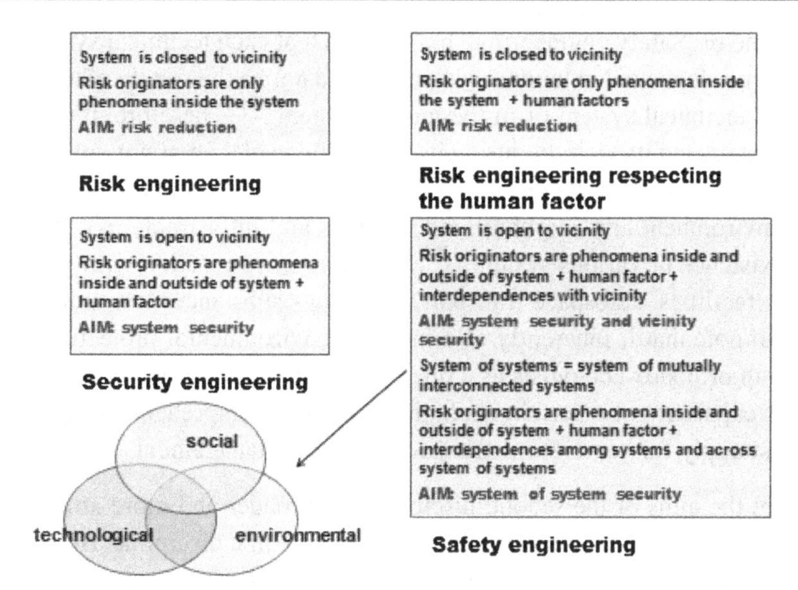

Figure 7.8: Engineering Types Considering the Risk.

case corresponds better to reality, and therefore it is considered as the groundwork preparation for strategic management with regard to safety. Various scenarios for risk realization and the probability of occurrence are determined; from these the mean value and its dispersion are determined by clear mathematical procedures (the given approach is in norms and standards based on a stochastic approach). The actual reality, however, is more complicated because, as was stated above, the data contain uncertainties and vagueness that are connected with the variability of conditions in time and space. At present exactly defined heuristic procedures[3] are used.

Advanced risk engineering disciplines, i.e. original engineering disciplines related to safety, considered the consequences of human errors but they did not investigate their causes. The systematic elimination of human errors was included in engineering disciplines from the mid-1980s as a reaction to the Chernobyl accident[20].

During the last 30 years, two engineering disciplines were produced based on risk management and aimed to ensure the security of a system and later the security of a system and its vicinity:

1. The discipline of "security engineering" has the aim that each individual system (technical or of another nature) is in a state of security with regard to the sources of internal and external risk[29]. The security of the vicinity is outside its area of interest. It is applied, for example, in the provision of bank information systems, boundaries, attacks against cyber networks, etc.

2. The discipline of "safety engineering" has the aim that each technical system produced by humans and implemented in human systems should not be the source of unacceptable risks either in the technical system or in the human system[30]. Therefore, it deals with a technical system and its vicinity during its whole life cycle, i.e. it not only solves technical system problems, but also respects public assets (human lives and health, welfare, property, environment and neighborhood facilities and infrastructures). It inherently includes coexistence of various systems[31]. It is used in connection with nuclear, chemical and similar facilities, aerospace transport, hazardous substances transport, etc. It is necessary to note that it inherently also includes environmental protection[20], and its conflicts with orthodox ecologists as a rule come from knowledge deficiency and from the insufficient capability of a certain ecological group to comprehend priorities that the human must apply for survival in strategic territorial management.

It is evident that the aims of the second discipline are broader and more ambitious, and therefore their achievement is a substantial challenge to the first discipline. Both disciplines in their recent form include risks connected with the human factor.

The key concepts of present engineering directed towards safety are:

1. The approaches are based on risk – the work intensity and documentation is adequate for the risk level.
2. The professional approach is based on the reality that only the critical attributes of quality and the critical parameters of the process are considered.
3. Problem solution is oriented to critical items – the critical aspects of technical systems ensuring the consistency of system operations are followed and managed.
4. Verified quality parameters are included in the project proposal.
5. The emphasis is on quality engineering procedures – the accuracy of selected procedures under given conditions must be proved.
6. The aim of a safety upgrade – permanently improving the processes with the analysis of the root causes of malfunctions and failures.

From the given facts it follows that the considered engineering types are multidisciplinary and interdisciplinary, and therefore they use various methods, tools, and techniques because the safety management targets cannot be reached only technically and/or by skill, but methods, tools, and techniques respecting the data logic, technological, financial, managerial, and decision-making aspects must be used, because they form an integral part of decision-making for technical problems, human factors, costs, and time planning.

In practice system security (the security of a system) is achieved by tool security engineering[29]. A high-powered tool represented by the engineering of safety called "safety engineering" deals not only with technical problems, but also respects public assets in the vicinity of the system. It applies methods, tools, and techniques based on engineering and management approaches in order that the system can be safe for all public assets during its whole life cycle[30]. The predecessor of both engineering types was risk engineering, the standards and norms of which began to be developed in the middle of the twentieth century[3,15,29,30].

In the original risk engineering, risk determination used the following given principles: risk was determined based on the design of the system; risk determination was directed at a certain level of the system and its components, i.e. there was no consideration of the outer vicinity and the protection of public assets; only knowledge of the system and processes was required, i.e. knowledge of the outer vicinity and protection of public assets was not required and if the risk existed then it was determined and solved, but without the option to remove the risks connected with an inappropriate solution for a given site and system.

Risk engineering depends on risk management and it looks for a solution in such a way that it considers disasters individually and includes all risks where the probability of occurrence is equal to or greater than 0.05. Usually it only includes disasters whose sources are within the system and hence it very often only solves technical aspects of the problem[20].

The understanding of safety management is particularly marked from the risk management viewpoint by the following characteristics: design and construction of projects with risk reduction; operation with the integration of early warning systems and with procedures for the management of acceptable levels of risk; and overcoming the abnormal, emergency, and critical conditions of the operation[29,30].

Advanced safety engineering uses the following principles in risk determination:

- Risk is determined during the whole life cycle of the given system, i.e. from design, building, operation, and putting into operation
- Risk determination is directed toward user's demands and to the level of provided services
- Risk is determined according to the criticality of impact on the processes, provided services and on assets that are determined by public interest
- Unacceptable risks are mitigated using tools for risk management, i.e. according to technical and organizational proposals, by standardization of operating procedures or by automatic check-up.

Safety engineering encompasses risk management regarding all possible disasters at a stroke and it searches for an optimum problem solution applying the All Hazard Approach[32] (i.e. it considers all possible disasters without respecting whether their sources are within or outside a given system and it uses the precautionary principle). It uses "safety management", i.e. risk management supporting human system security, which also includes sustainable system development. In technical slang we say that safety management creates the inherent safety of a human system against design disasters and by implementation of the precautionary principle we upgrade resistance against unacceptable impacts beyond design disasters, the occurrence of which is so low in probability that it is unforeseeable[20]. In practice, principles are introduced as fail safe, and carry out only determined functions, i.e. if you cannot fulfill an aim, do not do anything.

The key concepts of safety engineering are as follows:

1. The approach to a problem is based on the risk with the rule that the intensity of work and documentation are adequate for the risk level.
2. In the professional procedure, in respecting the solved problem logic we must consider the critical attributes of quality and the critical parameters of the process.
3. The problem solution is directed at critical items, i.e. the topics are monitored and the management of critical aspects of technical systems ensuring the operational consistency of systems is performed.
4. Certified parameters of quality must be included in the project proposal for problem solving.
5. The emphasis is on quality engineering procedures, which means that the correctness of selected procedures in given conditions must be demonstrated.
6. During the whole life cycle there is the aim of a safety upgrade (with the help of safety management systems), i.e. we go on continually improving the processes with the use of analysis of the root causes of defects and failures.

In respecting the principles given above, relevant data sets and only verified methods that provide outputs with a designated testified competence must be used. Because there are cases that do not cope well with vagueness in data, in practice we use procedures designated as good practice procedures/good engineering practice procedures. Modus operandi procedures in individual domains based on experience lead to good results. The given procedure is used in cases in which there is no approved unified procedure. This is often used in measurements in laboratories, in negotiations with humans, etc.

There are many factors, including the human factor, that influence problem solving in real-world conditions, and these factors are not only random but also epistemic. The measures, activities, and procedures denoted as good engineering practice are typical for engineering disciplines.

Good engineering practice (good engineering procedure) is then defined as a set of engineering methods and standards that are used during the life cycle of a technical system with the aim of reaching an appropriate and cost-efficient solution. It is supported by appropriate documentation (conceptual documentation, diagrams, charts, manuals, test reports, etc.).

In a given context, engineering expertise is the expression of capabilities as follows:

- Apply the knowledge of mathematics, science, and engineering
- Propose and realize experiments

- Analyze and interpret data
- Propose components or the whole system according to requirements and in the framework of realistic limitations identify, formulate, and solve engineering problems
- Ensure effective communication
- Understand the impact of engineering solutions in a broader context
- Use advanced tools and methods in engineering practice
- Adhere to professional and operational responsibilities and ethics
- Lead an interdisciplinary team.

Most of the demands given above are directed at correcting negative manifestations of the human factor.

Based on the knowledge of the past decade, it is necessary to admit when considering risk realization that, in addition to random uncertainties, there also exist knowledge (epistemic) uncertainties, i.e. vagueness in the data. By admitting that this additional uncertainty exists, we *de facto* admit the existence of significant changes in the process of risk realization, which go significantly beyond the simple effects of random changes. Thus, in recent years approaches from the theory of *possibilities*, i.e. Dempster–Shafer theory[25,26], have been introduced in practice for modeling safety and reliability. This assumes that the available data and our knowledge have vagueness, i.e. they contain knowledge (epistemic) uncertainties in addition to random uncertainties. Using this theory, variants corresponding to different processes are modeled – what is possible due to knowledge shortcomings. Of these, the optimum variant is selected. To select the options, the service of experts is used and calculations are combined with the best practices. Experience has shown that one expert is not enough, but that it is necessary to combine the knowledge of several experts. Such a combination can be ensured by analytical methods or heuristics, such as the DELPHI panel discussion[4].

7.4 Conclusion

For human safety and for human systems safety (i.e. territory, organization, plant) we must manage the integral risk including the human factor, i.e. find the path of cross-section risks management and concentrate the investigation on interdependences and critical spots with the potential to start the system cascade failures, and on the basis of such site knowledge prepare measures and activities to ensure the continuity of limited infrastructure operation and human survival.

Considering the present critical knowledge evaluation, we recognize that one of the causes of interdependence inducing failure cascades in a human system or in its parts is human error (intentional or unintentional) in management. Therefore, in both management activities and engineering activities we must carry out all procuration with the aim of averting human failure, especially in decision-making. Because the consequences of decision-making are often huge, the causes of human failure at management level are given in detail above.

References

[1] I. Matoušková, The decision in situations with threat, in: Riešenie krízových situácií v špecifickom prostredí, FŠI, Žilina, 2008, pp. 533–540 (in Czech).
[2] UN, Human Development Report, UN, New York, 1994, <www.un.org>.
[3] D. Procházková, Analysis and Management of Risks, CVUT, Prague, 2011 (in Czech).
[4] D. Procházková, Methods, Tools and Techniques for Risk Engineering, CVUT, Prague, 2011 (in Czech).
[5] J. Trpiš, Probabilistic assessment of reliability of human factor in industry, in: Bezpečnost a ochrana zdraví při práci, VŠB a SPBI, Ostrava, 2010, pp. 281–287 (in Czech).
[6] D.A. Wiegmann, S.A. Shappell, A Human Error Approach to Aviation Accident Analysis: The Human Factors Analysis and Classification System, Ashgate Publishing, 2003, pp. 48–49.
[7] J. Reason, Human Error, Cambridge University Press, 1990.
[8] D. Procházková, J. Bumba, V. Sluka, B. Šesták, Dangerous Chemical Substances and Industrial Accidents, PA ČR, Prague, 2008 (in Czech).
[9] A.M. Malkin, C. Winder, Applying the safe place, safe person, safe systems framework: case study findings across multiple industry sectors, in: Reliability, Risk and Safety. Theory and Applications, CRC Press/Balkema, Leiden, 2009, pp. 697–704.
[10] I.A. Papazoglou, et al., Occupational risk management for contact with moving parts of machines, in: Reliability, Risk and Safety. Theory and Applications, CRC Press/Balkema, Leiden, 2009, pp. 713–720.
[11] P.A. Bragatto, et al., The impact of the occupational safety control programs on the overall safety level in an industrial cluster, in: Reliability, Risk and Safety. Theory and Applications, CRC Press/Balkema, Leiden, 2009, pp. 745–752.
[12] D. Procházková, Human System Safety, SPBI, Ostrava, 2007 (in Czech).
[13] D. Procházková, Strategy of Management of Safety and Sustainable Development of Territory, PA ČR, Prague, 2007 (in Czech).
[14] J.F. Gustin, Disaster, Recovery Planning: A Guide for Facility Managers, Fairmont Press, Lilburn, 2002.
[15] R. Briš, C.G. Soares, S. Martorell (Eds.), Reliability, Risk and Safety: Theory and Application, CRC Press/Balkema, Leiden, 2009.
[16] D. Procházková, System of management of organisation safety, in: Bezpečnost a ochrana zdraví při práci, VŠB-TU, Ostrava, 2009, pp. 223–232 (in Czech).
[17] E. Hanáková, Basic terms used in work hygiene, Internal Working Document, VÚBP, Prague, 2007 (in Czech).
[18] Team of authors from ČMKOS a ASO: Work Safety – Integral Part of Life, ČMKOS, Prague, 2008 (in Czech). <www.cmkos.cz, www.esfcr.cz>.
[19] L. Foldyna, Reliability of human factors of crisis managers of regional offices, in: Bezpečnost a ochrana zdraví při práci, VŠB-TU, Ostrava, 2009, pp. 121–131 (in Czech).
[20] D. Procházková, Strategic Safety Management of Territory and Organization, ČVUT, Prague, 2011 (in Czech).
[21] D. Procházková, Protection of Humans and Property, ČVUT, Prague, 2011 (in Czech).
[22] US Project Management Institute, A Guide to the Project Management Body of Knowledge, US Project Management Institute, Washington, 2004.
[23] EU, Risk assessment and mapping guidelines for disaster management. Working Paper SEC (2010) 1626, EU, Brussels, 2010.
[24] D. Procházková, Methodology for Estimation of Costs for Renovation of Property in Territories Affected by Natural or Other Disasters, SPBI Spektrum XI, Ostrava, 2007 (in Czech).
[25] G.A. Shafer, Mathematical Theory of Evidence, Princeton University Press, Princeton, 1976.

[26] A.P. Dempster, Upper and lower probabilities induced by a multivalued mapping, Annals of Mathematical Statistics 38 (5) (1967) 325–339.
[27] OECD, Assessing Societal Risks and Vulnerabilities, OECD Studies in Risk Management, Paris, 2006.
[28] J.S. Armstrong, Review of corporate strategic planning, Journal of Marketing 54 (1990) 114–119.
[29] R. Anderson, Security Engineering – A Guide to Building Dependable Distributed Systems, John Wiley, 2008.
[30] H.E. Roland, B. Moriarity, System Safety Engineering and Management, John Wiley, 1990.
[31] H. Bossel, Systeme, Dynamik, Simulation – Modellbildung, Analyse und Simulation komplexer Systeme, Books on Demand, Norderstedt/Germany (2004). <www.libri.de>.
[32] FEMA, Guide for All-Hazard Emergency Operations Planning. State and Local Guide (SLG) 101, FEMA, Washington, 1996.

CHAPTER 8

Robust Road Environment Perception for Navigation in Challenging Scenarios

Chunzhao Guo*,†, Seiichi Mita†, David McAllester‡
*Toyota Central R&D Labs., Inc., †Toyota Technological Institute
‡Toyota Technological Institute at Chicago

Chapter Outline

8.1 Introduction

In order to adequately navigate an autonomous vehicle through its environments, or increment road safety with Advanced Driver Assistance Systems (ADAS), the host vehicle

Advances in Intelligent Vehicles. http://dx.doi.org/10.1016/B978-0-12-397199-9.00008-2

must perceive the structure of that environment, modeling world features relevant to navigation. One of the primary perception tasks is to provide a description of the road so that the intelligent navigation system can plan safe as well as legal actions. A comprehensive situational road detection system is paramount to the effectiveness of proprietary navigational and higher-level functions. The detected free road space and lane boundaries can provide significant context information to reduce the region of interest (ROI), re-weight hypotheses, and remove false positives for other functions such as vehicle and pedestrian detections. Road detection has been extensively studied for several decades and dramatic developments have been accomplished. A number of approaches focused on the use of vision exclusively [1,2], whereas others utilized LIDAR (Light Detection and Ranging) [3], sometimes in combination with vision [4]. While the LIDAR-based systems can perform extremely well in certain situations regarding range measurement, they are not capable of acquiring visual traffic information, such as traffic signs, traffic signals, on-road text, road colors, etc. The detection and recognition of such information are crucial for the successful deployment of future intelligent transportation systems in mixed traffic, in which the intelligent vehicles have to share the road environment with all road users, such as pedestrians, motorcycles, bicycles, and other vehicles. Vision can deliver a great amount of information over a long range, making it a powerful means for sensing the structure of the road environment, identifying the on-road objects, and recognizing the visual traffic information. Moreover, such passive sensing systems have no risk of inter-vehicle interference or pollution to the environment. Therefore, vision is necessary and promising for road environment perception and other applications related to intelligent vehicles and ADAS.

Generally, the task of road detection is difficult for computer vision, as the road appearance is affected by a number of factors that are not easily measured and change over time and space, such as road materials, illumination and weather conditions, etc. Examples of the variety of road environments can be seen in Figure 8.1, which shows the environmental variability even within small regions. Along with the various types of the roads, weather conditions and time of day can have a great impact on the visibility of the road surface. These challenges produce a roadmap of the vision-based road detection system for vehicular autonomy, as shown in Figure 8.2. It reflects the relationship between the degrees of autonomy that the vehicle can operate and the robustness of the vision system in different complexities of road environments. The more road scenarios where the vision system can provide reliable road information, the more frequently the vehicle can operate in the automated driving mode and the higher degree of autonomy the vehicle can achieve. In other words, the degree of a vehicle's autonomy is proportional to the robustness of the vision system. Most vision systems proposed nowadays only produce reliable results in well-structured roads and good conditions so that they can assist the human driver occasionally. Some systems can show good performance in relatively well-structured roads and some challenging illumination conditions, and reach the "semi-automated" phase. Most of the current systems still have

Figure 8.1: Example Road Images in Challenging Scenarios.
(a–i) Different types of roads with varying colors, textures, shapes, and conditions. (j–l) Images of the same stretch of road shown in the daytime, nighttime, and a day with heavy rain. (a–f) and (j–l) were taken by our stereo camera systems from roads near our campus (less than 1 km in total), which show the environmental variability even within small regions.

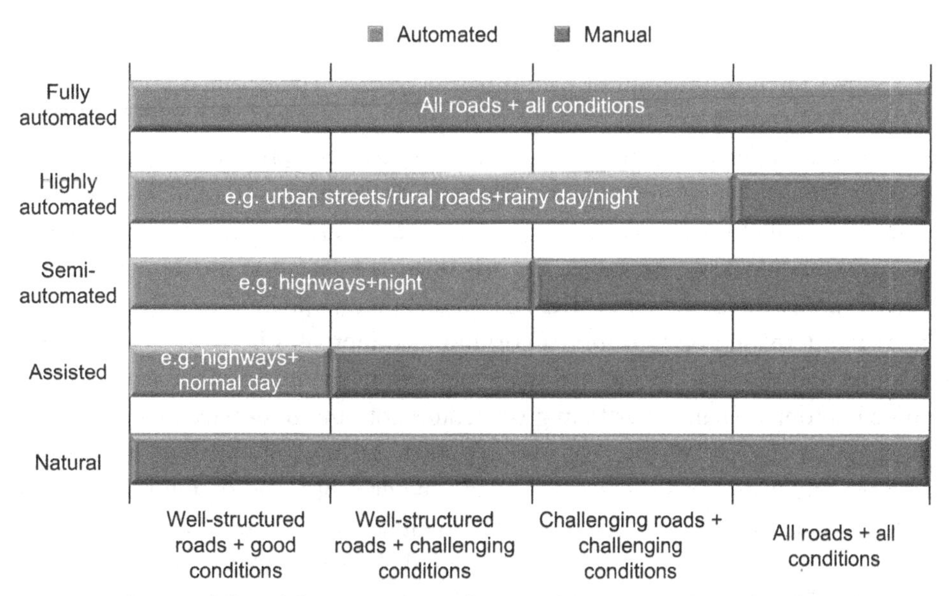

Figure 8.2: Roadmap of the Vision-Based Road Detection System for Vehicular Autonomy over Road Scenarios.

difficulties in detecting reliable road information in challenging roads and conditions; however, it is necessary to overcome these difficulties to achieve "highly automated" and "fully automated" intelligent vehicles. This is the motivation of the proposed work, in which we focus on the challenging scenarios, and develop a sophisticated robust stereovision-based system that can overcome these challenges.

From Figure 8.1 we can see that roads often have heterogeneous surfaces in challenging scenarios, and their appearances may change over time and space due to a number of factors. Therefore, it is not reliable to find such roads by utilizing/modeling their appearance descriptors. Actually, what constitutes a road being drivable is not its appearance but its physical properties. For car-like vehicles, the road is drivable as long as it is solid and reasonably flat, and the painted lane markings are also constructed on a reasonably flat road surface. The solid property of the road should be examined by using object recognition methods with *a priori* knowledge. Here, we only consider the geometric properties of the road, and define a drivable road region as a reasonably flat region in front of the vehicle on the ground plane where a vehicle can pass safely, under the flat-road assumption. Therefore, in the proposed approach, we utilize the reasonably flat geometric property of the road as the exclusive criterion for drivable road detection in the image, and the visual features are used only to provide the evidence for what regions the image pixels belong to. Therefore, such judgment criteria, as well as the road classification, is independent of the road's appearance, and invariant to road environment transitions and condition variations, thereby solving the challenges indicated in Figure 8.1. In this case, the entire heterogeneous road portions in the image can be contained in the detected drivable region no matter how dramatically the road's appearance changes in challenging scenarios.

Figure 8.3 shows the flow diagram of the proposed system. The geometric relationships between the stereo camera and the road are dynamically estimated and calibrated for each input image pair to utilize the underlying geometry cues more precisely in the subsequent steps, so as to enhance the accuracy and robustness. Subsequently, modules 1 and 2 are implemented in parallel for detecting and tracking the drivable road and lane boundaries respectively. In module 1, we formulate the drivable road detection problem as binary labeling in a Markov Random Field (MRF) based on homography induced by the road plane. We develop an alternating optimization algorithm that alternates between computing the binary labeling and learning the optimal parameters from the input image pair itself. The proposed automatic parameter tuning procedure not only improves the accuracy of road detection, but also makes the proposed approach adaptive to changing environments without having any prior knowledge of the road. The drivable road region is subsequently obtained after removing the effect of textureless regions, which often cause problems in stereovision systems. In module 2, an Artificial Neural Network (ANN) classifier is applied to the left and right Inverse Perspective Mapped (IPM) images in parallel to detect painted lane markings. Next, Normalized Cross Correlation (NCC) is employed to reveal the geometric

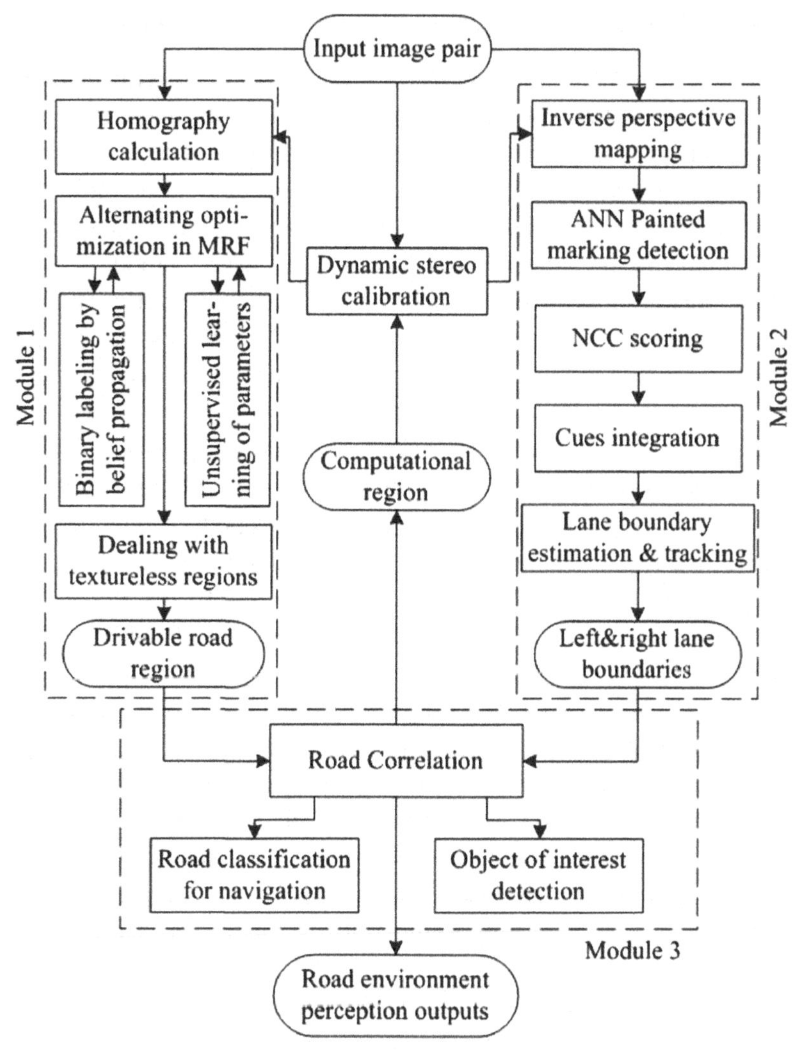

Figure 8.3: Flow Diagram of the Proposed System.

cue between the preserved left and right lane markings, which is then integrated with the intensity cue by constructing a weighted graph, reflecting the belief of each pixel as a true lane feature. The left and right lane boundaries are estimated in a tracking process using a particle filter. In module 3, the detected drivable road region and lane boundaries are correlated for a number of high-level functions, such as the road classification for navigation and object of interest detection, etc. Here, several common road users, including vehicles, pedestrians, motorcycles, and bicycles, are considered as particular objects of interest, which are then identified using a multi-class object detector with deformable part-based models and the Histogram of Oriented Gradient (HOG) features. The road and lane correlation also provides the computational region for the dynamic stereo calibration.

The proposed system is distinguished from previous related work in the following ways:

1. We formulate the road detection problem as a Maximum A Posteriori (MAP) problem in MRF based on stereo with homography. The proposed MRF model results in high classification performance due to the concentration on the relevant challenging road scenarios with a well-defined energy function that exploits the essence of drivable roads.
2. We focus on the challenging scenarios and develop a road detection approach that can overcome the challenges. Compared with appearance-based methods, the proposed exclusive judgment criterion based on the reasonably flat geometric property of the drivable road is more accurate and robust, since it reflects the essential characteristics of the road being driven, and is able to contain all heterogeneous roads in the image even when they are changing over time and space.
3. We place strong emphasis on the robustness of the proposed approach, since this is the key issue for vision-based road detection systems, as explained previously. On the basis of the exclusive judgment criterion based on the geometric property of the drivable road, we further enhance the robustness of the proposed approach in the following four ways. First, the use of MRF allows that the road classification to be determined jointly with the spatial interactions present at the scene, thereby ensuring local consistency in the detection results. Second, the region tracking makes use of the temporal support from the consecutive images. In cases where the image is too noisy or blurred due to particular factors, the image support for matching will become weak, while the strong temporal support can prevent false detections. Third, the extrinsic camera parameters are re-estimated for each input image pair to retrieve the precise geometric property of the drivable road so as to overcome the challenges brought about by dynamic vehicle movements. Fourth, the MRF parameters are learned online by applying a hard Expectation Maximization (EM) algorithm to maximize a conditional likelihood so as to overcome the difficulties due to the challenging and changing environment.
4. We construct a weighted graph by integrating the intensity and geometric cues in a "soft" way, reflecting the belief of each pixel as a true lane feature. This can reveal the uncertainty of lane feature detection in challenging scenarios, and bring it to the grouping level.
5. We identify particular objects of interest based on the context information obtained by the road correlation, which can improve both the accuracy and efficiency of road environment perception.

8.2 Use of Geometric Properties of the Road

In the proposed system, in order to utilize the geometric property of the drivable road while maintaining a manageable computational load, we employ IPM for lane boundary detection and homography induced by the planar road plane for drivable road detection.

8.2.1 Inverse Perspective Mapping

We employ IPM for lane boundary detection for the following three reasons. Firstly, IPM remaps the road points in the left and right images into points in the same world coordinate ($Z = 0$) and the resulting image represents a top view of the road region in front of the vehicle, revealing the geometric relationships of the road between the stereo pair and the road plane. Secondly, IPM normalizes the size of the lane markings and reduces the range of lane boundaries, which is beneficial for the use of the ANN classifier. Thirdly, the IPM image is in accordance with the global coordinates so that the detection results in the IPM image can easily be used to update the navigation map.

IPM can be performed with the aid of a camera, such as a viewpoint position, viewing direction, aperture, resolution, etc. An alternative, simpler way to conduct IPM is a linear transformation on homography represented by a 3×3 matrix $H_{\mathrm{I}i}$, as shown in Figure 8.4, which can correspond each road point u_i on the image plane to a point x_i on the road plane π using the following equation:

$$x_i = H_{\mathrm{I}i} u_i, \tag{8.1}$$

in which i indicates the left or the right image. $H_{\mathrm{I}i}$ can then be derived from a simple external camera calibration with four reference points [5].

8.2.2 Homography Induced by the Road Plane

The underlying geometric properties of the drivable road between the image pair can be revealed by using planar homography, which can relate points of images on a plane to corresponding image points in a second view [5]. For the road plane π, there is an IPM, $x = H_{\mathrm{Il}} u_{\mathrm{l}}$, between the left image plane and π; and an IPM, $x = H_{\mathrm{Ir}} u_{\mathrm{r}}$, between the right image

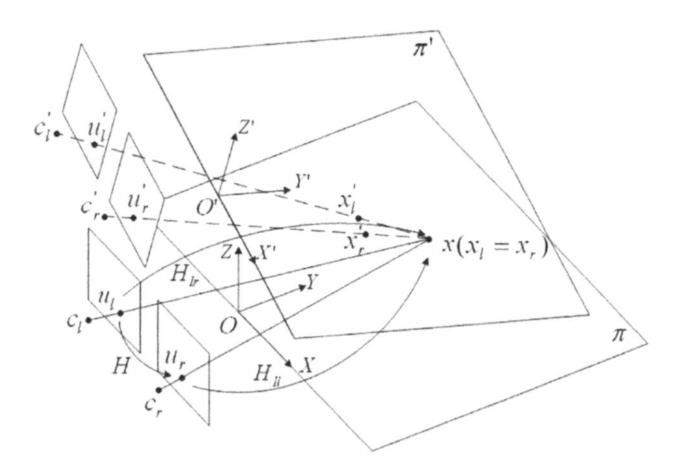

Figure 8.4: Geometry Relationships Between Cameras and the Road Plane.

plane and π. The composition of the two inverse perspectivess is a homography induced by the plane π,

$$H = H_{\mathrm{Ir}}^{-1} H_{\mathrm{Il}}, \tag{8.2}$$

between the two images. As for road detection, when we employ H to find correspondences between the image pair, only the road points that can comply with the homography will show a good match while the other non-road points will not. Therefore, based on H, all the regions belonging to the same planar road plane should be contained in the matched area and classified as the drivable road region no matter how dramatically the road appearance changes in challenging conditions. The reason that we use original left and right images with the homography induced by the road plane instead of the IPM images for drivable road region detection is that objects outside the road plane are distorted in the IPM images. Part of the distorted objects may be matched so as to increase the possibility of false detections.

8.2.3 Dynamic Stereo Calibration

Many vision-based road detection methods assume that the cameras are calibrated beforehand and the extrinsic camera parameters are known and fixed. However, these assumptions are not practical in real-world applications since the vehicle may tilt and the cameras may vibrate. Therefore, the geometric relationships, e.g. the homography in the proposed system, should be re-estimated online.

As shown in Figure 8.4, c_{l}, c_{r}, and π are the normal positions of the left and right cameras and the road plane respectively. The inverse perspective matrices H_{Il}, H_{Ir} obtained from the camera calibration can map the image points u_{l}, u_{r} to x_{l}, x_{r} in global coordinates and $x_{\mathrm{l}} = x_{\mathrm{r}}$ since u_{l}, u_{r} are from the same road point x in the same coordinates. Once the position and pose of the cameras change and move to c_{l}', c_{r}' due to vehicle tilt or camera vibration, the global coordinates that the pre-calculated H_{Il}, H_{Ir} correspond to also change along with the cameras from O to O'. In this case, H_{Il}, H_{Ir} will map u_{l}', u_{r}' to x_{l}', x_{r}' and $x_{\mathrm{l}}' \neq x_{\mathrm{r}}'$.

We assume the transformation from O' to O needs to be translated by δ and rotated by θ about the X-axis, ϕ about the Y-axis, and ψ about the Z-axis. The relationship between O' and O can be described by

$$x_i = R x_i' + T, \tag{8.3}$$

where R and T are the rotation and translation matrices respectively. Subsequently, the inverse perspective mapping from a road point u_i' on the shifted image plane to a point x_i on the road plane can be represented by the following equation:

$$x_i = A H_{\mathrm{I}i} u_i', \tag{8.4}$$

where $H_{\mathrm{I}i}$ is the pre-calculated matrix in homogeneous coordinates and

$$A = \begin{pmatrix} \cos\varphi\cos\psi & \sin\theta\sin\varphi\cos\psi - \cos\theta\sin\psi & \cos\theta\sin\varphi\cos\psi + \sin\theta\sin\psi & \delta_x \\ \cos\varphi\sin\psi & \sin\theta\sin\varphi\sin\psi + \cos\theta\cos\psi & \cos\theta\sin\varphi\sin\psi - \sin\theta\cos\psi & \delta_y \\ -\sin\varphi & \sin\theta\cos\varphi & \cos\theta\cos\varphi & \delta_z \\ 0 & 0 & 0 & 1 \end{pmatrix}. \tag{8.5}$$

In the present case, in order to estimate current homography H' from current inverse perspective matrices H'_{Il}, H'_{Ir} that can map u'_{l}, u'_{r} to x where $x'_{\mathrm{l}} = x'_{\mathrm{r}}$, we estimate the shift vector $S = (\theta, \varphi, \psi, \delta_x, \delta_y, \delta_z)$ by minimizing the total error function $E(S)$ in (8.6) between the stereo image pair, using the Levenberg–Marquardt Algorithm (LMA) [6]:

$$E(S) = \sum_{u'_{\mathrm{l}}, u'_{\mathrm{r}} \in R_{\mathrm{c}}} \sum_{k} \omega_k \left(\Phi^k_{\mathrm{l}}(AH_{\mathrm{Il}}u'_{\mathrm{l}}) - \Phi^k_{\mathrm{r}}(AH_{\mathrm{Ir}}u'_{\mathrm{r}})\right)^2, \tag{8.6}$$

where $\Phi_{\mathrm{l}}(\ )$ and $\Phi_{\mathrm{r}}(\ )$ are the feature vectors we constructed for the image pair, and ω_k is the weight of the kth element. For color images, the vector is nine-dimensional, consisting of three color values plus a six-dimensional color gradient vector in the x and y directions. For grayscale images, it is a three-dimensional vector that consists of the intensity value plus a two-dimensional gradient vector. R_{c} is the computational region, which is actually the planar road region of the image.

8.3 Drivable Road Region Detection and Tracking

During the last several decades, many vision-based road detection systems have been proposed. Some methods use a monocular camera to extract the road region by employing specific features based on the road appearance. For example, Nieto and Salgado [7] compute the texture orientation for each pixel, then seek the vanishing point of the road by a voting scheme, and finally localize the road boundary using the color cues. McCall and Trivedi [2] find the roads by extracting the painted lane boundaries. Since the imaged road texture varies greatly with the distance to the camera, color analysis is used exclusively by some methods, which often model the road appearance in different color spaces. For example, the Hue–Saturation–Intensity (HSI) color features are used to model the road pattern [8]. The Gaussian distribution and a mixture of Gaussians in the Red–Green–Blue (RGB) color space are used to describe the road appearance [4]. Such appearance-based methods can work very well in certain environments. However, they are characterized by lack of effectiveness in cases where the roads do not correspond sufficiently to the models of the prior defined features. Moreover, it is difficult to represent all of the possible road patches in predefined models in challenging traffic scenes. Therefore, they usually detect a portion of the drivable roads that complies with their models in such scenarios. Some other methods work on a sequence of temporally consecutive camera images of the scene based on a monocular camera, and make use of the displacement of pixels between two consecutive images. For example, the reverse optical flow technique is used to provide an adaptive segmentation of the road region on multiple scales [9]. The Structure From Motion (SFM) technique is used to estimate a map-based road boundary model [10]. These motion-based methods can provide generic detection of the drivable roads and give information about the displacement of the target and structure/depth of the scene. However, they cannot work well on chaotic roads when the camera is unstable and the estimation of the optical flow is not robust enough. Recently, methods for geometry estimation from a single still image have seen a revival. These methods typically segment the image and infer the most likely 3D configuration of each segment based on monocular cues, such as texture gradient, aerial perspective, blurring, etc. Even though impressive results have been demonstrated [11], these methods are still not accurate enough to directly support applications like autonomous driving and ADAS.

The above monocular-based methods can produce good results when the extracted features are discriminative and robust. However, when the features are noisy or changing, these methods may produce very poor results. Generally, the stereovision-based methods are more robust than the monocular-based ones, since they have more information such as triangulate feature points in 3D, and are more robust to loss of scale and dynamic vehicle movements. Given a stereo image pair, stereo matching-based methods extract the 3D structure of the scene by solving the corresponding problem and computing the disparity map. 3D urban reconstruction has been demonstrated [12]. However, stable reconstruction of the 3D structure by computing the disparity map is computationally expensive due to the requirement of solving the corresponding problem for every pixel. Others utilize the stereo cues without calculating the disparities. The homography induced by the ground plane is employed to find the agreement of the road pixels between the image pair [13]. Compared with the stereo matching-based methods, although these methods are convenient only under the flat-road assumption, they are simpler and more effective with respect to a dedicated drivable road detection system for autonomous driving applications. The proposed approach belongs to this type of system.

8.3.1 MRF Model for Road Detection

In the proposed approach, we present road detection as a binary labeling $L: x \rightarrow \{0,1\}$ for each pixel x at pixel location p in the reference image, and the assigned label denoted by a random variable f is

$$f(p) = \begin{cases} 1 & p \in \text{Drivable road region} \\ 0 & \text{Otherwise.} \end{cases} \tag{8.7}$$

The MRF framework is employed for the labeling because it models the spatial interactions present in the scene so that the labels of the points are determined jointly. Our goal is to find the correspondence that matches pixels of similar intensity and gradients between the left and right images based on the homography induced by the road plane while minimizing the number of discontinuities, since the labels should not vary in either road or non-road regions but just change at pixel locations along the boundary between the two regions. We accomplish this by minimizing the following energy function, which describes the quality of labeling:

$$E = E_{\mathrm{D}} + E_{\mathrm{S}}, \tag{8.8}$$

where E_{D} is a data-dependent energy term containing the costs of assigning the labels to the pixels. E_{S} enforces smoothness by penalizing the discontinuities.

Under the proposed formulation, we define the data term E_{D} as the sum of a matching energy E_{M} and a tracking energy E_{T} to utilize both image evidence and temporal support of the road. We take the left image as the reference image, and the matching energy is defined as follows:

$$E_{\mathrm{M}} = \sum_{p \in P} \left(\sum_{k} \lambda_k (\Phi_{\mathrm{l}}^k(p) - \Phi_{\mathrm{r}}^k(Hp))^2 \cdot f(p) + \lambda_m (1 - f(p)) \right), \tag{8.9}$$

where $\Phi_{\mathrm{l}}(p)$ and $\Phi_{\mathrm{r}}(p)$ are the feature vectors described in the previous section, and λ_k are weights for each element of the vector. The use of the gradient vector is aimed at improving the performance by encouraging the labeling discontinuities to be aligned with the intensity edges. H is the homography matrix induced by the road plane and thus Hp is the pixel in the right image that corresponds to the pixel location p in the left image under H. The weighted sum of the squared difference of the two vectors is utilized to measure the compatible cost, and λ_m is a thresholding factor designed to adjust the influence of the compatible cost on the labeling. In order to minimize the matching energy, $f(p)$ will be assigned as 1 if the compatible cost is smaller than λ_m. Otherwise the variable will be assigned as 0.

The tracking energy is very important for road detection in changing environments since most of the road points are detected repeatedly. Considering the situation that image noises, such as raindrops and backlights, are different between the image pair, the road points may get weak image support from the matching energy but will still have strong temporal support from the tracking energy so as to prevent false detections. Therefore, more robust detection can be expected. Here, we define the tracking energy using the difference in labeling between the current and adjacent previous reference image based on the vehicle's motion:

$$E_{\mathrm{T}} = \sum_{p \in P} \min\left(\tau_{\mathrm{t}},\ \lambda_{\mathrm{t}} I_{\mathrm{l}}(p) \left| f^t(p) - f^{t-1}(T(p)) \right| \right), \tag{8.10}$$

where λ_t is the weight of the tracking energy and τ_t is the truncation threshold. f^t and f^{t-1} are the labeling in the reference image at time t and $t-1$. T indicates the transformation according to the motion of the host vehicle from $t-1$ to t, and the transformed labeling map serves as a prediction of the new labeling. The transformation can be derived from either the ego-motion information of the host vehicle or the homography between sequential images.

In stereo problems, the smoothness term is generally based on the difference between labels, rather than on their actual values. Therefore, we define the smoothness term as

$$E_S = \sum_{(p,q)\in G} \min(\tau_s, \lambda_s|f(p) - f(q)|), \tag{8.11}$$

where λ_s is the weight of the smoothness term and τ_s is the truncation threshold.

Therefore, the energy function in (8.8) can be rewritten as follows, by summing (8.9), (8.10), and (8.11), which is used for the inference based on belief propagation (BP) in the MRF:

$$E = E_M + E_T + E_S. \tag{8.12}$$

8.3.2 Unsupervised Learning of the Parameters

Most current vision-based road detection systems for autonomous driving typically rely on hand-crafted algorithms manually tuned for particular classes of road environments [14]; however, performance falls short in changing environments. To deal with this problem, an adaptive approach is necessary by applying learning techniques to the vision-based system. A number of systems that incorporate learning have been proposed. Some systems were trained offline using hand-labeled data based on supervised learning [15], which has two major disadvantages: (1) labeling requires a lot of human effort; (2) offline training limits the scope of the host vehicle's expertise to environments seen during training. More recently, self-supervised systems have been developed by employing near-to-far learning [4], in which a reliable detector (such as LIDAR) that detects the drivable region in a short range of the current road environment provides labels for inputs to another long-range detector for online supervised training. In this way, such systems can adapt to changing environments; however, a reliable short-range detector is necessary and the learning extends the detection range without improving the detection accuracy itself. In the proposed system, we develop an adaptive MRF-based road detection approach by applying unsupervised learning of the input image pair itself.

For the problem of estimating MRF parameters, Cheng and Caelli [16] proposed a Bayesian approach to estimate four parameters with a restricted MRF model for stereo matching use, Markov Chain Monte Carlo (MCMC). Zhang and Seitz [17] proposed a method to compute three parameters, including the truncation thresholds for both data and smoothness terms and a regularization weight between the two terms, for general MRF stereo matching based on the mixture models for the histograms of pixel matching errors and neighboring disparity differences. Both of these approaches model the joint distribution of cues over the left and right images with various independence assumptions, which are sometimes bad assumptions. Our approach models the conditional probability distribution of the right image given the left image so that there is no danger of corrupting the model by modeling the distribution over both of the images poorly. Moreover, our approach is simpler and more efficient, which is very important for real-time applications.

The parameters of our energy function were learned using a hard conditional EM algorithm applied to the stereo data to maximize a conditional likelihood. The truncation thresholds τ_{t} and τ_{s} will not be estimated since they are insensitive to changing environments under our formulation of road detection. Therefore, we consider a general conditional probability model $P_{\beta}(I_{\mathrm{r}}|I_{\mathrm{l}})$ over the input image pair I_{l} and I_{r}, defined in terms of a parameter vector $\beta = \{\lambda_k, \lambda_m, \lambda_{\mathrm{t}}, \lambda_{\mathrm{s}}\}$ and the latent variable f,

$$P(I_{\mathrm{r}}|I_{\mathrm{l}},\beta) = \sum_{f} P(I_{\mathrm{r}},f|I_{\mathrm{l}},\beta). \tag{8.13}$$

The hard conditional EM is an algorithm for locally optimizing the parameter vector β so as to maximize the probability of I_{r} given I_{l}:

$$\beta^* = \arg\max_{\beta} \ln P(I_{\mathrm{r}}|I_{\mathrm{l}},\beta) \tag{8.14}$$

with the following two updates:

$$\text{Hard E step}: \quad f := \arg\max_{f} P(I_{\mathrm{r}},f|I_{\mathrm{l}},\beta) \tag{8.15}$$

$$\text{Hard M step}: \quad \beta := \arg\max_{\beta} \ln P(I_{\mathrm{r}},f|I_{\mathrm{l}},\beta). \tag{8.16}$$

In the proposed system, the hard E step is implemented using an efficient belief propagation algorithm which computes f by minimizing the proposed energy function [18]. The implementation of the hard M step relies on a factorization of the probability model into two conditional probability models:

$$P(I_{\mathrm{r}},f|I_{\mathrm{l}},\beta) = P(f|I_{\mathrm{l}})P(I_{\mathrm{r}}|I_{\mathrm{l}},f,\beta), \tag{8.17}$$

where $P(f|I_{\mathrm{l}})$ is independent of β. Therefore, (8.16) can be written as the following update:

$$\beta := \arg\max_{\beta} \ln P(I_{\mathrm{r}}|I_{\mathrm{l}},f,\beta). \tag{8.18}$$

Here, we present an alternating algorithm to find the optimal binary labeling of the road image and learn the parameters from the stereo pair itself. Given f, we apply the hard conditional EM algorithm to compute the parameters in β using (8.18). Using the learned parameters, we then implement BP to find the assignment of the labeling that minimizes (8.12). The alternating optimization procedure runs until convergence or a fixed number of iterations.

(1) Computing λ_k, λ_m given *f*. As mentioned previously, the points in the road region correspond between the stereo image pair based on the homography while other points do not. Therefore, considering the matching energy (8.9), the conditional probability model $P(I_r|I_l, f, \lambda_k)$ for the hard M step is defined as

$$P\left(\Phi_r^k(p') = x|\Phi_l^k, f, \lambda_k\right) = \begin{cases} \varsigma \exp(-(\Phi_l^k(p) - x)^2/(2\sigma_k^2)) & p' = Hp \text{ and } f(p) = 1 \\ 1/N & \text{Otherwise,} \end{cases} \tag{8.19}$$

where σ_k^2 is the variance for the Gaussian distribution and ς is a normalization factor. N is the number of possible intensity/gradient values for the feature vector. Substituting (8.19) into (8.18) results in an equation with a closed form solution, i.e. $\lambda_k = 1/(2\sigma_k^2)$, which can be used to update λ_k. σ_k^2 can be obtained from the input stereo image pair as follows:

$$\sigma_k^2 = \frac{1}{N_R} \sum_{p \in R} (\Phi_l^k(p) - \Phi_r^k(Hp))^2, \tag{8.20}$$

where N_R is the size of road region R.

As described previously, λ_m is a thresholding factor to discriminate good matches from bad matches. If we take the compatible cost term as a graph, the good matches should be dark (background) and the bad matches should be bright (foreground). Inspired by Otsu's thresholding method [19], we determine λ_m by maximizing the inter-class variance σ_b^2 defined as

$$\sigma_b^2 = \omega_1'(c)\omega_2'(c)(\mu_1(c) - \mu_2(c))^2, \tag{8.21}$$

where the weights ω_i' are the probabilities of the two classes separated by a threshold c, and μ_i are class means. In our hard M step, the inter-class variance is calculated using (8.21) given f, and the result is written as $\sigma_{b,f}^2$. Subsequently, λ_m iterates through all the possible threshold values to calculate the inter-class variance σ_{b,λ_m}^2 and is set to the value that minimizes the absolute difference between $\sigma_{b,f}^2$ and σ_{b,λ_m}^2.

(2) Computing λ_t given f^t,f^{t-1}. For the use of tracking, the conditional probability model is actually $P(I_r|I_l, f^t, f^{t-1}, \beta)$, which can be factorized as

$$P(I_r, |I_l, f^t, f^{t-1}, \beta) = P(f^t|I_l, f^{t-1}, \lambda_t)P(I_r, |I_l, f^t, \beta_{\overline{\lambda_t}}), \tag{8.22}$$

where $\beta_{\overline{\lambda_t}}$ indicates the parameters, apart from λ_t. $P(I_r, |I_l, f^t, \beta_{\overline{\lambda_t}})$ is irrelevant to the tracking energy. Considering the tracking energy (8.10), we define $\Delta f_t(p) = |f^t(p) - f^{t-1}(T(p))|$ and then the conditional probability for the tracking step is defined as

$$P(\Delta f_t(p) = x|I_l, f^{t-1}, \lambda_t) = \begin{cases} \eta \exp(-\mu x) & f^t(p) = 1 \\ 1/2 & \text{Otherwise,} \end{cases} \tag{8.23}$$

where μ is the rate parameter for the exponential distribution and η is a normalization factor. Substituting (8.23) into (8.18) results in an equation with a closed form solution to update λ_t, which can be obtained from the adjacent left image pair as follows:

$$\lambda_t = \frac{1}{\left(\frac{1}{N_R} \sum_{p \in R} (I_l(p)|f^t(p) - f^{t-1}(T(p))|) \right)}. \tag{8.24}$$

(3) Computing λ_t given *f*. λ_s can be calculated in the same way as λ_t if we define $\Delta f_s(p, q) = |f(p) - f(q)|$. However, in our formulation of road detection, we force the detected drivable road to be a connected region in front of the vehicle. Therefore, in the experiments we just set λ_s to a constant value and let the other weight parameters in the energy function be changeable.

(4) Computing *f* given β. As mentioned previously, we implement an efficient belief propagation approach for inference to find the assignment of the labeling that minimizes (8.12).

8.3.3 Dealing with Textureless Regions

Stereo vision algorithms typically compute erroneous results in regions where there is little or no texture in the scene due to the inherent ambiguities. In the proposed approach, which is based on a formulation of stereo with homography, textureless regions that connect to the road region, such as black cars, white walls, and the blue sky, may frequently cause errors in detection, since the feature vectors do not change as part of such textureless regions despite their pixel positions changing when transforming the image from one view to the other.

Figure 8.5 illustrates such a problem with a real example. T_1, T_2 are the textureless regions in the reference image and T_1', T_2' are the corresponding textureless regions in the transformed image. The proposed approach will classify all of the matched pixels into the drivable road region. Apparently, a portion of pixels in T_1' matches well, thereby resulting in such an error. Therefore, the road classification for textureless regions should be

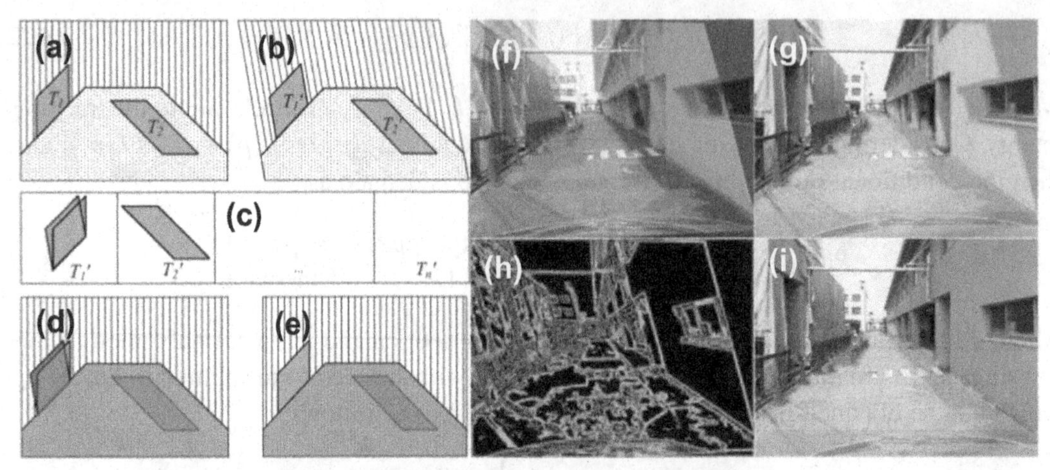

Figure 8.5: Dealing with Textureless Regions.
(a, b) The reference image and the transformed image respectively. Dots indicate the road region, strips indicate the non-road region, and gray indicates the textureless regions. (c) A list of textureless regions in the transformed image. Blue indicates the matched regions. (d) The detection result before dealing with textureless regions. (e) The detection result after dealing with textureless regions. (f) A comparison between the transformed image and the reference image. (g) The road detection result before dealing with textureless regions. (h) The textureless regions extracted from the transformed image. (i) The road detection result after dealing with textureless regions, where the textureless regions outside the road plane have been removed and the preserved road region was used for the unsupervised learning of the parameters. (For Interpretation of the references to colour in this figure legend, the reader is referred to the online version of this book.)

determined over the entire regions rather than single pixels inside them. As shown in Figure 8.5, a textureless region belongs to the drivable road if and only if the entire pixels inside its contour coincide well, such as T_2, T_2'. For the other textureless regions, such as T_1, T_1', the disagreements of their contours indicate that they do not comply with the homography induced by the road plane so that they are outside the road plane. In this case, there must be some pixels in T_1' that lie outside the detected road region, as indicated in Figure 8.5.

In the proposed approach, we perform the following procedures to deal with the problem caused by the textureless regions:

1. Extract the textureless regions in the transformed image. The textureless regions can be defined as regions where the squared horizontal intensity gradient averaged over a square window of a given size is below a given threshold. In the proposed approach, we take the gradient of the image in four directions (horizontal, vertical, and the two diagonals) to avoid connecting different textureless regions into one component. Subsequently, the textureless regions are obtained by thresholding and dilating the image gradient result.
2. Label and record each connected textureless region.

3. Compute the percentage of the outlier pixels that lie outside the detected road region in every textureless region.
4. Remove the entire textureless region from the detected road region if its outlier percentage is greater than a given threshold.

We judge the textureless regions by examining the outlier percentage of the regions rather than matching the regions between the reference and transformed images due to its simplicity and robustness. In this case, we just need to extract and record the textureless regions in one image, and this can also avoid the unpaired regions resulting from problems of occlusions and camera gain/bias difference between stereo pairs. As shown in Figure 8.5, parts of the textureless walls and pillars were classified erroneously into the detected road region, since the displacement of the pixels did not cause an apparent difference in the feature vector. After extracting the textureless regions and examining the outlier percentage, the textureless regions that are outside the road plane were removed.

8.3.4 Results

The objective of the proposed work is to develop an accurate and robust drivable road detection approach without using any prior knowledge of the road. The proposed formulation of MRF-based road detection with unsupervised learning not only improves the accuracy of road detection but also makes the proposed approach adaptive to changing environments to achieve robustness. In the experiments, we evaluate the optimality and adaptability of the proposed approach. Here, optimality means the learned parameters and the optimal detection results should correspond to each other. Adaptability means the learned parameters should be able to adapt to the current input image pair and correspond to the optimal performance in the current scene, rather than the optimal parameters that were learned with a predefined training set.

In the proposed hard conditional EM-based unsupervised learning, the optimal parameters are learned from the current input stereo pair itself. First, we fix all the other parameters with the learned values but vary each of the λ_1, λ_2, λ_3, λ_m values singly in a certain range and implement BP to find the binary labeling by minimizing the proposed energy function. We use three input image pairs as examples, and plot the error rates as a function of each evaluated parameter in Figure 8.6(a) and (b). The vertical line indicates our learned values, whose corresponding error rates are quite close to the minimum error rates of the graphs. From Figure 8.6 we also can see that the learned parameters adapt with respect to each of the road images and correspond to the optimal performances in the corresponding scene. Therefore, the use of hard conditional EM learning in the proposed approach makes the parameters well adapted to the optimal values automatically so as to achieve optimality and adaptability.

Next, we also evaluate the significance of the functions of online extrinsic parameter re-estimation and tracking for road detection in challenging environments, by comparing the road

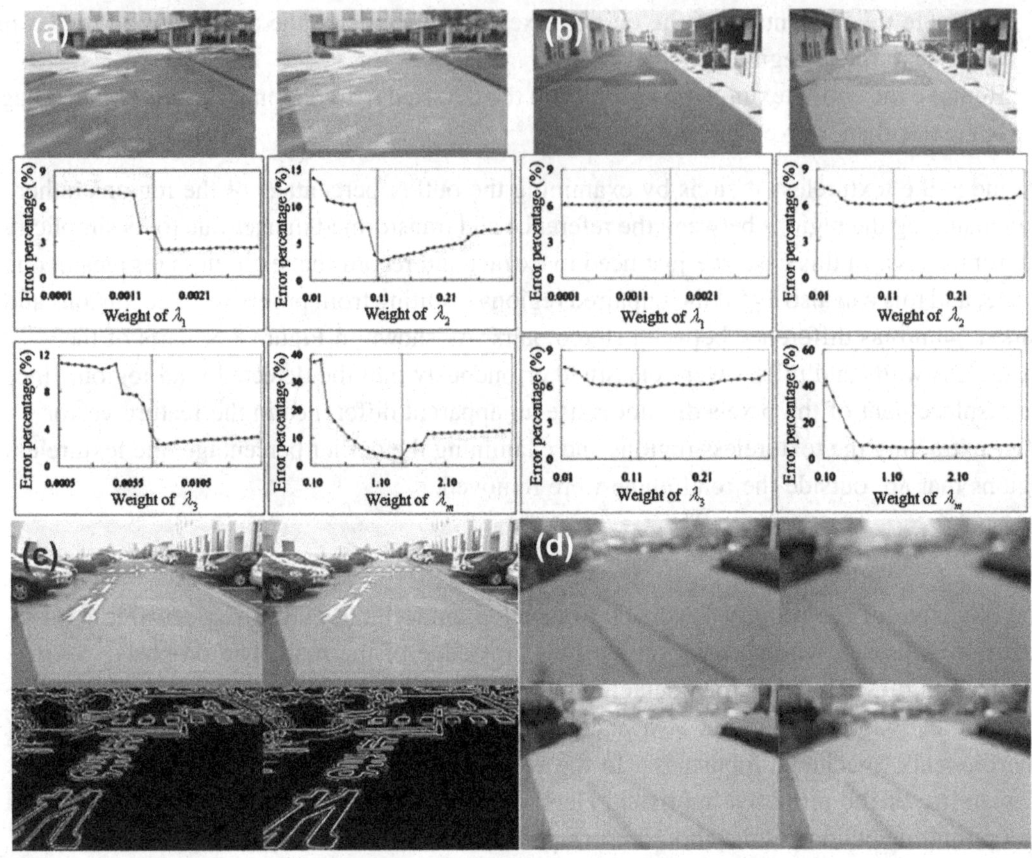

Figure 8.6

(a, b) Optimality and adaptability evaluations of the proposed approach. Detected road region is indicated by blue. The vertical line indicates the learned values. (c) Road detection results with the function of online extrinsic parameter re-estimation deactivated and activated respectively. (d) Road detection results with the tracking function deactivated and activated respectively. (For Interpretation of the references to colour in this figure legend, the reader is referred to the online version of this book.)

detection performances with these functions being activated and deactivated. Figure 8.6(c) shows an example of the comparison of the results without online extrinsic parameter re-estimation (the left column) and with it (the right column). Without the re-estimation, the road points around the pedestrian crossing are not contained in the detected road region because the two images do not coincide well. When dynamic estimation was applied, the homography is optimized. Subsequently, the match of the road points becomes very good, and the road points around the road markings are contained in the detected road region.

For evaluating the tracking function, we intentionally cause the stereo camera to be out of focus occasionally. In Figure 8.6(d), the upper row shows the input out-of-focus image pair. The difference between the reference image and the transformed image becomes smaller, since everything is blurry in the image. In this case, the matching energy is small for both the road and non-road regions. Without the tracking function, the detected road region will be larger than the actual one. In the case of activating the tracking function, the tracking energy will play a more important role when computing the data term in the MRF. Therefore, some of the non-road region will not be contained in the detected road even if their matching energy is small. As shown in the image, the false positives are eliminated when the tracking function is active.

Furthermore, we give more example results of the proposed system in various road scenarios with different challenges, such as heterogeneous surfaces, sloped/rough terrains, lighting/weather variations, crowded traffic, small on-road obstacles, etc., as shown in Figure 8.7. From the figure we can see that, the proposed system can overcome these challenges, and the detected drivable road regions are in good agreement with the real situations and can give a safe path for the host vehicle to facilitate autonomous driving behaviors such as road following, obstacle avoidance, and off-road navigation.

Figure 8.7: Example Results in Challenging Traffic Scenes.
(a–d) Results of the challenging scenarios shown in Figure 8.1. The person riding a bicycle is detected as non-road. (e,k) Night detections with poor image quality. (f,l) Rough road surfaces and textureless buildings. (g) An uphill slope with shadows. (h) Raining day with influence of wipers. (i) Rural roads with withered leaves. (j) An uphill slope with strong backlight and shadows. (m) An uphill slope with heterogeneous surfaces due to different materials and patterns. (n) Multicolored roads with multiple lanes and other vehicles. (o) Complex road markings and a small obstacle surrounded by the drivable roads. (p,q) Crowded scenarios with road markings and other vehicles.

For quantitative evaluation, we tested our system on a video clip with a total of 3031 frames, which records roads over 2.5 km in challenging and changing environments, including downhill/uphill slopes, worn pavements, multicolored pavements, bending roads with consecutive inclined surfaces, strong backlight, heavy shadows, etc. The quantitative evaluations for drivable road detection are based on the following three ratios: the false positive ratio FPR $= N_{FP}/N_P \times 100\%$, the false negative ratio FNR $= N_{FN}/N_N \times 100\%$, and the error rate $= (N_{FP} + N_{FN})/(N_P + N_N) \times 100\%$. Here N_{FP} is the number of non-road pixels erroneously labeled as road (false positives) in the reference image, N_P is the number of road pixels in the ground truth labeling, N_{FN} is the number of road pixels erroneously labeled as non-road (false negatives) in the reference image, and N_N is the number of non-road pixels in the ground truth labeling. The ground truth was obtained by a human operator. We plot the quantitative evaluation results every 10 frames in Figure 8.8 and give the statistical results in Table 8.1. For comparison, we also give the results using two different matching methods for road detection. One is based on the Sum of Absolute Difference (SAD) values [20], and the other is based on the minimization of the energy function (8.12) in MRF with fixed manually tuned parameters. As we can see, both the accuracy and robustness of the proposed approach are much better than those of the reference methods.

For an inclined road, the proposed system can detect a portion of it as drivable in practice, since our system also measures the local consistency, and the inclined road is connected with the flat road and they share similar textures. An example of the detection results of inclined road is given in Figure 8.9. In (a), the inclined road is a low-textured region when it is far from the host vehicle. It is detected as drivable; however, it is not removed as a textureless region, since it is connected with the road region that has the same texture. In (b), the inclined road is getting closer, and a portion of it is labeled as drivable since the proposed approach ensures local consistency in the vicinity of the flat road and there are no boundaries elevated out of the ground in between. However, curbs will not be labeled as drivable since they generate boundaries that are elevated out of the ground. In (c), the inclined road becomes the flat ground to the host vehicle, thereby being detected as drivable. In future, a non-planar road detection system will be developed within the proposed framework, in which non-planar roads will be modeled by a succession of planar road sections, and the multiple homographies induced by the planar sections will be estimated and used for road labeling section by section.

8.4 Painted Lane Boundary Detection and Tracking

Roadways are typically marked with painted boundaries to assist safe and efficient transportation, especially in urban streets. In such traffic scenes, vehicles must drive not only safely but also legally. The problem of finding lane boundaries can be divided into three sub-problems: lane feature detection, lane boundary estimation, and lane tracking. In the proposed

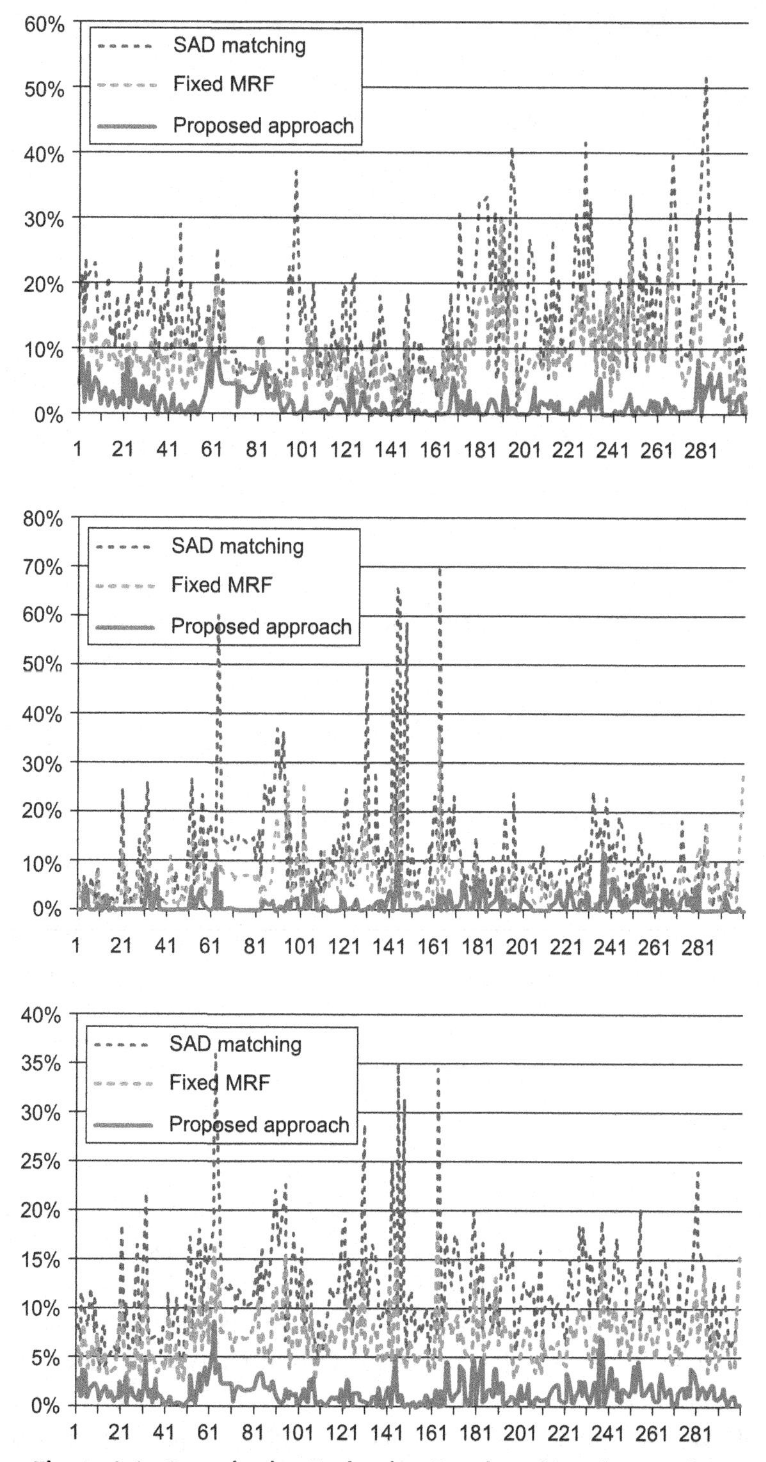

Figure 8.8: Quantitative Evaluation Results of Road Detection.
Top: false positive rate; middle: false negative rate; bottom: error rates.

Table 8.1: Quantitative Evaluations and Comparisons of Road Detection

		SAD Matching	Fixed MRF	Proposed Approach
Average value	FPR (%)	14.2906	9.0306	1.6717
	FNR (%)	10.6785	5.6831	1.1688
	Error rate (%)	11.5767	6.8079	1.3567
Standard deviation	FPR (%)	8.6448	4.8014	1.9916
	FNR (%)	10.0083	5.0849	1.7408
	Error rate (%)	5.1032	2.7223	1.1891

Figure 8.9: Example Detection Result for an Inclined Road.
The inclined road is connected with the flat road and has similar texture.

system, we apply an ANN classifier to detect painted lane markings, and the left and right lane boundaries are estimated in a tracking process using a particle filter.

Vision-based lane detection problems have been studied extensively. McCall and Trivedi [2] provide an excellent survey. Recently, there has been much work on modeling uncertainty problems in lane detection and tracking. Sehestedt et al. [21] described the use of a particle filter for boundary tracking. Kim [22] presented a system that uses an ANN classifier, RANdom SAmple Consensus (RANSAC) spline fitting, particle filtering, and a combination of dynamic Bayesian network and maximum-likelihood estimation. Huang and Teller [23] proposed a probabilistic model of lane curvature called the lateral uncertainty model. These systems are concerned with the sub-problem of lane boundary estimation given a set of noisy observed lane features.

The proposed approach differs from previous related work in the following aspects. Most previous algorithms detect lane features in the "hard" way by classifying the points in the image into two groups of "road markings" and "non-road markings" on the basis of whether they have some property of the lane boundary, such as color, shape, gradient, etc. In this case, the false positives will be used and treated equally for lane boundary estimation just like the true positives, which will increase the complexity and difficulty of boundary estimation and decrease the accuracy. In the same manner, the abandoned false negatives may result in misdetection of the lane boundary. This is a serious problem especially in challenging conditions since road markings are usually poor and change with noises in such scenarios.

Here, we utilize geometric cues of the road scene, which are little affected by the challenging conditions, together with intensity cues to detect and represent the lane features in a "soft" way by constructing a weighted graph with integrated cues, reflecting the confidence of each pixel as a true lane feature. Therefore, the uncertainties are modeled at the very beginning of the detection procedure, and then brought to the grouping level of the lane boundary estimation.

8.4.1 ANN Painted Lane Marking Detection

We apply machine learning to the left and right IPM images in parallel for lane marking detection, since the painted lane boundaries are constructed not randomly but with limited patterns. A comparative study of both classification performance and computation time on various painted markings classification methods has been presented [22]. Based on that, we chose to use an ANN classifier with two layers and seven hidden nodes since it is faster, yet the performances are still good. For training, we have gathered image patches of 100 painted lane markings and 100 non-painted markings. For detection, the ANN classifier is applied on a small image patch of 9×3 windows around each pixel of the IPM images. Pixels whose ANN scores exceed a loose threshold will be preserved as the painted lane markings. Non-maximum suppression is then applied to the preserved pixels to ensure well-localized positions.

8.4.2 NCC Scoring Between IPM Images

We compute the NCC value between the left and right processed IPM images to measure the similarity of corresponding pixel locations. The aim of this step is to utilize the underlying geometric cue, since the entire lane markings lie on the road plane and all the road points are mapped into the same global coordinates. For each detected lane marking pixel in the left IPM image, the NCC in (8.25) is computed with the pixel at the same location in the right IMP image:

$$\mathrm{NCC} = \frac{\sum_{(i,j)\in W}\left[f_1(i,j) - \overline{f_1}\right]\left[f_2(i,j) - \overline{f_2}\right]}{\sqrt{\sum_{(i,j)\in W}\left[f_1(i,j) - \overline{f_1}\right]^2 \sum_{(i,j)\in W}\left[f_2(i,j) - \overline{f_2}\right]^2}}, \tag{8.25}$$

where W is the computational window, and $f_1(i, j)$ and $f_2(i, j)$ are the image blocks in the left and right IPM images respectively. $\overline{f_1}$ and $\overline{f_2}$ are the average values of the blocks. Pixels whose NCC values exceed a loose threshold will be further preserved as the lane features.

8.4.3 Cue Integration

Sensing the real world is an inherently uncertain process. Many previous approaches model uncertainty for lane estimation based on noisy observations of binary classified lane features, in which false positives are treated equivalently to the true positives. In the proposed system, we intuitively model the uncertainty in lane feature detection since the uncertainty occurs at the very beginning. In particular, we construct a weighted graph by integrating the intensity and geometry cues, reflecting the belief of each pixel as a lane feature, which assures that each pixel has a weight so as to play a different role when estimating lane boundaries using a particle filter. In this way, the uncertainty of lane feature detection and the uncertainty of lane boundary estimation can be integrated for the probabilistic reasoning to deal with challenges due to the varying appearances of lane boundaries.

The weighted graph $G = (V, E, B)$ is shown in Figure 8.10, where each node $v^j \in V$ represents a pixel, associated with the ANN score S_j as the node weight. Each edge $e^j \in E$ connects pixels at the same location between the left and right IPM images, associated with the NCC value NCC^j. The integrated weight $b^j \in B$ of a node in the reference layer, which is constructed from the left IPM image, is computed as follows:

$$b^j = \alpha_1 \exp(k(1 - NCC^j)) + \alpha_2 S^j / S^{max}, \tag{8.26}$$

where k is a convergence parameter, α_1, α_2 are retuning parameters for the cues integration, and S^{max} is the maximum value of S^j.

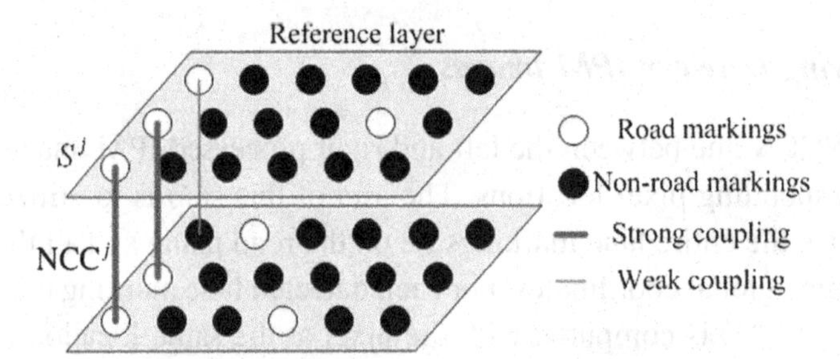

Figure 8.10: Weighted Graph with Integrated Cues.

8.4.4 Boundary Estimation Using a Particle Filter

The following task is to estimate lane boundaries by utilizing the multiple evidence on the weighted graph. In order to deal with the challenge of complex road geometry, we employ Catmull-Rom splines [24] to represent the left and right road boundaries separately, allowing more flexibility and less restriction so as to give more weight to the observed evidence. This representation has been shown to be able to deal with challenging scenarios such as lane curvature, lane changes, and emerging, ending, merging, and splitting lanes [22].

A particle filter is a very versatile and robust stochastic filter. In the proposed system, we sample the set of control points to generate the left and right lane boundary hypotheses based on T (described in the previous section). To choose the best pair of left and right hypotheses as the detected lane boundaries, we define the likelihood function $S_{\mathrm{l,r}}$ as follows:

$$S_{\mathrm{l,r}} = \left(S_{\mathrm{LF}}^{\mathrm{l}} - \gamma S_{\mathrm{PC}}^{\mathrm{l}}\right)\left(S_{\mathrm{LF}}^{\mathrm{r}} - \gamma S_{\mathrm{PC}}^{\mathrm{r}}\right)S_{\mathrm{PW}}^{\mathrm{l,r}}, \tag{8.27}$$

where $S_{\mathrm{LF}}^{\mathrm{l}}$, $S_{\mathrm{PC}}^{\mathrm{l}}$ and $S_{\mathrm{LF}}^{\mathrm{r}}$, $S_{\mathrm{PC}}^{\mathrm{r}}$ are the lane-feature support score and curve penalty score of the left and right hypotheses respectively, and γ is a weight parameter. $S_{\mathrm{PW}}^{\mathrm{l,r}}$ is the correlation penalty score of the hypothesis pair. The first two scores are simply for the measurement of a single lane boundary and the last one is to handle the lane width.

(1) Lane-feature support. The lane-feature support is composed of the belief and distance supports, which can be calculated as follows:

$$\mathrm{S_{LF}} = \sum_{p} \exp(\mu \cdot b^{p})\mathrm{Distance\ Score}(p), \tag{8.28}$$

where p is a pixel on the lane boundary hypothesis and μ is a weight parameter. The distance support is defined as

$$\mathrm{Distance\ Score} = \begin{cases} 1 - \mathrm{Distance}(p)/K & \mathrm{Distance}(p) < K \\ 0 & \mathrm{otherwise,} \end{cases} \tag{8.29}$$

where K is a threshold, and Distance(p) is defined as

$$\mathrm{Distance}(p) = ||p - \mathrm{Nearest\ Lane\ Feature}||, \tag{8.30}$$

where $||\quad||$ indicates the calculation of Euclidean distance.

(2) Curve penalty. In addition to the lane-feature support, a curve-penalty score is used to prevent overfitting. A penalty is imposed when the direction of the curve is changed while the lane-feature support score is very small. The penalty is calculated as [22]:

$$\mathrm{S_{PC}} = \sum_{p} \left| (y(p_2) - y(p_1))\left(\frac{\mathrm{d}x}{\mathrm{d}y}(p_2) - \frac{\mathrm{d}x}{\mathrm{d}y}(p_1)\right) \right| \tag{8.31}$$

for all point pairs p_1 and p_2, where the lane-feature support score is smaller than an empirical threshold.

(3) Correlation penalty. We assume that the lane width can be increasing or decreasing linearly at a small ratio. This assumption is reasonable, since dramatic changes are very rare in the real world and the IPM image, in agreement with the local navigation map, only contains the lane information in the near range. Given a pair of left and right lane boundaries, the lane width is sampled at various distances and fit into a linear equation to obtain the average lane width and the change of the lane width. The lane boundaries penalty $\mathrm{S_{PW}}$ is then calculated as the maximum residual distance.

8.4.5 Results

Figure 8.11 shows a number of example detection results in several typical but challenging environments, such as a curved road with turnouts (a), a dark road with worn painted lane boundaries (b), raindrops (c), and strong backlights (d). In each image block (a)–(d), the first column shows the original left images, the second column shows the detected lane boundaries superimposed on the original images together with the road detection results, and the third column shows the results for the IPM images, which can be used for the update of the local navigation map. These results illustrate our method's ability to overcome noise and localize road markings of different road shapes, raindrops, wipers, strong backlight, dark roads, and poor lane markings. The proposed framework of the weighted graph and integrated cues enables the proposed approach to maintain accurate and robust detection

Figure 8.11: Example Results in Challenging Traffic Scenes.

even in these challenging lighting and weather conditions. The effects of noise are largely eliminated, due to the use of geometric cues and the "soft" detection method. As we can see from the figures, the detected drivable road regions are in good agreement with real situations and the detected painted lane boundaries can provide positional as well as curvature information of the lanes.

8.5 Road Correlation

8.5.1 Road Classification

In the proposed system, we develop two parallel modules for the detection and tracking of drivable road region and painted lane boundaries. We did not use the intermediate results of one module for the other but studied these two modules with the intensity and geometric cues independently in order to avoid arbitrary results. However, as mentioned previously, road and painted lane boundaries are not constructed randomly but under certain rules and regulations to assist in safe and legal transportation. Therefore, as shown in Figure 8.12, we correlate the results of the two modules and then label each pixel of the reference image as belonging to one of the following three categories, which serve as an input to the navigation system:

1. **Super-drivable.** Pixels in the drivable road region between the left and right lane boundaries, indicating safe and legal areas, which can be used for behavior such as road

Figure 8.12: Example Classification Results by the Correlation of the Detection Drivable Road and Painted Lane Markings.
(a) Situations with two painted lane boundaries. (b) Situations with a single lane boundary. Purple indicates the super-drivable region, blue indicates the drivable region, and the other regions are the non-drivable ones. (For Interpretation of the references to colour in this figure legend, the reader is referred to the online version of this book.)

following. In the case of a single lane boundary, we estimate the super-drivable road region with a default safe space based on the position of the detected lane boundary.
2. **Drivable.** Pixels in the drivable road region outside the lane boundary pair, indicating safe areas, which can be used for behavior such as overtaking, lane changing, turning, etc.
3. **Non-drivable.** Pixels outside the drivable road region, indicating dangerous areas, which can be used for behavior such as obstacle avoidance.

Since the road points are detected repeatedly in multiple frames, we use a histogram of the labels over five frames to accumulate evidence for a particular label of each point in the global map during navigation. Although we do not use the intermediate results of one module for the other, the results of the two modules really help each other to improve the accuracy as well as the robustness by providing a better computational region through correlation. As mentioned previously, the precise geometric cues induced by the road plane are crucial for the proposed system, which are derived from dynamic stereo calibration over the computational region. The more precise planar road region, the computational region, is the best performance that the system can achieve. Usually the detected drivable road region is used as the computational region. However, it may contain some false positives in the roadsides due to textureless regions without clear contours or false negatives in the middle due to complexity of the road surface, particularly in some challenging conditions and situations, which means the geometric cues are not purely or precisely induced by the planar road plane. To overcome these problems, we expand the super-drivable road region by moving the left and right boundaries to the roadsides separately until the overlap between each boundary and the drivable road region is smaller than an empirical threshold. The expanded super-drivable road region is then used as the computational region.

8.5.2 Object of Interest Detection

As mentioned previously, intelligent vehicles have to share the road environment with all road users, such as pedestrians, motorcycles, bicycles, and other vehicles. Therefore, in the proposed system, such particular objects of interest, including vehicles, pedestrians, motorcycles, and bicycles, are identified within the context of the information obtained by road correlation.

Multi-Class Object Detector

Object identification is challenging in that objects present dramatic appearance changes according to camera viewpoints and environmental conditions, and also have intra-class variability. There are two common solutions to tackle this challenge. One is to employ robust features, since the overall performance of the system depends on the discriminative power of features used in the detection algorithm. For example, the HOG feature is considered to be one of the strongest features, which captures the shape information of an object, and is robust for local shape variations. The other is to establish a part-based model for an object of interest. Rather than trying to capture a global pattern of an object with one template, part-based models focus on parts of an object and, in consequence, provide more flexible and robust representations. Recently, Felzenszwalb et al. [25] demonstrated a deformable part-based model that outperformed the single template model by using a latent Support Vector Machine (SVM) formulation in combination with a variation of HOG features. In this chapter, we adopt this work for object of interest detection.

The structure of the deformable part-based model is shown in Figure 8.13, where we use vehicle detection as an example. The large rectangle represents the root filter, and the eight smaller rectangles express part filters. Assume we have K object models corresponding to the objects of interest, and let M^j be the jth model with a root filter v_0^j and eight additional part filters $v_1^j, \ldots, v_8^j$. Each filter keeps the texture information of the object and object parts images as HOG features, as shown in Figure 8.14. To calculate the HOG features, the input image is divided into many 8×8 cells. Each pixel in the cell influences the orientation of its gradient with a strength that depends on the gradient magnitude for the cells. The features for the part filters are computed at twice the resolution of the root filter to keep the local fine features.

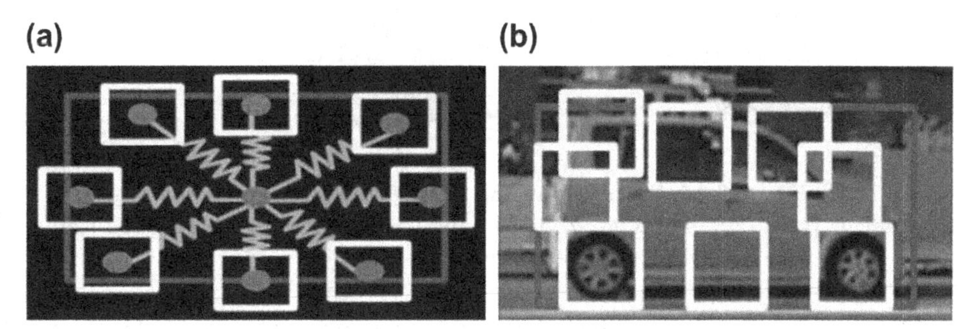

Figure 8.13: A Deformable Part-Based Model for Vehicle Detection.
(a) The model structure, where the springs represent deformation cost functions. (b) A detection example.

The HOG features and deformation cost functions can be learned by using latent SVM from the training data set. The detained learning algorithm can be found in Ref. [25]. In order to fix

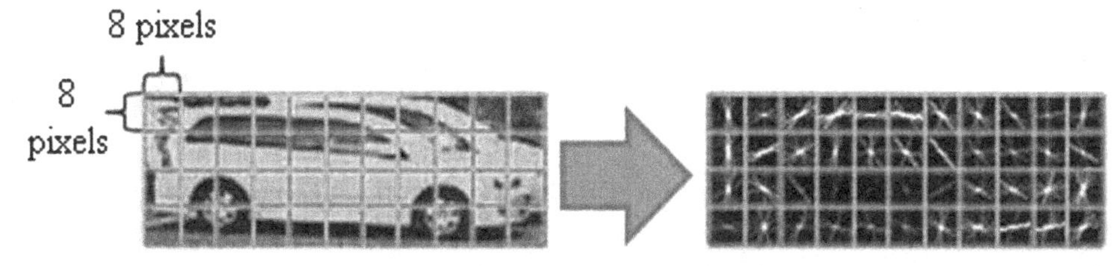

Figure 8.14: HOG Feature Calculation.

each filter size, the HOG pyramid H consists of a multi-scale resized image, which has λ layers, and the image is resized by a ratio of $2^{1/\lambda}$, as shown in Figure 8.15. Normally, the HOG features have to be calculated in each layer. However, in the proposed system, the size information of the object hypotheses can limit the HOG pyramid levels.

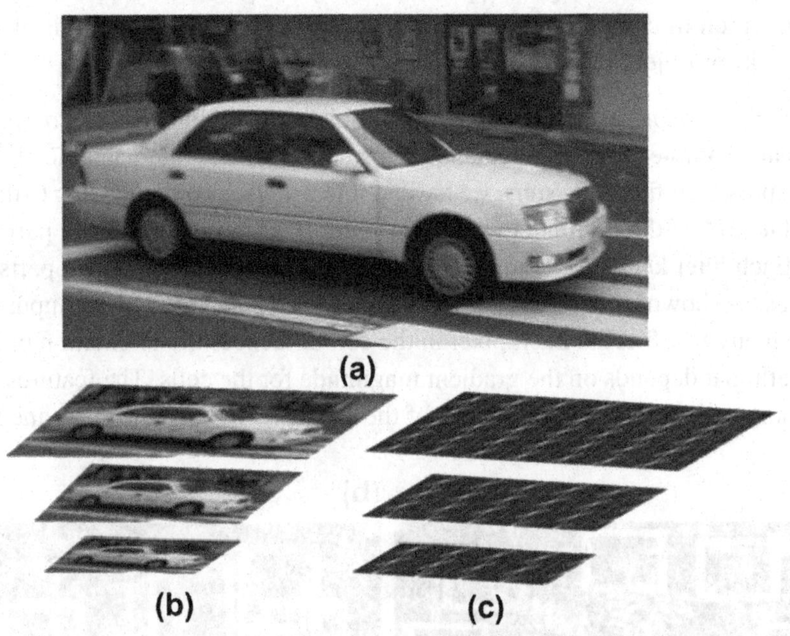

Figure 8.15: Procedure of the HOG Pyramid Calculation.
The input image (a) is resized and an image pyramid (b) is constructed. The HOG features are calculated in each layer, as shown in (c).

Let Ω be a space of location for each part within an image, and $p(u_i, v_i, l_i) \in \Omega$ specify a position (u_i, v_i) and scale l_i in the pyramid. Let $m_i^j(p)$ be the score for placing v_i^j in location p. For a non-root part, let $a_i^j(p)$ specify the ideal location for v_i^j as a function of the root location. Let Δ be a space of displacements, and let $\oplus : \Omega \times \Delta \rightarrow \Omega$ be a binary operation taking a location and a displacement to another location. Let $d_i^j(\delta)$ specify a deformation cost for a displacement of v_i^j from its ideal location relative to the root. We can define an overall score corresponding to model M^j for a root location based on the maximum score of a configuration rooted in location p:

$$\text{score}^j(p) = m_0^j(p) + \sum_{i=1}^{8} \max_{\delta_i \in \Delta} \left[m_i^j\left(a_i^j(p) \oplus \delta_i \right) - d_i^j(\delta_i) \right]. \tag{8.32}$$

Here we calculate the maximum scores for parts filters, over displacements of the part from its ideal location, of the part score minus the deformation cost associated with the displacement. Example detection results of such multi-class object identification are shown in Figure 8.16.

Figure 8.16: Example Results of Multi-Class Object Detection.
Vehicles (a), pedestrians (b), motorcycles (c), and bicycles (d) are indicated by blue, azure, green, and dark green rectangles respectively. (e) and (f) show the detections of multiple objects in one image. (For Interpretation of the references to colour in this figure legend, the reader is referred to the online version of this book.)

Object of Interest Detection with Road Correlation

A common object detection approach is the sliding window method, in which a classifier is applied at all positions and scales of an image. However, testing all points in the search space with a non-trivial classifier can be very slow, especially for multi-class object identification. In the proposed approach, we perform the object of interest detection with the context information obtained by the results correlation in the following ways:

1. **Region-of-interest reduction.** A common region of interest for object detection/identification is usually set based on a predefined template or prior knowledge of the targets. In the proposed system, we reduce the ROI greatly based on the correlation with the road detection module. For the example image shown in Figure 8.17, the ROI was reduced by 70%.
2. **Hypothesis re-weight.** The drivable road detected by the proposed system also provides important contextual information that can be used to improve object detectors. For example, road users, including the four objects of interest, are more likely to be connected with or close to the drivable road. Therefore, we re-weight the scores as follows by adding a road geometric contextual term to (8.32):

$$\text{score}^j(p) = m_0^j(p) + \sum_{i=1}^{8} \max_{\delta_i \in \Delta} \left[m_i^j\left(a_i^j(p) \oplus \delta_i\right) - d_i^j(\delta_i) - \min\left(\tau_i^j, \rho_i^j D_i^j\right)\right], \tag{8.33}$$

Figure 8.17
(a) A common ROI set by a predefined rectangle template. (b) The ROI with the correlation results. (c) The detection result. Green indicates the ROI, blue indicates the drivable road. (For Interpretation of the references to colour in this figure legend, the reader is referred to the online version of this book.)

where D_i^j measures the distances of the filters to the drivable road boundary, ρ_i^j is the weight parameter, and τ_i^j is a truncation parameter. Figure 8.18 shows such an example, in which our hypothesis re-weight successfully reduced the false positives of the object of interest detection.

Figure 8.18
(a) Example result with a false positive of pedestrian detection. (b) The false positive was removed due to the hypothesis re-weighting.

8.6 Conclusions

In this chapter, we presented a vision-based road environment perception system designed for challenging scenarios, which is notable for the following contributions:

1. The formulation of road detection as an MAP problem in MRF and a well-defined energy function that incorporates geometry information, image evidence, and temporal support of the road.
2. The unsupervised learning of the parameters from the stereo pair itself using a hard conditional EM algorithm.

3. A weighted graph constructed with integrated intensity and geometric cues based on a "soft" detection of the lane features, reflecting the belief of each pixel as a lane feature. This framework can be expected to deal with more complex problems since it has the ability to differentiate nodes, thereby reducing the complexity of the problem.
4. The implementation of multi-class object identification by using deformable part-based models with HOG features based on the road correlation.

All of these contributions improve the accuracy as well as the robustness of road environment perception in challenging scenarios without any prior knowledge. Future works will focus on the modeling/understanding of traffic scenes based on the outputs of the proposed system.

References

[1] M. Bertozzi, A. Broggi, GOLD: A parallel real-time stereo vision system for generic obstacle and lane detection, IEEE Transactions on Image Processing 7 (1) (1998) 62–81.
[2] J. McCall, M. Trivedi, Video-based lane estimation and tracking for driver assistance: Survey, system, and evaluation, Transaction on Intelligent Transportation Systems 7 (1) (2006) 20–37.
[3] M. Buehler, K. Lagnemma, S. Singh (Eds.), The DARPA Urban Challenge, Springer Tracts in Advanced Robotics, vol. 56, Springer, 2010.
[4] S. Thrun, et al., Stanley: The robot that won the DARPA grand challenge, Journal of Field Robotics 23 (9) (2006) 661–692.
[5] R. Hartley, A. Zisserman, Multiple View Geometry in Computer Vision, Cambridge University Press, Cambridge, UK, 2000.
[6] P. Gill, W. Murray, Algorithms for the solution of the nonlinear least-squares problem, SIAM Journal on Numerical Analysis (1978) 977–992.
[7] M. Nieto, L. Salgado, Real-time vanishing point estimation in road sequences using adaptive steerable filter banks, in: Proc. Advanced Concepts for Intelligent Vision Systems, LNCS, 2007, pp. 840–848.
[8] M. Sotelo, et al., Virtuous: Vision-based road transportation for unmanned operation on urban-like scenarios, Transaction on Intelligent Transportation Systems 5 (2) (2004) 69–83.
[9] A. Lookingbill, et al., Reverse optical flow for self-supervised adaptive autonomous robot navigation, International Journal of Computer Vision 74 (3) (2007) 287–302.
[10] M. Darms, et al., Map based road boundary estimation, in: Proc. IEEE Intelligent Vehicles Symposium, 2010, pp. 609–614.
[11] A. Gupta, A. Efros, M. Hebert, Blocks world revisited: Image understanding using qualitative geometry and mechanics, in: Proc. of 11th European conference on Computer Vision, Part IV, 2010, pp. 482–496.
[12] D. Gallup, et al., Piecewise planar and nonplanar stereo for urban scene reconstruction, in: Proc. IEEE Conf. ComputerVision and Pattern Recognition, 2010, pp. 1418–1425.
[13] A. Seki, M. Okutomi, Robust obstacle detection in general road environment based on road extraction and pose estimation, in: Proc. IEEE Intelligent Vehicles Symposium, 2006, pp. 437–444.
[14] N. Vandapel, et al., Natural terrain classification using 3D ladar data, in: Proc. Conf. on Robotics and Automation, 2004, pp. 5117–5122.
[15] A. Huertas, et al., Stereo-based tree traversability analysis for autonomous off-road navigation, in: IEEE Workshop of Applications of Computer Vision, 2005, pp. 210–217.
[16] L. Cheng, T. Caelli, Bayesian stereo matching, in: Proc. IEEE Conf. Computer Vision and Pattern Recognition, 2004, p. 192.
[17] L. Zhang, S. Seitz, Parameter estimation for MRF stereo, in: Proc. IEEE Conf. Computer Vision and Pattern Recognition, 2005, pp. 288–295.

[18] P. Felzenszwalb, D. Huttenlocher, Efficient belief propagation for early vision, in: Proc. IEEE Conf. Computer Vision and Pattern Recognition, 2004, pp. 261–268.
[19] N. Otsu, A threshold selection method from gray-level histograms, IEEE Transactions on Systems, Man and Cybernetics (1979) 62–66.
[20] C. Guo, S. Mita, Drivable road region detection using homography estimation and efficient belief propagation with coordinate descent optimization, in: Proc. Conf. on Intelligent Vehicles Symposium, 2009, pp. 317–323.
[21] S. Sehestedt, et al., Robust lane detection in urban environments, in: Proc. IEEE Int. Workshop on Intelligent Robots and Systems, 2007. San Diego, CA.
[22] Z. Kim, Robust lane detection and tracking in challenging scenarios, IEEE Trans. Intelligent Transportation Systems 9 (1) (2008) 16–26.
[23] A. Huang, S. Teller, Lane boundary and curb estimation with lateral uncertainties, in: Proceedings of the 2009 International Conference on Intelligent Robots and Systems, 2009, pp. 1729–1734. St Louis, MO.
[24] J.D. Foley, et al., Computer Graphics: Principles and Practice, second ed., Addison-Wesley, 1990.
[25] P. Felzenszwalb, D. McAllester, D. Ramanan, Object detection with discriminatively trained part-based models, IEEE Pattern Analysis and Machine Intelligence 32 (9) (2010) 1627–1645.

CHAPTER 9

V2I-Based Multi-Objective Driver Assistance System for Intersection Support

Jianqiang Wang, Jiaxi Liu, Xiaohui Qin, Keqiang Li

State Key Laboratory of Automotive Safety and Energy, Tsinghua University, Beijing, China

Chapter Outline

Advances in Intelligent Vehicles. http://dx.doi.org/10.1016/B978-0-12-397199-9.00009-4

9.1 Introduction

With the use of automobiles increasing over the years, traffic accidents have become an increasingly severe problem. Data from the MTPRC (Ministry of Transport of the People's Republic of China) shows that intersection crashes account for 20% of all the accidents happening in China annually [1], and have caused many casualties and property loss due to the high vehicle density at the intersections. Many efforts have been made to improve the situation through infrastructure construction and traffic light optimization, but these attempts still could not solve the problem completely [2–4]. In recent years, the emergence of ADAS (Advanced Driving Assistance Systems) based on VIC (Vehicle Infrastructure Cooperation) offers a new way to address this matter [5].

As an important branch of ADAS, IDAS (Intersection Driver Assistance System) can reduce traffic violations greatly, and can also reduce accidents or even prevent them from occurring at intersections, and has attracted much attention worldwide.

The main idea behind IDAS is to prevent drivers from incurring traffic violations, violating red lights, for instance, through indicators, warnings or auto-braking. Also, IDAS perceives the timing of traffic lights through communication with the infrastructure, thus recommending to drivers an optimal driving speed to go through the intersection without unnecessary stopping on the basis of safety. IDAS can also inform drivers in advance if the green light is going to time out, which will give drivers more time to take the proper action, and thus improve driving comfort [6,7].

Much work has been conducted on IDAS, and can be divided into three categories: (a) V2I (Vehicle to Infrastructure) wireless communication; (b) vehicle positioning; (c) IDAS safety algorithms.

9.1.1 V2I Wireless Communication

Two kinds of wireless communication methods are used in V2I systems: (a) DSRC (Dedicated Short Range Communication) [8]; (b) beacon-based wireless communication technology.

DSRC is developed based on "IEEE 802.11 Wi-Fi". The advantages of DSRC include a high data transfer rate, low time delay, high stability, strong anti-jamming capability, and concentrated signal coverage. Therefore, DSRC is suitable for V2I systems that require stable communication for a specific section only [9,10].

DSRC has become the first choice for Intelligent Transportation Systems (ITS) based on V2I in Europe and the USA. However, the technical standard for DSRC is still under discussion. Therefore, many researchers apply some wireless communication technology that can be easily replaced by DSRC. In INTERSAFE conducted in Europe and VII directed in America, "IEEE 802.11b" was used, and an update to DSRC was planned [11–15].

Japan selected a beacon-based wireless communication method in its ITS. The practical system VICS (Vehicle Information and Communication System) applied radio wave and infrared beacons [16,17].

DSRC and beacons are two main wireless communication approaches used in ITS. However, each has its own advantages and disadvantages. Compared with beacon-based wireless communication, DSRC has a higher transfer rate, larger data processing capability, and broader coverage. However, it has a more stringent requirement on the hardware and is more sensitive to jamming. Beacon-based wireless communication adopts a fixed operating region and emission direction, so many beacons have to be installed around interested sections and thus the cost is a weak point, but its requirement for hardware is lower.

9.1.2 Vehicle Positioning Technology

Since IDAS can only work at intersections, a relative small area, we have to utilize a precise positioning method to provide accurate information, otherwise the performance of the assistance system cannot be guaranteed. The positioning precision should be high enough to distinguish different lanes, e.g. 1 m [11].

E-map-aided GPS positioning is now the most common way for positioning in the automobile industry. However, stand-alone GPS positioning cannot satisfy the precision demand of IDAS. In order to settle the problem, the BMW platform in Europe (INTERSAFE) and the American ICAS (Intersection Collision Avoidance System) adopted DGPS (Differential Global Position System) [11,18]. DGPS can reach a precision of 1 cm [19], which is accurate enough for IDAS. However, a large number of base stations have to be deployed so as to apply DGPS, and an accurate E-map is required.

The test vehicle VW in Europe INTERSAFE combined radar, machine vision, and detailed E-map to achicvc accurate positoning.

Toyota used infrared beacon-based wireless communication to develop its IDAS. The vehicle could receive messages from the infrared beacon only in a specific small area and this determined the relative position between the vehicle and the intersection. After the beacon, the remaining distance between the host vehicle and the intersection was short enough for INS (Inertial Navigation System) to guarantee the precision [7].

9.1.3 IDAS Control Algorithm

The American IDAS adopted a fixed acceleration bound. IDAS monitored the vehicle speed and the remaining distance up to the intersection, and the real-time acceleration needed to stop the vehicle right at the intersection was calculated. Once the acceleration exceeded $0.35g$, a warning was generated [18].

IDAS in the literature [6] can get the phase and timing of the intersection in advance, and TTI (Time to Intersection) was calculated according to the current speed and the remaining distance to the intersection. A comparison between the TTI and the timing at the intersection would then be made in order to tell the driver the phase of the traffic light when the host vehicle arrived at the intersection.

On the basis of statistical data about speed change at the intersection, Ref. [20] proposed a warning algorithm based on speed threshold. To keep a low false alarm rate, the trends of speed change 1 s before warning should be taken into consideration. If the vehicle was slowing down, then the warning would be canceled so as not to distract the driver. This algorithm included driver characteristics, which made it more acceptable to drivers. However, the author did not put it into practice.

The IDAS created in Europe, INTERSAFE, not only gave out a warning on traffic light violations, but also provided recommended driving speeds based on current vehicle status, environmental information, and traffic lights. For instance, if the green light was changing and there was no other vehicle in front, then the system would suggest that the driver accelerate to pass the intersection before the green light changed.

In Ref. [7], Toyota's IDAS provided TSVW (Traffic Signal Violation Warning) and GLDS (Green Light Driving Support).

IDAS should take enough driver characteristics into consideration in designing the IDAS algorithm to enhance the acceptability of the system. Similarly, the information fusion of vehicle status and driver operation can reduce false warnings and missed warnings. A well-defined IDAS algorithm should take full advantage of I2V. Besides providing warnings against potential traffic signal violations, IDAS should also offer assistance information in non-dangerous situations to help the driver pass intersections without stopping and therefore improve traffic efficiency.

9.2 Structure of IDAS

The functions of IDAS include violation warning and driving comfort promotion. The following section describes the basic functions of IDAS and then the architectural structure of IDAS.

9.2.1 Function Definition of IDAS

The primary function of IDAS is to prevent drivers from incurring traffic signal violations. The On-Board Unit (OBU) of IDAS can receive the phase and timing of traffic signals as well as vehicle position, and therefore can provide other helpful information in those situations that are not dangerous and help the driver pass intersections without having to stop. Furthermore, rear-end collisions are a frequent occurence at intersections, so a rear-end collision warning should also be merged into IDAS. Hence, the functions of IDAS can be defined as follows:

1. **Passing Support (PS).** PS provides speed recommendations in non-dangerous situations and helps the driver to pass intersections without stopping or even without decelerating. PS aims at increasing intersection passing rates and reducing unnecessary stops, and therefore improves driving smoothness and comfort at intersections.
2. **Traffic Light Violation Warning (TLVW).** TLVW avoids traffic light violations through informing, warning the driver, or even automatically braking. In another aspect, when a green signal is about to finish, TLVW can also inform the driver in advance and avoid the sudden amber phase surprise for the driver that can lead to hard, uncomfortable braking. Therefore, TLVW improves driving safety and comfort at intersections.
3. **Rear-End Collision Warning (REW).** REW warns the driver in cases of imminent crash danger to provide him/her with a sufficient time margin to react. The range sensor (Radar or Lidar) of IDAS can detect/assess the various target vehicles ahead of the host vehicle and measure the kinematic attributes of each target. The speed sensor (e.g. ABS sensor) can measure the speed of the host vehicle. Alerts will be given to the driver in anticipation of an existing potential rear-end collision based on a warning algorithm.

9.2.2 Architecture of IDAS

Based on the function definition, the architecture of IDAS is designed as shown in Figure 9.1. The hardware of IDAS includes a Road-Side Unit (RSU) and OBU. RSU is a traffic light equipped with a wireless communication device, which broadcasts the signal phase and timing information for the intersection area. Dedicated Short-Range Communication (DSRC) is investigated worldwide as a wireless communication technology used in ITS, and will be the best solution for this application.

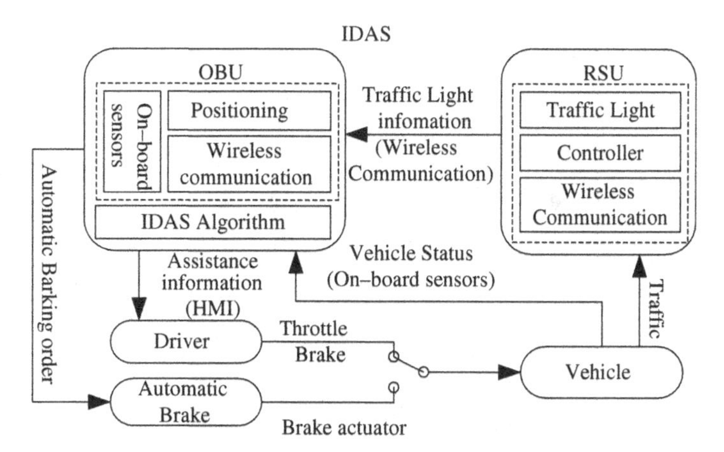

Figure 9.1: Architecture of IDAS.

In the OBU, the wireless receiver receives signals coming from the traffic light; positioning systems provide vehicle positioning at intersections; on-board sensors collect vehicle status information such as velocity, acceleration, and so on. According to the above information, the IDAS algorithm assesses the vehicle's situation, and determines the most suitable driving operation, whether to pass for driving smoothness and traffic efficiency or to yield to the signal and stop for safety.

In non-emergent situations, the OBU presents TLVW, PS or REW information to the driver through the Human–Machine Interface (HMI). When TLVW or REW is on, HMI gives multi-level visual and auditory warnings. When PS is on, HMI presents speed recommendations through a visual image and auditory information signal. In emergent situations when the driver fails to respond to the warning and the vehicle is going to violate the traffic signal, the OBU controls the vehicle to stop by automatic braking.

According to the system architecture, the layer configuration of OBU is shown in Figure 9.2. Except for the OBD described above, some infrastructure needs to be installed on the road side, such as a digital transceiver, traffic light equipped with wireless communication device, and a Radio-Frequency Identification (RFID) tag.

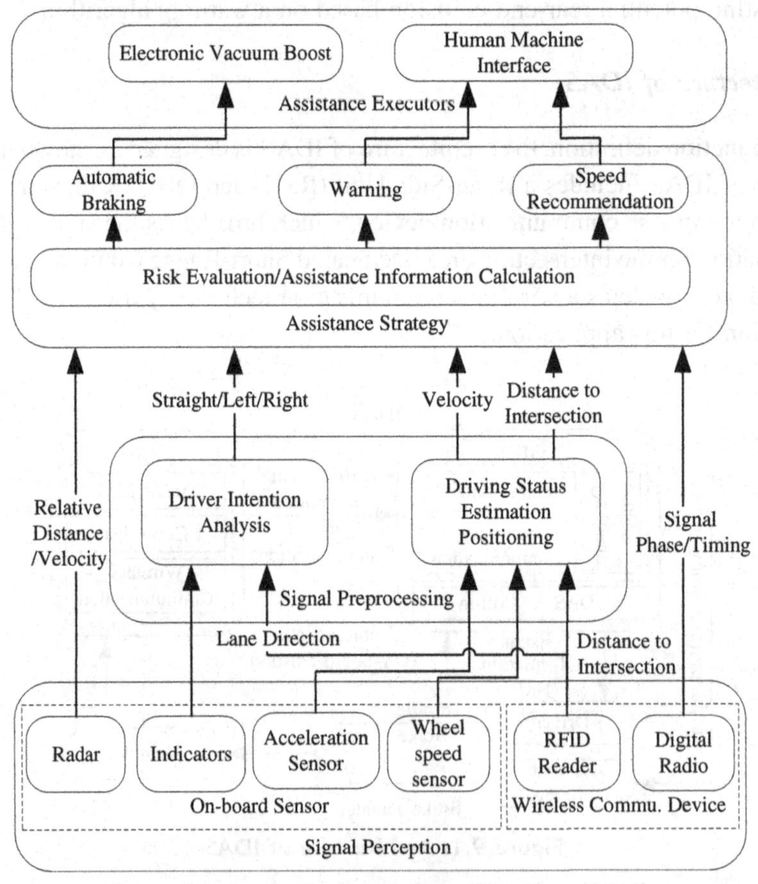

Figure 9.2: Layer Configuration of On-Board Unit.

Also, the following four key technologies are needed to develop a complete IDAS:

1. Accurate positioning with respect to intersection
2. Vehicle status estimation
3. IDAS algorithm fully accepted by drivers
4. Vehicle control technology.

9.3 IDAS Algorithm

Based on the architecture described above, the following section describes signal perception level algorithms of IDAS, including vehicle status estimation based on a Kalman filter, vehicle positioning technology based on RFID, and driver intention recognition. Then algorithms of PS and TLVW are demonstrated below.

9.3.1 IDAS Signal Perception Level Algorithms

Signal perception level algorithms process raw signals to determine vehicle speed, relative distance to the intersection, and driver's intention, which are necessary for core algorithms.

Vehicle Status Estimation

Vehicle status including vehicle speed and acceleration are important inputs to IDAS. The signal wheel speed sensor is not reliable in driving conditions due to tire slippage. Additionally, noise pollutes the signal from the accelerometer and renders it unusable.

Based on the literature [21–23], an algorithm is designed to estimate vehicle speed and filtered acceleration as shown in Figure 9.3 [24], where Bias Correction is used to compensate the zero drift of the accelerometer as shown in Eq. (9.1). Slip estimation is explained in Eq. (9.2) and the Kalman filter is depicted in Eq. (9.3):

$$a_{\mathrm{m}}^{*}(k) = a_{\mathrm{m}}(k) - a_{\mathrm{bias}} \tag{9.1}$$

$$\begin{aligned} \lambda &= f(F_x/F_z) \\ F_z &= M/L\cdot(l_{\mathrm{a}}\cdot g + H\cdot a) \\ F_x &= \begin{cases} M\cdot g\cdot f & F_{\mathrm{b}} = 0 \\ -a\cdot M\cdot \beta & F_{\mathrm{b}} > 0 \end{cases} \end{aligned} \tag{9.2}$$

$$\begin{aligned} \begin{bmatrix} v(k) \\ a(k+1) \end{bmatrix} &= \begin{bmatrix} 1 & \tau \\ 0 & 1 \end{bmatrix} \begin{bmatrix} v(k) \\ a(k) \end{bmatrix} + \begin{bmatrix} 0 \\ \tau \end{bmatrix} w_{\mathrm{at}} \\ \begin{bmatrix} v_{\mathrm{wm}}(k) \\ a_{\mathrm{m}}^{*}(k) \end{bmatrix} &= \begin{bmatrix} 1-\lambda & 0 \\ 0 & 1 \end{bmatrix} \begin{bmatrix} v(k) \\ a(k) \end{bmatrix} + V. \end{aligned} \tag{9.3}$$

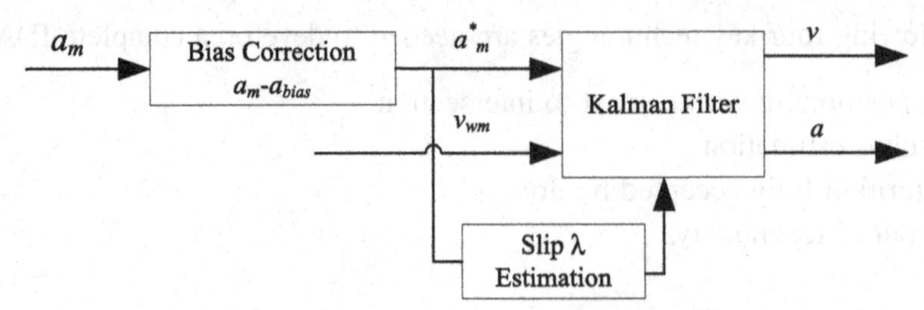

Figure 9.3: Kalman Filter for Velocity and Acceleration Estimation.

In Eq. (9.1), a_{m} is the raw signal from the accelerometer and a^*_{m} is the modified signal. K is time tag and a_{bias} represents the zero drift of the accelerometer.

In Eq. (9.2), λ stands for tire slip and F_z denotes the load on the driven axle, F_x represents the longitudinal force from the driven axle, L is the wheel base, l_{a} is the distance between the front axle and the mass center, H is the height of the mass center, M stands for the overall mass of the vehicle, f is the coefficient of rolling resistance, a represents vehicle acceleration, g is the acceleration of gravity, and β is the ratio of braking force on the rear axle, which accounts for the total braking force of the vehicle.

In Eq. (9.3), v represents the real velocity of the vehicle, v_{wm} stands for the rotation speed of the driven wheel, *wat* means system noise, V is the two-dimensional observation noise, and τ is the time step of calculation. The filtering results are shown in Figure 9.4.

Vehicle Positioning

Vehicle positioning method

RFID is a technology that uses radio waves to transfer data from an electronic tag, called an RFID tag or label, through a reader attached to an object for the purpose of identifying and tracking the object. Some RFID tags can be read from several meters away and beyond the line of sight of the reader. RFID has many applications. Logistics and transportation are major areas of implementation for RFID technology. For example, yard management, container shipping, and freight distribution use RFID tracking technology. Transportation companies around the world pay great attention to RFID technology due to its impact on business value and service efficiency.

Here, a real-time RFID-based vehicle positioning method is described. A series of passive RFID tags are mounted in the middle of lanes on the road surface, and store position information, distance information to intersection, lane number, lane direction and road curvature, gradient, etc. When the vehicle passes over an RFID beacon, the RFID reader and antenna carried by the vehicle activate the tag and read the information. Then the

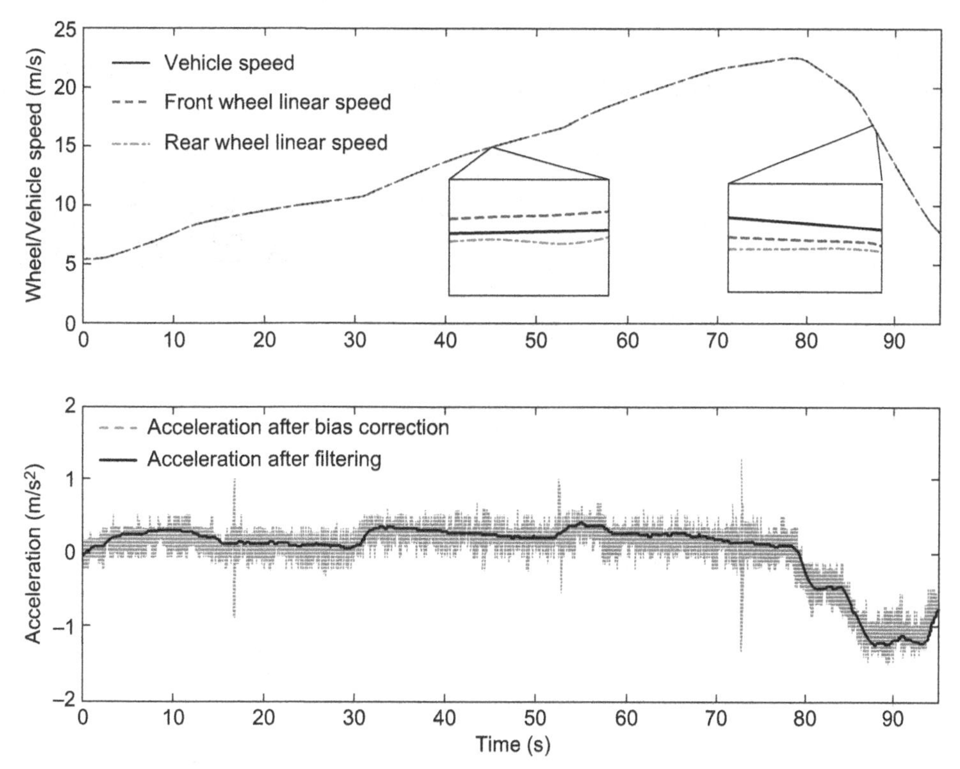

Figure 9.4: Filtering Results of Acceleration and Speed.

On-Board Unit (OBU) uses the information acquired from the RFID beacon to identify the vehicle's current position. Such a configuration is opposite to what is commonly used – the reader is fixed and the beacon is on the vehicle. It is this new configuration that brings significant advantages for much more accurate vehicle positioning that many other sensor technologies cannot provide.

The layout of the RFID beacons and the reader in the vehicle is depicted in Figure 9.5. Although more tags in the road will improve the positioning accuracy, only two RFID tags in each lane near the intersection are assumed to reduce costs (e.g. one is at the stop line and the other one is 150 m upstream).

Positioning data read and distance estimate

RFID beacons are distributed discretely in lanes. As a result, the input data acquired from RFID beacons are discontinuous. IDAS should provide the driver with continuous and reliable assistance information. To generate continuous positioning data with adequate accuracy, vehicle kinematics is fused with the RFID beacons data, as depicted in Figure 9.6.

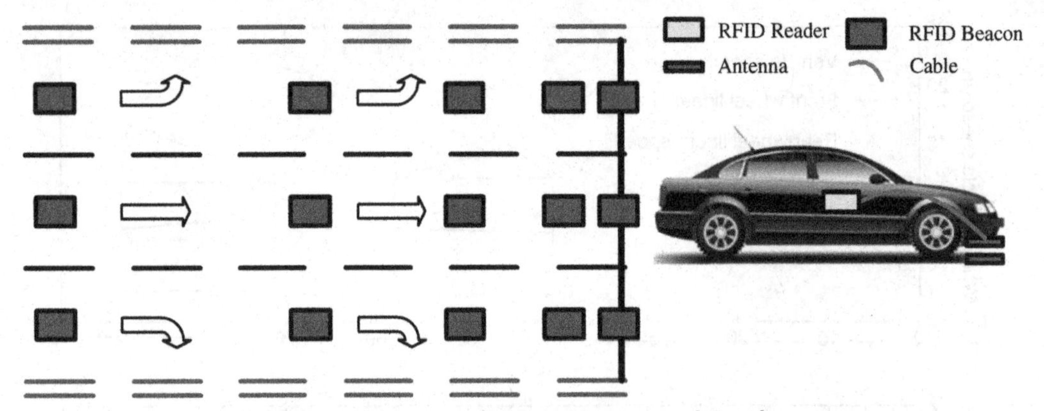

Figure 9.5: Layout of RFID Beacons and Reader.

The vehicle position can then be calculated by

$$d_{2\text{inter}} = d_{\text{tag}} - d_{\text{integration}}$$

$$d_{\text{integration}} = \begin{cases} 0 & |F_{\text{tag}}| = 1 \\ \sum\limits_{k=0} v(k) \cdot \tau + \dfrac{1}{2} \sum\limits_{k=0} a(k) \cdot \tau^2 & F_{\text{tag}} = 0, \end{cases} \quad (9.4)$$

where $d_{2\text{inter}}$ is the current distance to the intersection; d_{tag} is the stored distance information obtained from the RFID tag at the last time; $d_{\text{integration}}$ is the driving distance integrated from the speed; F_{tag} is the road position flag with value 1 when the system can get the information from the RFID tag, and 0 otherwise; k is the data sequence number, starting the count when the system reads the tag and reset to 0 before reading the next tag; and v and a are vehicle speed and acceleration respectively.

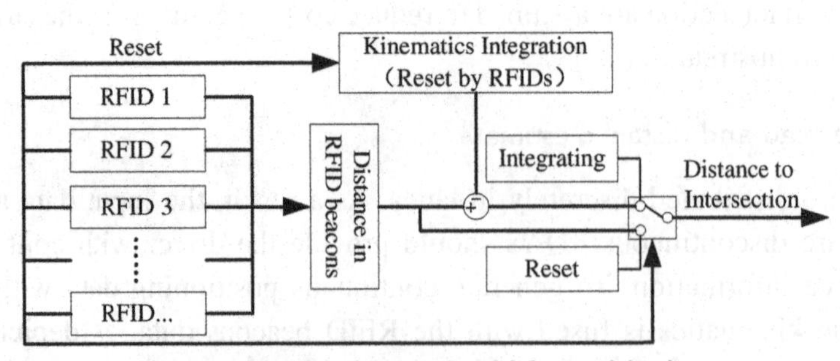

Figure 9.6: Algorithm of Vehicle Positioning.

On going a step further, position methods based on GPS and RFID can be fused. When the signal is of good quality, GPS is used to yield the accurate position, and thus fewer RFID beacons need to be installed. When the signal quality is limited, RFID is applied to calculate the accurate position. Fusion of these applications will make it possible to achieve better results.

Experimental verification

As mentioned above, the vehicle positioning approach is based on RFID. When the vehicle is passing over the tag, the vehicle position is given by the accurate position information previously stored in the tag (ground truth). Otherwise the vehicle position is estimated by kinematics integration. The lane information can also be obtained by the tag.

The position accuracy can be ensured by two aspects. (1) RFID is a short-range-oriented wireless communication set. The RFID readers on the vehicle can only receive the tag's information within 2 m. Also, the antenna facing direction would prevent the reader from receiving strong signals from the tag further away, which the antenna is not directed to. (2) Although the cumulative error of the distance estimation from the speed may increase with the moving distance, the known distance between two adjacent tags is relatively short and can be used to correct the error. In order to validate the method and the accuracy of the positioning, one test method is proposed by radar measurement below.

The main idea of this experiment is to use the calibrated radar distance measure to check the accuracy of the proposed positioning approach. The experimental vehicle and the test yard are shown in Figure 9.7. The experimental vehicle is equipped with a millimeter-wave radar set and an RFID reader. The radar is installed on the vehicle's front bumper. The radar wave beam faces the vehicle's running direction. The antenna is installed below the front bumper facing the ground. The tags installed on the test road have the same configuration as at an intersection (one at the stop line and the other at a certain distance, such as 100 m upstream). A fixed target is mounted on the driveway of the vehicle, via which the on-board radar can detect its distance.

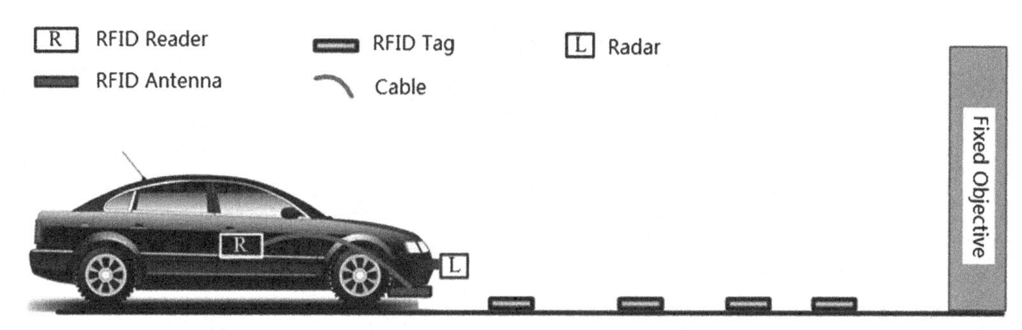

Figure 9.7: Position Approach Experimental Validation.

As is shown, the accuracy of the radar is within 1 meter, which is adequate for the validation. The radar measured distance and the distance estimated from the RFID are compared to check the accuracy, which can be conducted either online or offline.

The experimental scenarios include different vehicle speeds: constant, acceleration, and deceleration over several periods of time. The on-board computer estimates the distance between the test vehicle and the tag at the stop line. Meanwhile, the radar measures the distance between the vehicle and the fixed target. The data are recorded by CANoe through the CAN Bus. CANoe is a development and test software tool for a single ECU or for ECU networks. One of the test results is shown in Figure 9.8.

From the figure, the continuous curved line is obtained based on the position information stored in the first tag (top left of Figure 9.8) and the kinematics integration from the speed. It coincides with the dashed curved line obtained by radar detection, which shows that the estimated distance to the stop line using the proposed approach is equivalent to the radar measured distance. Similarly, the '×' marks on the continuous curved line mean that the estimated distance is equivalent to the accurate distance at tag locations. The experimental results demonstrate the feasibility of the proposed vehicle positioning approach.

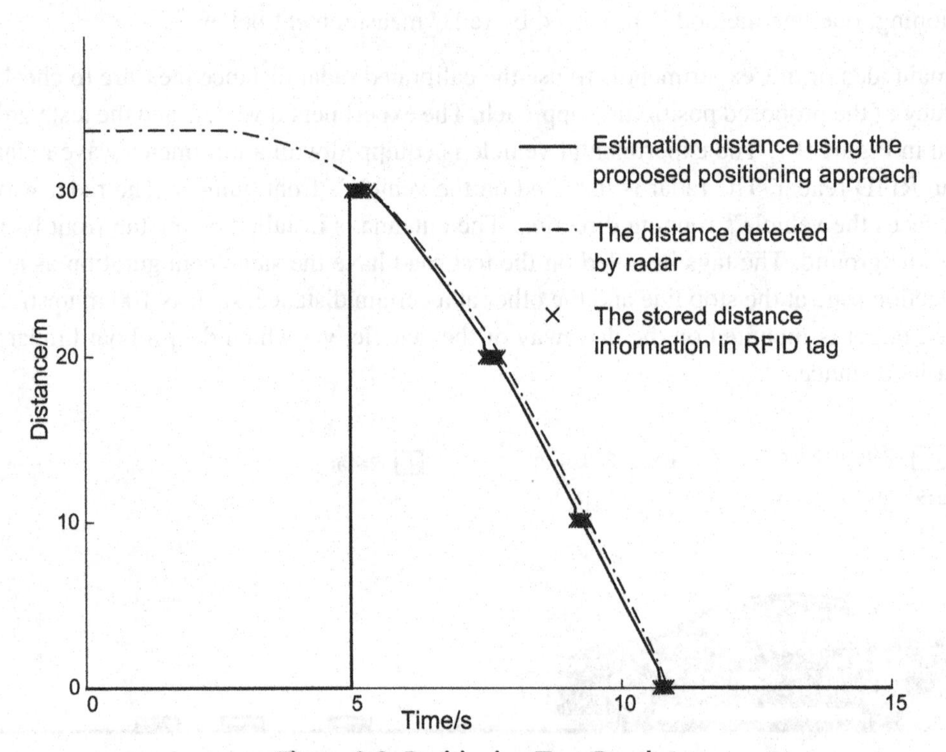

Figure 9.8: Positioning Test Result.

9.3.2 Driver Intention Recognition

The driver's intention is the driver's choice at an intersection, i.e. in which direction they are about to go, including straight ahead, left turn, and right turn.

Two messages can be used to judge the driver's intention when approaching an intersection: (a) steering light; (b) the lane in which the host vehicle is in now. The judging law can be explained using Table 9.1.

Table 9.1: Driver Intention Judgment Table

Lane Direction	Steering Light	Driver's Intention
Straight ahead	–	Straight ahead
Left turn	–	Left turn
Right turn	–	Right turn
Straight ahead or left turn	Off	Straight ahead
	Left turn light	Left turn
Straight ahead or right turn	Off	Straight ahead
	Right turn light	Right turn light
All directions	Off	Straight ahead
	Left turn light	Left turn light
	Right turn light	Right turn light

9.3.3 IDAS Assistance Algorithm

When driving at intersections, vehicles may be in various scenarios and situations that can be classified into 19 typical types according to three factors: traffic control signal (green, amber, red or stop sign); driver's intended direction (straight ahead, right turn or left turn); presence of leading vehicle (LV; present or not).

These 19 typical scenarios have one thing in common, that the final behavior of the vehicle will be either passing or stopping. An algorithm of Passing Support (PS) is designed for scenarios in which the vehicle can pass the intersection, whereas the algorithm of Traffic Light Violation Warning (TLVW) is designed for scenarios in which the vehicle has to stop at the stop bar. The additional Rear-End collision Warning (REW) algorithm is for the possibility of potential rear-end collisions. The final hybrid IDAS algorithm is formed by matching the three algorithms with every typical scenario.

Overall Structure of IDAS Assistance Algorithm

The overall structure of the IDAS assistance algorithm is shown in Figure 9.9. The assistance algorithm is composed of two modules: Assistance Function 1 and Assistance Function 2. Input signals determine which module will be executed.

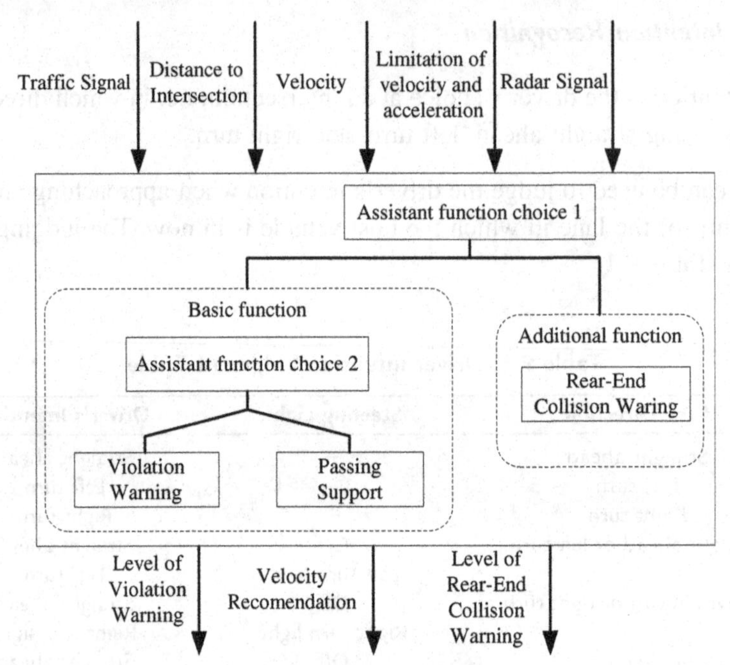

Figure 9.9: Structure of IDAS Assistance Algorithm.

RCW has a higher priority in Assistance Function 1. Whenever there is a possibility of a rear-end collision, a warning will be released first. Assistance Function 2 is based on critical speed. Critical speed is defined as the lowest final speed the host vehicle must reach in order to go through the intersection under the condition of a comfortable acceleration or deceleration. If the critical speed does not violate the traffic laws, then PS will be carried out. Otherwise, TLVW will be applied.

Rear-End Collision Warning Algorithm

REW is an additional function of IDAS. Currently, much research work has been done about rear-end collision warning. According to a literature review and field tests, an algorithm with multiple levels based on Time to Collision (TTC) has the advantages of simple architecture and variable sensitivity for different types of drivers. Therefore, it possesses a higher level of practicability and driver acceptance.

TTC is defined as the ratio of relative distance and velocity between the leading vehicle and host vehicle as shown in Eq. (9.5), and two warning thresholds are designated as TTC_{warn1} and TTC_{warn2}:

$$\mathrm{TTC} = \frac{d_{2\mathrm{lead}}}{v_{\mathrm{rel}}}, \tag{9.5}$$

where d_{2lead} is the relative distance between the leading vehicle and host vehicle, and v_{rel} is the relative speed.

The REW algorithm is designed as follows:

1. $TTC > TTC_{warn1}$: No danger of collision, inform the driver of the leading vehicle's existence.
2. $TTC_{warn2} < TTC < TTC_{warn1}$: Certain danger of collision exists, issue warning level 1.
3. $TTC < TTC_{warn2}$: Great danger of collision exists, issue warning level 2.

The values of TTC_{warn1} and TTC_{warn2} represent the sensibility of the algorithm and can be adjusted by the driver as shown in Table 9.2. The more aggressive the driver, the smaller the values.

Table 9.2: REW Threshold

Sensibility Level	Speed <50 km/h	Speed >50 km/h
1	1.5 s	2 s
2	2.5 s	6 s
3	4 s	6 s
4	6 s	8 s
5	8 s	10 s

PS Control Algorithm based on Critical Speed

The PS algorithm calculates the speed recommendation and at the same time forms a reference for switching between the two functions of PS and TLVW. A PS algorithm is proposed based on a Critical Passing Speed (CPS), which is defined as the speed to which the vehicle should adapt at a limited acceleration that is comfortable for the driver, if the vehicle passes the stop bar at the very time point in which the traffic light changes from green to amber or from red to green. The two main parts of the PS algorithm are the calculation of CPS and the feasibility evaluation of CPS respectively, as described in Figure 9.10.

Three questions should be considered in the calculation of CPS. (1) If the CPS is presented to the driver as a speed recommendation, the driver needs a reaction time to interpret and respond. After the driver's response, the vehicle powertrain system also needs a certain amount of time to react to the driver's operation. (2) The vehicle will experience a speed-changing process from its current speed to the recommended one. (3) After the speed-changing process, the vehicle may still need to drive a distance to pass the stop bar.

Calculation of critical speed and its feasibility in green light and red light situations are discussed separately.

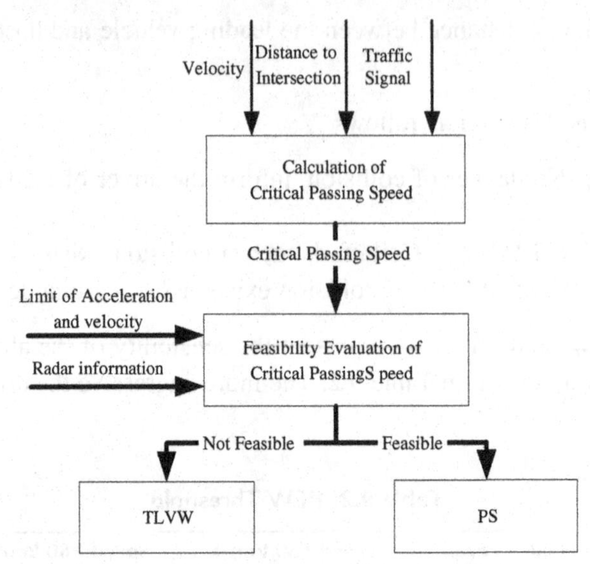

Figure 9.10: Structure of PS Algorithm.

For a green light, ignoring the previous motion state, suppose that the host vehicle will maintain a constant speed from the current position. Then the critical speed to get through the intersection in the remaining time of the green light can be calculated by

$$v_{\text{threshold,app}} = \frac{d_{\text{2inter}}}{t_{\text{left,g}}}, \tag{9.6}$$

where d_{2inter} is the distance between the vehicle and the stop bar, $t_{\text{left,g}}$ stands for the remaining time of the green light, and $v_{\text{threshold,app}}$ represents the critical speed.

If $v_{\text{threshold,app}}$ satisfies inequality (9.7), then no further calculation is needed:

$$v_{\text{threshold,app}} \leq v_{\text{cur}}. \tag{9.7}$$

If the inequality (9.7) is not satisfied, it means more acceleration to get through the intersection is required, and we have to re-calculate the critical speed.

Three hypotheses were proposed in order to simplify the calculation: (1) During the reaction time for the driver and powertrain, the vehicle speed is constant. (2) In the speed-changing process, the vehicle speeds up or slows down at a constant acceleration whose value is the upper or lower limit from the driver's comfort perspective. (3) After the speed-changing process, the vehicle drives constantly at the recommended speed. According to the hypothesis above, the calculation of CPS can be expressed as

$$\begin{cases} d_{\text{re}} = v_{\text{cur}} \cdot t_{\text{re}} \\ t_{\text{acc}} = \dfrac{v_{\text{threshold}} - v_{\text{cur}}}{a_{\max}} \\ d_{\text{acc}} = \dfrac{v_{\text{threshold}}^2 - v_{\text{cur}}^2}{2a_{\max}} \\ d_{\text{const}} = v_{\text{threshold}}\left(t_{\text{left,g}} - t_{\text{re}} - t_{\text{acc}}\right) \\ d_{\text{2inter}} = d_{\text{re}} + d_{\text{acc}} + d_{\text{const}}, \end{cases} \tag{9.8}$$

where t_{re} is the sum of driver and vehicle powertrain reaction time; v_{cur} is the current vehicle speed; d_{re} is driving distance under hypothesis (1); $v_{\text{threshold}}$ is the CPS; $a_{\max}$ is the maximum acceleration or minimum deceleration; t_{acc} is the acceleration or deceleration time under hypothesis (2) and d_{acc} is the driving distance under hypothesis (2); $t_{\text{left,g}}$ is the traffic light expected changing time; d_{const} is the driving distance under hypothesis (3); and d_{2inter} is the current distance to intersection.

Equation (9.8) is a quadratic formula that is not easy to solve by a controller in practice. Numerical methods are better solutions.

After yielding $v_{\text{threshold}}$, its feasibility will be checked based on two limits:

1. $v_{\text{threshold}}$ does not violate traffic laws. If $v_{\text{threshold}}$ exceeds the limit of the traffic laws, then it is unfeasible and TLVW algorithms should be applied.
2. There is a possibility of rear-end collision. The situation can be divided into two sub-situations: (a) the leading vehicle cannot get through the intersection; (b) the leading vehicle can pass the intersection, but the host vehicle will be too close to it after acceleration in order to go through the intersection. These two situations will be discussed separately below.

In the first situation, the leading vehicle cannot pass the intersection in the remaining time of the green light. So the host vehicle must stop as well to avoid collision. Suppose the leading vehicle maintains a constant speed, then if Eq. (9.9) is met, the leading vehicle will have to stop, which makes PS unfeasible. Then the assistance algorithm should turn to TLVW.

$$\begin{aligned} d_{\text{2inter,lead}} &= d_{\text{2inter}} - d_{\text{rel}} \\ v_{\text{cur,lead}} &= v_{\text{cur}} - v_{\text{rel}} \\ \frac{d_{\text{2inter,lead}}}{v_{\text{cur,lead}}} &> t_{\text{left,g}}, \end{aligned} \tag{9.9}$$

where $d_{\text{2inter,lead}}$ is the distance between the leading vehicle and intersection, d_{rel} is the relative distance between the host vehicle and leading vehicle, $v_{\text{cur,lead}}$ stands for the speed of the leading vehicle, and v_{rel} represents the relative speed with respect to the leading vehicle.

In the second situation, although the leading vehicle is able to pass the intersection, the host vehicle has to accelerate to a high speed in order to get through the intersection too. Thus, the relative distance between the two vehicles may be too small considering the safety factors. Here, we define a Critical Dangerous Relative Distance (CDRD) as explained by Eq. (9.10). Whenever CDRD is reached, the situation is classified as dangerous.

$$d_{danger} = \begin{cases} TTC_{limit} \cdot (v_{threshold} - v_{cur,lead}) + d_{offset} & v_{threshold} > v_{cur,lead} \\ d_{offset} & v_{threshold} < v_{cur,lead} \end{cases}, \tag{9.10}$$

where d_{danger} is the CDRD defined above and d_{offset} is introduced into the equation to make the algorithm more conservative.

If critical speed is smaller than the speed of the leading vehicle, a rear-end collision will not be possible, then the critical speed can be recommended to drivers. Otherwise, a further calculation is needed.

Acceleration is needed to reach the critical speed. Whether there will be a possible collision or not depends on the relative distance. If the CDRD is reached, it is considered as a dangerous situation. Displacement of the host vehicle during the acceleration process can be calculated using Eq. (9.11), and the danger criterion is expressed by the inequality (9.12). If the relative distance between the two vehicles after acceleration is smaller than CDRD, i.e. inequality (9.12) is satisfied (collision may happen), the critical speed calculated above should not be applied.

$$d_{acc} = \begin{cases} \dfrac{v_{threshold}^2 - v_{cur}^2}{a_{max}} & v_{threshold} > v_{cur} \\ 0 & v_{threshold} < v_{cur} \end{cases} \tag{9.11}$$

$$d_{lead,acc} = v_{cur,lead} \cdot t_{acc}$$

$$d_{rel} + d_{lead,acc} - d_{acc} < d_{danger}. \tag{9.12}$$

If inequality (9.12) is not satisfied, further judgment is needed.

If the critical speed is larger than the speed of the leading vehicle, there is still a potential danger of collision. The possibility of collision is determined by t_{danger}, the time during which the host vehicle accelerates from its original position till the moment its relative distance reaches CDRD. The relationship between t_{danger} and CDRD can be represented by

$$d_{danger} = d_{rel} + v_{cur,lead} t_{danger} - \left[d_{acc} + v_{want}\left(t_{danger} - t_{acc}\right)\right]. \tag{9.13}$$

If inequality (9.14) is satisfied, the danger will occur before the stop bar and therefore the critical speed should not be applied. Otherwise, collision avoidance can be achieved after the intersection.

$$t_{\text{danger}} = \frac{d_{\text{rel}} - d_{\text{acc}} + v_{\text{want}} t_{\text{acc}} - d_{\text{danger}}}{v_{\text{want}} - v_{\text{cur,lead}}} < t_{\text{left,g}}. \tag{9.14}$$

The PS problem is addressed while the traffic light is red. Again, there are two situations. (a) Critical speed is larger than the current vehicle speed. This means that when the vehicle reaches the stop bar, the traffic light is already green. So there is no need to pause at the stop bar. (b) Critical speed is smaller than the current vehicle speed. The vehicle should decelerate so as to avoid stopping before the stop bar when the traffic light changes from red to green. Suppose the host vehicle maintains a constant speed and arrives at the stop bar when the traffic light turns green, then we have an equation as expressed by

$$v_{\text{threshold,app}} = \frac{d_{2\text{inter}}}{t_{\text{left,r}}}, \tag{9.15}$$

where $d_{2\text{inter}}$ is the distance between the host vehicle and stop bar, $t_{\text{left,r}}$ is the remaining time of the red light, and $v_{\text{threshold,app}}$ denotes an initial calculated critical speed.

If $v_{\text{threshold,app}}$ satisfies inequality (9.16), in that the CPS $v_{\text{threshold}}$ equals $v_{\text{threshold,app}}$, the calculation of critical speed is terminated. Otherwise, more calculations are necessary to yield a new useful critical speed.

$$v_{\text{threshold,app}} \geq v_{\text{cur}}. \tag{9.16}$$

In order to simplify the calculation of critical speed, three hypotheses are proposed:

1. The vehicle maintains a constant speed during the reaction time of the driver and powertrain.
2. The vehicle adopts a constant deceleration in order to decelerate to a recommended speed, and the deceleration is within the range in which the driver feels comfortable.
3. After reaching the recommended speed, the vehicle will maintain that speed.

Based on these hypotheses, we have the following formulae to calculate critical speed:

$$\begin{cases} d_{\text{re}} = v_{\text{cur}} \cdot t_{\text{re}} \\ t_{\text{decc}} = \dfrac{v_{\text{cur}} - v_{\text{threshold}}}{d_{\max}} \\ d_{\text{decc}} = \dfrac{v_{\text{cur}}^2 - v_{\text{threshold}}^2}{2d_{\max}} \\ d_{\text{const}} = v_{\text{threshold}} \left(t_{\text{left,r}} - t_{\text{decc}} - t_{\text{re}} \right) \\ d_{2\text{inter}} = d_{\text{re}} + d_{\text{decc}} + d_{\text{const}}, \end{cases} \tag{9.17}$$

where d_{re} is the displacement of the host vehicle in reaction time, d_{max} is the maximum deceleration that makes the driver feel comfortable, t_{decc} is the time it takes for the host vehicle to decelerate to the recommended speed from its original speed, d_{decc} is the corresponding displacement, and $t_{left,r}$ is the remaining time of the red light.

If the real root of Eq. (9.17) does not exist, it means the host vehicle cannot pass the intersection without pausing, and TLVW should be applied.

The feasibility should be checked after yielding $v_{threshold}$, and two factors are worthy of consideration:

1. If the $v_{threshold}$ obtained from Eq. (9.17) is too low, the host vehicle may cause a negative effect on other vehicles. Also, it would not make any sense if the critical speed is too low. Therefore, TLVW should be applied in this situation.
2. If LV exists, the motion of the host vehicle will be limited, and under such conditions TLVW should be applied.

Algorithm of TLVW

When critical speed is not feasible and the vehicle has to stop at the intersection, the TLVW is active. When the TLVW perceives that the driver is about to violate the traffic signal, it gives visual and auditory warnings, or automatically brakes the vehicle when necessary. Furthermore, TLVW informs the driver about the upcoming signal change in advance, and avoids the situation where a sudden change surprises the driver, which results in hard braking.

A dynamic TLVW algorithm with many levels is proposed based on velocity thresholds, which are defined as velocity profiles of the vehicle when it stops at the stop bar with constant decelerations of -1.5, -3, and $-5\ m/s^2$. These deceleration values are selected based on driver behavior characteristics such as stop behavior at intersections [20], traffic light violation possibility [18], and comfortable acceleration [25,26]. The velocity vs. distance-to-intersection coordinate is divided into four warning zones, as described in Figure 9.11.

According to Figure 9.11, a static TLVW is defined as follows:

1. $v < v_{thre,warn1}$: Informing
2. $v_{thre,warn1} < v\ v_{thre,warn2}$: TLVW warning level 1
3. $v_{thre,warn2} < v < v_{thre,inter}$: TLVW warning level 2
4. $v > v_{thre,inter}$: Automatic braking.

According to statistics of velocity profiles of vehicles that stop at intersections, even though drivers yield to the traffic signal and stop, more than 50% of velocity profiles will exceed the warning threshold during deceleration [20]. Therefore, a static algorithm will result in a high false warning rate. To solve this drawback, the history of velocity profiles should be taken into account when designing TLVW algorithms. A TLVW algorithm checks vehicle

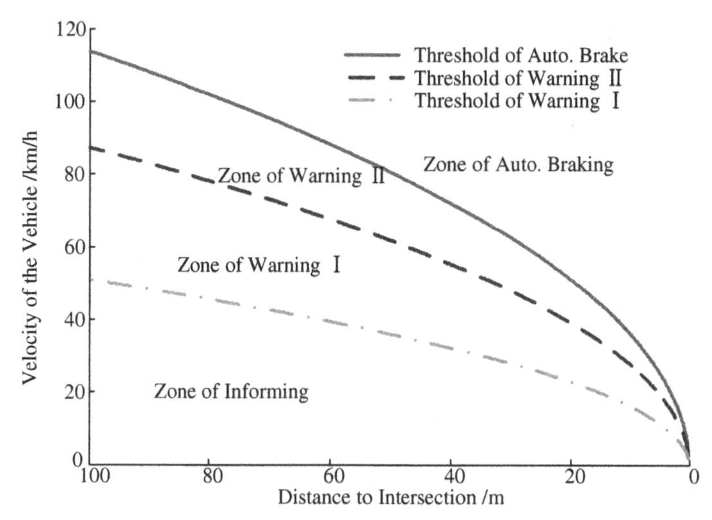

Figure 9.11: Warning Zones of TLVW.

acceleration at the last second before the velocity crosses the warning thresholds. If vehicle acceleration shows that the driver has already yielded to the traffic light, the warning will be canceled. Otherwise TLVW gives a warning according to the current warning zone.

Matching IDAS Algorithms with Driving Scenarios at Intersections

As described above, a vehicle may be in 19 typical scenarios when driving toward an intersection. For each scenario, the traffic control signal and detail regulation will be different, so the above three algorithms cannot simply be combined, and certain adjustments should be made to match the algorithms to each scenario, as described in Table 9.3.

Table 9.3: Matching ITLAS algorithm with driving scenarios at intersections

			REW	PS	TLVW
Green	Straight	LV	×	√	No braking
		Non-LV	√	√	No braking
	Left turn	LV	×	×	No braking
		Non-LV	√	×	No braking
	Right turn	LV	×	×	Only inform
		Non-LV	√	×	Only inform
Amber	Straight and left turn	LV	×	×	√
		Non-LV	√	×	√
	Right turn	LV	×	×	Only inform
		Non-LV	√	×	Only inform
Red	Straight and left turn	LV	×	√	√
		Non-LV	√	×	√
	Right turn	LV	×	×	Only inform
		Non-LV	√	×	Only inform
Stop sign			√	×	√

9.4 Simulation of IDAS

9.4.1 Simulation Model based on Simulink

In order to verify the IDAS algorithm and access its effect on intersection driving, a micro-simulation model of intersection traffic is built based on Simulink. This simulation model includes a model of OBU and a model of the intersection environment, which includes models of road, traffic light, Subjective Vehicle (SV), Leading Vehicle (LV), and drivers of SV, as described in Figure 9.12.

The intersection environment model includes:

1. **Driver Decision Model.** To simulate the driver's perception, decision, and operation.
2. **SV Model.** To simulate the kinematic motion of SV.
3. **LV Model.** To simulate the kinematic motion of LV.
4. **Model of Traffic Light and Wireless Communication.** To simulate the traffic light and its communication with vehicles.

An OBD Model is used to simulate the real IDAS controller mounted on the SV, including data acquisition and assistance algorithm.

9.4.2 Simulation and Results

Two kinds of simulations are designed: (a) Fixed Initial Condition Simulation (FICS); (b) Random Condition Simulation (RCS). The difference between the two types of simulation models is based on whether the initial conditions of the simulation are fixed or random.

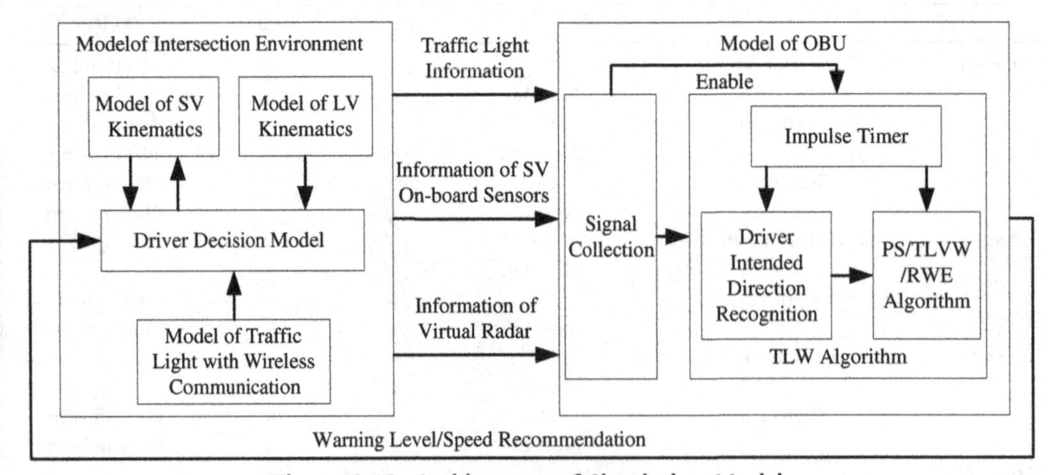

Figure 9.12: Architecture of Simulation Model.

FICS is designed to verify the rationality and function of IDAS. Such typical initial traffic conditions as initial position and initial speed are configured. IDAS should cope with these conditions. The simulation results show that the IDAS algorithm is working properly.

RCS aims at accessing the impact of this IDAS algorithm on intersection driving. The initial velocity of the SV and initial time point of the traffic signal is random. The velocity of the SV is distributed normally around a mean value, and the initial time point of the traffic signal is distributed evenly in its cycle. For every mean velocity of 30, 40, and 50 km/h, 10,000 simulations are conducted, and four statistical indices are used to compare the behavior of vehicles with and without IDAS, and to access the effect of IDAS on intersection driving. The results are shown in Figure 9.13.

Unintended traffic light violation is a scenario in which it is not possible for the driver to brake the vehicle to a full stop in the limited distance to the stop bar when the traffic light suddenly turns to amber from green. Large deceleration means the deceleration of the vehicle exceeds a value of -1.5 m/s^2. The rate of unintended violation represents driving safety at the intersection, and the rate of large deceleration reveals riding comfort. As described in Figure 9.13, when the vehicle is not equipped with IDAS, i.e Traffic Light

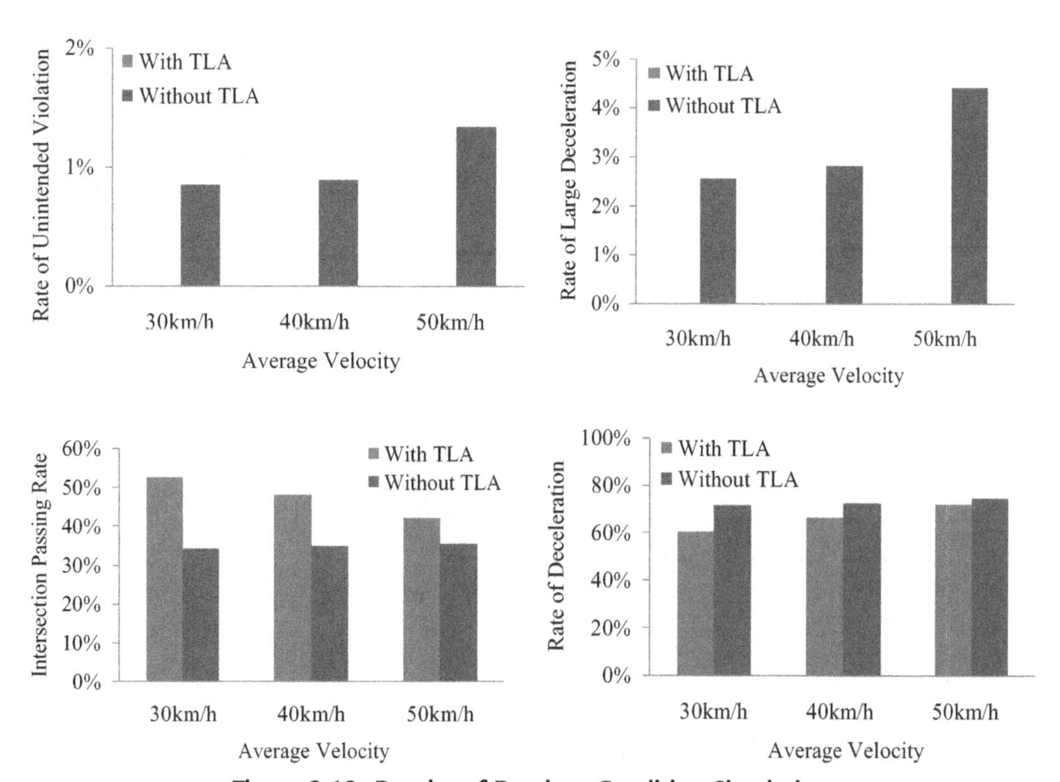

Figure 9.13: Results of Random Condition Simulation.

Assistance (TLA), the rates of unintended traffic light violations and large deceleration increase with velocity, but IDAS can avoid these two situations. Intersection passing rate is the ratio of the times that a vehicle passes the intersection without stopping to overall simulation times; rate of deceleration is the ratio of times that a vehicle experiences deceleration resulting in a traffic light to overall simulation times. These two indices reveal driving smoothness at the intersection. From Figure 9.13, it can be seen that IDAS increases the passing rate and reduces the rate of deceleration, and therefore IDAS improves driving smoothness.

9.5 Real Vehicle Test

In order to further test the performance of IDAS, a real vehicle test is needed. Here, the IDAS platform is composed of OBD and RSU, and a wireless communication is required to transmit data between the two parts. OBD is based on xPC technology [27]. Many experiments have been carried out concerning intersection assistance. Results show that IDAS is beneficial to drivers and can improve safety.

9.5.1 Design of IDAS Prototype

Architecture of an IDAS Prototype

The prototype consists of RSU and OBU, as shown in Figure 9.14. RSU is composed of a traffic light with a wireless communication device and RFID beacons, and the OBU contains an RFID reader, radio receiver, xPC-based controller and assistance executors, including HMI and Electronic Vacuum Boost (EVB).

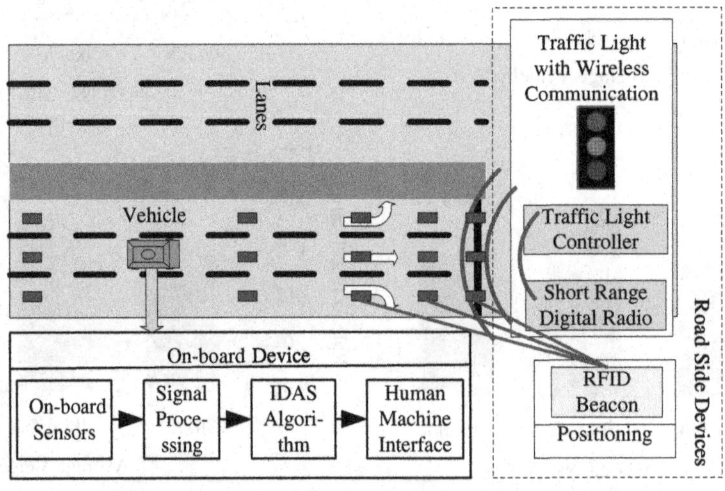

Figure 9.14: Architecture of IDAS Prototype.

The traffic light with a wireless communication device is modified from an ordinary mobile one. The controller can control the traffic light shift cycle and simultaneously broadcast its phase and timing through a wireless communication transmitter. The wireless communication device in this prototype is a digital radio, which can easily be replaced by a DSRC when it becomes available.

The RFID beacons used in this prototype are passive RFID tags with high reading speed.

Design of Test Vehicle

Architecture of test vehicle

The overall architecture of the IDAS test vehicle is shown in Figure 9.15. The features of this test vehicle are as follows:

1. Most of the information such as vehicle status, which is acquired by an intersection driving assistance system, can be obtained from the original CAN bus of the vehicle. Therefore, a complex signal collection network is avoided.
2. The original CAN bus and added CAN bus are connected through a CAN gate. This design ensures that the added system can obtain information from the original CAN bus and avoid possible influence on the vehicle's original system.
3. The testbed is designed based on modularization and is convenient for extensions.
4. To meet the requirements of IDAS, some devices are added and connected to the CAN bus, which includes the receiver of a short-range communication radio and its ECU, the reader and antenna of the RFID and its ECU, and the central controller of the IDAS OBU.

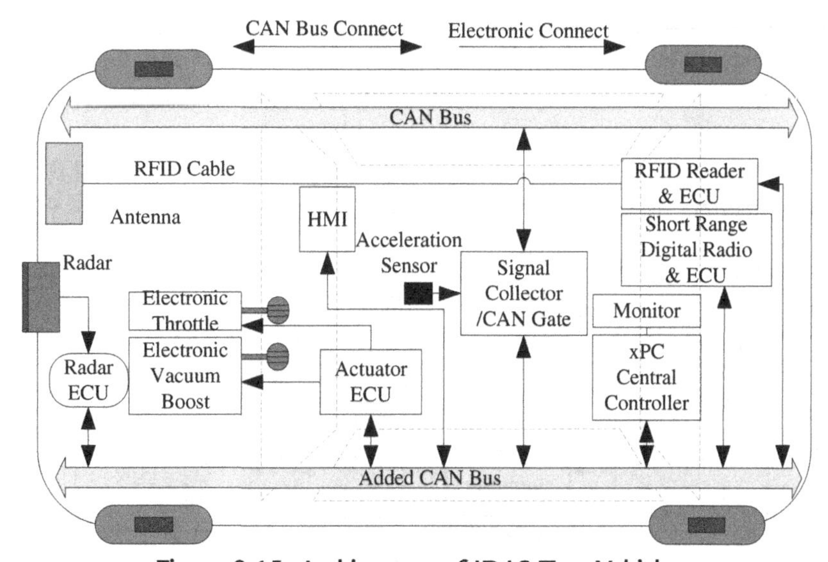

Figure 9.15: Architecture of IDAS Test Vehicle.

Electronic vacuum booster

Based on a normal vacuum booster, EVB is developed. The additional electromagnetic coil makes it possible for the booster to achieve braking via an electronic signal [28].

In order to build a dynamic model of the vehicle, the performance of the EVB has to be taken into consideration. In fact, what is really interesting is how the actual brake pressure will respond to the desired braking pressure. The controller adopts a PID algorithm to control EVB.

We utilize a sinusoidal scanning method to acquire the frequency response characteristics of EVB. Also, we determine the yield of the transfer function of EVB, as expressed by

$$G(s) = \frac{bs + 1}{a_1 s^2 + a_2 s + 1}. \tag{9.18}$$

Double-Mode Electronic Throttle (DMET)

Based on a normal electronic throttle system, a Double-Mode Electronic Throttle (DMET) system was designed as shown in Figure 9.16 [29]. Compared with the normal electronic throttle system, the DMET controller is a unique additional hardware. The acceleration pedal position sensor is connected to the DMET controller rather than the engine ECU. The output terminal of the DMET controller is linked to the engine ECU. Also, in order to carry out closed-loop control, the actual throttle angle must be accessible for the DMET controller.

Figure 9.17 shows the electric schematic diagram of the DMET controller. The DMET controller gets the throttle angle demand from the upper controller via a CAN bus and gets the actual throttle angle signal from the throttle angle sensor via an AD conversion module. Then the control variable will be figured out by the DMET controller with suitable control software. A DA conversion module is set to convert the control variable into an analog voltage signal that is equivalent to the output of the acceleration pedal position sensor. We mark the former as U1 and the latter as U2. Then U1 and U2 are in proportion to the desired throttle angle of the upper controller and driver respectively.

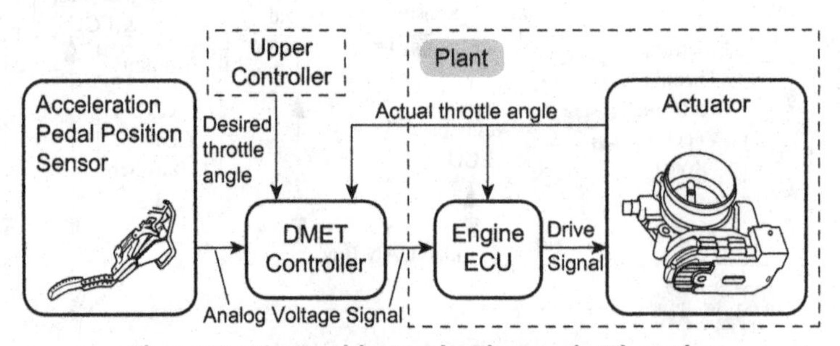

Figure 9.16: Double-Mode Electronic Throttle.

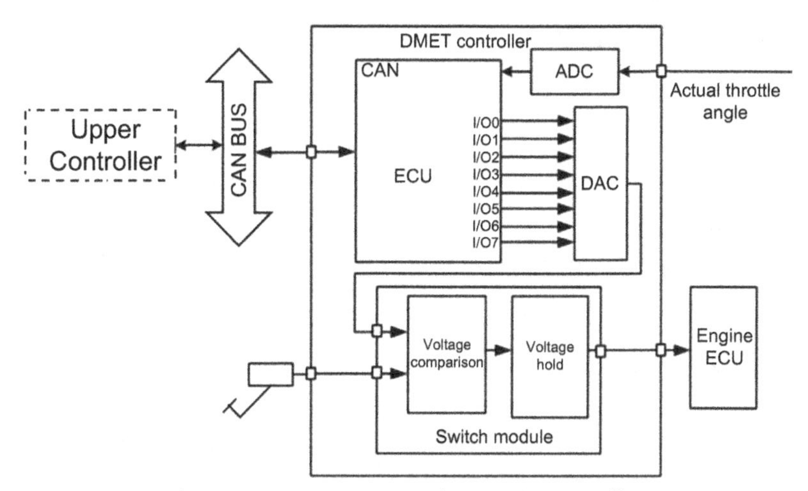

Figure 9.17: Structure of DMET Controller.

The DMET controller achieves seamless switching of the throttle angle demands between upper controller and driver via a switch module. The switch module compares U1 and U2 via a voltage comparison module and the output of the larger one. That is, the larger of the two voltage values determines the resulting throttle angle.

Because of the difficulty in establishing an accurate mathematical model, a PID control algorithm is used to design the feedback controller. The design of such a controller is independent of exact mathematical models and the controller parameters can be adjusted by experience to a certain extent.

In addition, hysteresis of the system results in some fluctuations of the controlled throttle angle, so a feedforward algorithm is added to the controller. This means that a controller is designed to carry out the command of the upper controller. Its diagram is shown in Figure 9.18.

In order to achieve seamless switching between the driver operating mode and automatic brake mode, a dead zone is also introduced into the PID controller.

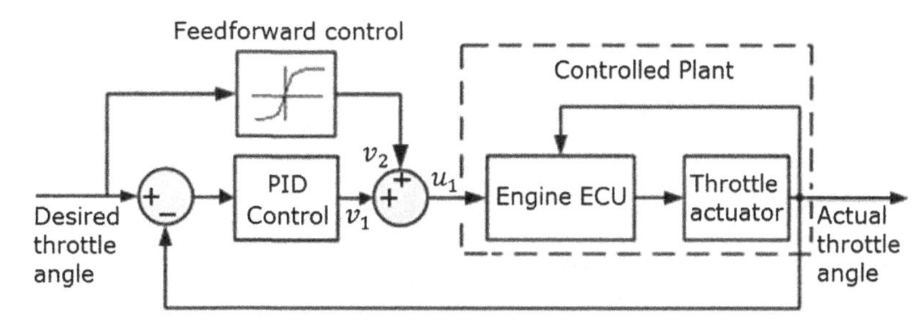

Figure 9.18: Architecture of the Software of DMET.

9.6 Field Test and Results

9.6.1 Test Scenario

In order to further test IDAS, three kinds of tests are carried out:

1. PS in green light
2. PS in red light
3. TLVW.

9.6.2 Results and Analysis

Field tests in a private intersection are conducted based on the developed prototype. The system function under conditions of PS and TLVW is verified. The test results are shown in Figure 9.19.

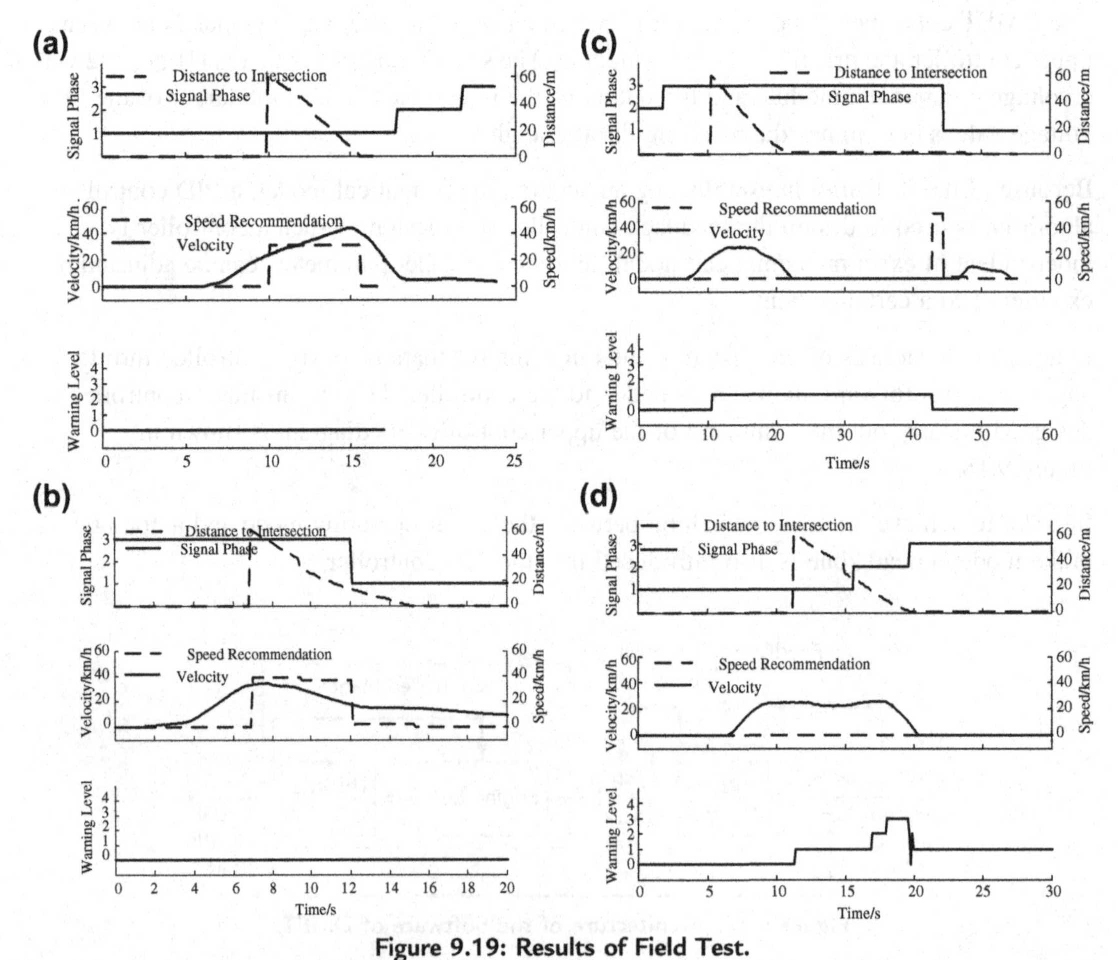

Figure 9.19: Results of Field Test.
(a) PS in green light. (b) PS in red light. (c) TLVW – informing. (d) TLVW – warning.

When the vehicle is driving in a normal situation at intersections, the hybrid IDAS algorithm is the function of PS and provides the driver with speed recommendations through HMI. Figure 9.19(a) describes PS under green light. The driver follows the speed recommendation and speeds up the vehicle, and then the vehicle passes the intersection without stopping. Figure 9.19(b) describes PS under red light. The driver follows the speed recommendation and slows down the vehicle, and then the vehicle passes the intersection without stopping after the signal changes.

The hybrid algorithm of IDAS shifts to the function of TLVW when the vehicle has the potential to violate the traffic light. Figure 9.19(c) describes the situation where the driver follows the traffic light and the TLVW informing signal. If the driver does not respond to the traffic light and informing signal, TLVW gives a stronger warning, as shown in Figure 9.19(d).

9.7 Conclusion and Discussion

In this chapter, an infrastructure–vehicle communication-based intersection driving assistance system is designed and developed. It is effective in solving problems of safety and congestion at intersections. In the field tests, the validity of the system is demonstrated and the following conclusions can be drawn:

1. The proposed IDAS system can make full use of the capability of infrastructure–vehicle communication systems in the way that it not only maintains the driving safety, but also simultaneously improves driving comfort and traffic efficiency at intersections.
2. The proposed hybrid IDAS algorithm can deal with various scenarios and presents appropriate speed recommendations, warnings or automatic braking.
3. The developed IDAS prototype can realize the designed functions of passing support, traffic light violation warning and rear-end collision warning, and demonstrate the advanced assistance system at intersections.

At level intersections where each lane only has one direction, i.e. straight ahead, right turn or left turn, the driver's intended direction can be determined from the lane direction and on-board indicator. However, as for intersections where some of the lanes have two or three possible directions, the driver's intention cannot be identified because the lane may have more than one possible direction and some drivers may not turn on the indicators until their vehicles are very close to the stop bars. Similar problems also exist when the intersection is a ring type or cloverleaf. Therefore, in the future, the IDAS algorithm should contain a module to identify the driver's intended direction by fusing multiple information such as lane direction, indicator, navigation, and driver operating sequences.

In addition, there are some special driver behavior characteristics at intersections, especially at the onset of the amber phase. However, these have not been fully considered in the proposed IDAS algorithm. In the future research, the IDAS system should be improved not only according to theoretical calculations, but also by taking the driver's behavior characteristics into account.

Acknowledgments

This work was supported by the National High Technology Research and Development Program of China ("863" Project, No. 2011AA110402).

References

[1] Transportation Department of the Ministry of Public Security of the People's Republic of China, The Annals of Road Traffic Accident Statistics of PRC (2010), Traffic Management Research Institute of the Ministry of Public Security, Wuxi, Jiangsu, 2010 (in Chinese).
[2] R.A. Retting, M.A. Greene, Influence of traffic signal timing on red-light running and potential vehicle conflicts at urban intersections. Transportation Research Record 1295 TRB, Nation Research Council, Washington, DC, 1998, pp. 23–26.
[3] P.T. Martin, V.C. Kalyani, A. Stevanovic, Evaluation of Advance Warning Signals on High Speed Signalized Intersection, University of Utah Traffic Lab, 2003, pp. 39–40.
[4] T.J. Smith, C. Hammond, M.G. Wade, Investigating the Effect on Driver Performance of Advanced Warning Flashers at Signalized Intersections. Human Factors Research Laboratory, University of Minnesota, Minneapolis, MN, 2001, pp. 101–114.
[5] M. Nekoui, H. Pishro-Nik, The effect of VII market penetration on safety and efficiency of transportation networks, in: Proceedings of the IEEE International Conference, 2009, pp. 1–5.
[6] I. Iglesias, L. Isasi, M. Larburu, et al., I2V communication driving assistance system: On-board traffic light assistant, in: Proceedings of the IEEE 68th Vehicular Technology Conference, 2008, pp. 1–5.
[7] IT and ITS Planning Division, Toyota Motor Corporation, Toyota's Approaches to ITS,: Toyota Motor Corporation, Toyota-Shi, Aichi, 2008, pp. 7–11.
[8] C.-Y. Chen, et al., California Intersection Decision Support: A Systems Approach to Achieve Nationally Interoperation Solutions, University of California, Berkeley, 2005, pp. 125–130.
[9] H. Oh, C. Yae, D. Ahn, et al., 5.8 GHz DSRC packet communication system for ITS services, vol. 4, Vehicular Technology Conference, 1999, 2223–2227.
[10] F. Bai, H. Krishnan, Reliability analysis of DSRC wireless communication for vehicle safety applications, in: Proceedings of the IEEE Intelligent Transportation Systems Conference, Toronto, 2006, pp. 355–362.
[11] K. Fuerstenburg, et al., Project Evaluation and Effectiveness of the Intersection Safety System, Sixth Framework Program, Hamburg, 2007, pp. 21–65.
[12] K.C. Fuersternburg, A new European approach for intersection safety – The EC-Project INTERSAFE, in: Proceedings of the 8th International IEEE Conference on Intelligent Transportation Systems, Vienna, 2005, pp. 343–347.
[13] J. Chen, S. Deutschle, K. Fuerstenburg, Evaluation methods and results of the INTERSAFE intersection assistants. Proceedings of the 2007 IEEE Intelligent Vehicles Symposium, Istanbul, 2007, pp. 142–147.
[14] C. Shooter, J. Reeve, INTERSAFE-2 architecture and specification, in: Proceedings of IEEE 5th International Conference on Intelligent Computer Communication and Processing, 2009, pp. 379–386.
[15] M. Nekoui, D. Ni, H. Pishro-Nik, et al., Development of a VII-enabled prototype intersection collision warning system. Proceedings of International Conference on Testbeds and Research Infrastructure for Development of Networks and Communities, 2009, pp. 1–8.
[16] K. Tamura, M. Hirayama, Toward realization of VICS – Vehicle Information and Communication System, in: Proceedings of the IEEE–IEE Vehicle Navigation and Information Systems Conference, 1993, pp. 72–77.

[17] S. Yamada, The strategy and deployment plan for VICS, Communications Magazine 34 (10) (1996) 94–97.
[18] J. Pierowicz, E. Jocoy, M. Lloyd, et al., Intersection Collision Avoidance Using ITS Countermeasures, US Department of Transportation, Washington, DC, 2000, pp. 5–46–5-48.
[19] G. Wang, DGPS Technology and Application, Electronic Industry Press, Beijing, 1996 (in Chinese).
[20] H. Berndt, S. Wender, K. Dietmayer, Driver braking behavior during intersection approaches and implications for warning strategies for driver assistant systems, in: Proceedings of the 2007 IEEE Intelligent Vehicles Symposium. Istanbul, Turkey, 2007, pp. 245–251.
[21] K. Kobayashi, K.C. Cheok, K. Watanabe, Estimation of absolute vehicle speed using fuzzy logic rule-based Kalman filter, in: Proceedings of the American Control Conference, Seattle, WA, 1995, pp. 3086–3090.
[22] C.K. Song, M. Uchanski, J.K. Hedrick, Vehicle speed estimation using accelerometer and wheel speed measurements, in: Proceedings of the 2002 SAE International Body Engineering Conference and Automotive and Transportation Technology Conference, Paris, SAE 2002-02-2229.
[23] K. Wantanabe, K. Kobayashi, Absolute speed measurement of automobile from noisy acceleration and erroneous wheel speed information, in: International Congress and Exposition, Detroit, MI, SAE 920644.
[24] J. Liu, S. Li, J. Wang, K. Li, A fast identification method of vehicle longitudinal dynamic parameters for intelligent cruise control, Transactions of the Chinese Society of Agricultural Machinery 41 (10) (2010) 6–10 (in Chinese).
[25] M. Canale, S. Malan, Analysis and classification of human driving behaviour in an urban environment, Cognition, Technology and Work 4 (3) (2002) 197–206.
[26] L. Zhang, A Vehicle Longitudinal Driving Assistance System Based on Self-Learning Method of Driver Characteristics. Doctoral dissertation, Tsinghua University, Beijing, 2009.
[27] L. Shengbo, W. Jianqiang, L. Keqiang, xPC technique based hardware-in-the-loop simulator for driver assistance systems, China Mechanical Engineering 18 (16) (2007) 2012–2015.
[28] J. Wang, D. Zhang, K. Li, J. Xu, L. Zhang, Design of vehicular integrated electronic vacuum booster systems, China Journal of Highway and Transport 24 (1) (2011) 115–121.
[29] D. Zhang, J. Wang, S. Li, L. Zhang, K. Li, Double-mode vehicular electronic throttle for driver assistance systems, 2009 IEEE Intelligent Vehicles Symposium, Xi'an, China, 3–5 June 2009.

CHAPTER 10

Letting Drivers Know What is Going on in Traffic

Fang Chen, Minjuan Wang, Lian Duan

Department of Interaction Design and Technology, Institute of Applied IT, Chalmers University of Technology, Sweden

Chapter Outline

10.1 Introduction

It is obvious that if someone is driving on the street, he or she should know what is going on in the traffic around them. However, that is not the case when looking at what happens with traffic accidents. Every year, there are about 1.2 million people killed in traffic accidents [1] and 95% of traffic accidents are caused by human factors. This indicates that accidents happened due to drivers' lacking awareness of what was going on around them in traffic.

Vehicles are an effective transportation tool, but they are also dangerous! When accidents happen, vehicles can kill or injure the driver and passengers, or even other innocent people. In

Advances in Intelligent Vehicles. http://dx.doi.org/10.1016/B978-0-12-397199-9.00010-0

the history of automobile development, safety issues have been taken into consideration and different technical solutions have been developed.

In the automobile industry, the target customers are the drivers, or the people inside the cars. Therefore, the first priority is to protect these people's lives and property. As a result, safety belts, air bags, etc., are well developed and have been installed in almost all cars. The weak part of these systems is that they work only when accidents happen, and they can only reduce the severity of injury, but cannot prevent accidents from happening.

These passive safety systems do not provide any help for other innocent people who are unfortunately involved in these accidents, and also they do not help to reduce the number of accidents. Nowadays, the focus of vehicle safety research has shifted from passive safety to active safety, and the technology in this area is developing very rapidly. Currently, what has been taken into safety consideration includes not only the issue of drivers, and the passengers inside the vehicles, but also other road users, and the innocent people who may have the potential to be involved in vehicle accidents.

Active Safety Systems or Advanced Drive Assistant Systems (ADAS) are technical systems that have the potential to assist drivers to both avoid and minimize the effects of a crash. These systems include, for example, brake assistance, electronic stability control systems, adaptive cruise control, and collision warning/avoidance/mitigation. Due to the development of computer process capacity, sensor technologies, and image analysis technologies, there are more and more such systems that have been developed, or are under development. Figure 10.1

Figure 10.1: A Bird's Eye View of ADAS Systems.

provides an overview of these systems. This figure does not cover every system that is being developed or under the development in the world.

Nilsson [2] divided these systems into three categories:

- **Information/warning.** These systems provide different traffic information to drivers. The information is provided through different modalities and by different emergency levels. Examples of presenting information are night vision showing a clearer view in the dark, and blind spot detection systems indicating that there are objects in the blind spot area. Examples of providing warnings are the forward/backward collision warning, lane departure warning, and curve warning that warns the driver of an impending collision. Most information presenting designs for the ADAS systems have the function of warning the driver of potential hazards.
- **Active assistance/semi-automation.** These systems attempt to assist drivers in their driving tasks, such as helping to control acceleration, braking, and/or in steering the vehicle. Typical examples are adaptive cruise control, intelligent speed adaptation, collision mitigation and avoidance, lane departure mitigation, etc.
- **Full/high automation.** This system takes over the control of driving and acts automatically during driving. A typical example is the city safety system.

If we consider the time perspective from normal driving to a crash, we can divide the driving process into three phases, as shown in Figure 10.2. Most of the time, the driving situation is normal and there is no potential hazard in the traffic. During this period, the driver feels confident and comfortable. We refer to this situation as the driver being in a comfort zone. As soon as any potential hazard appears, drivers will need to pay extra attention to the traffic. In most of these cases, the driver is able to do something to prevent an emergency from happening. We call this period the safety zone, when the driver can still keep his driving safely under control. If the driver does not notice the potential hazard in time to take any proper action to avoid the accident, this period before the crash happens is called the time zone of lost control. From the moment of the crash and after is called the post-crash period.

For different driving periods, the driver's needs for information support are very different. During driving, drivers would prefer to drive in the comfort zone and keep away from any pre-crash situation. During this period, the driver needs to have good traffic situation awareness. Therefore, the ADAS system should provide the driver with necessary traffic

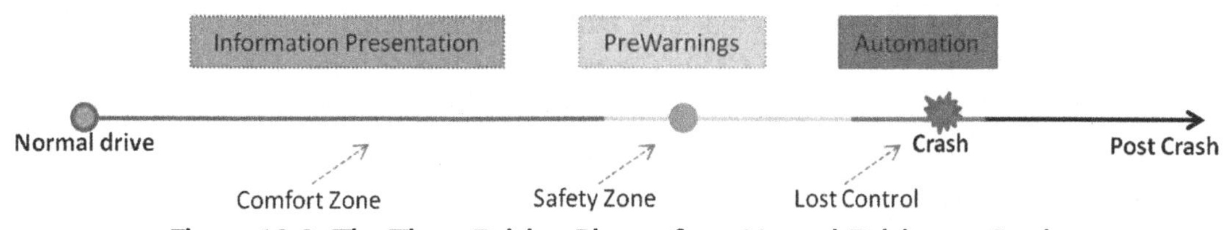

Figure 10.2: The Three Driving Phases from Normal Driving to Crash.

information to help the driver to stay inside the comfort zone for as long as possible. If any unpredictable event occurs, a pre-warning signal can be useful to help the driver focus attention on the potential traffic hazard, so that it can help the driver to stay within the safety zone during driving. If the driver does not manage to pay sufficient attention to the traffic hazard and take appropriate action to avoid the potential crash, this is the situation in which the driver loses control. In this period, the vehicle should take control and automatic action by the vehicle can take place.

With all these ADAS systems, it is possible to talk about letting the driver know what is going on in traffic, and let the driver stay within the comfort zone (Figure 10.2). Through ADAS systems, the driver can have good control of the vehicle, can make a good prediction of the traffic situation, and avoid accidents. However, we are still very far from that situation. The key problems are how to present the traffic information to drivers and how to design the driver–system interaction so that these systems can function to help drivers to avoid or minimize the effect of crashes. The bottleneck of ADAS technology applications is the Human–Machine Interface (HMI) design.

ADAS systems have the potential not only to protect drivers and passengers inside the vehicle, but also to protect other road occupants and pedestrians. Humans (drivers) play a vital role in the pre-crash phase. ADAS systems that protect other road occupants and pedestrians must be seen as distributed cognitive systems rather than as purely technical systems. The intelligibility of driving is no longer the property of the driver, but also of the ADAS systems. Knowledge of how drivers and vehicle systems function together is critical in achieving successful design of future vehicles. Hence, this presents a major challenge for HMI designers.

In the present automotive industry, HMI design for ADAS is mainly technically driven. Each technical system has a corresponding information presentation design. There is no communication and coordination between the systems. It depends on the driver's experiences on handling the in-vehicle information. This can cause a high visual and mental workload, as well as stress and frustration. For example:

- **Blind Spot Information System (BLIS).** The BLIS is a sensor or camera-based system used to provide the driver with information about vehicles, pedestrians or cyclists in areas not visible in the rear-view mirrors [3]. Information from these systems presented by a light on a pillar inside the car (normally these are mounted close to the side mirrors; see Figure 10.3 for an example) indicates other obstacles may be in the blind spot.
- **Forward Collision Warning (FCW).** The FCW uses sensor technology (laser or micro-wave radars) that measures distance, angular position, relative speed of the car, and any obstacles ahead. If there is a risk of crashing, the system provides the driver with a warning [3]. Warnings can be an optical warning like symbolic and text messages signaled by warning lamps or displays, or an acoustic warning like gongs, buzzers, or other sounds [4].

Figure 10.3: Blind Spot Indication.

- **Adaptive Cruise Control (ACC).** Using radar technology, the ACC automatically adjusts the vehicle's speed to maintain a safe distance from a vehicle ahead in the same lane. The ACC is expected to ensure that there is enough distance from the car in front, even if that driver unexpectedly reduces speed. The information presenting modality of the system, for example the BMW solution [5], is designed in this way: the ACC activation state (standby and active control) is indicated to the driver by an ISO symbol during the whole period of activation. In active control mode, the set speed is indicated by means of LEDs around the speedometer scale in a quasi-analog manner.
- **Lane Departure Warning (LDW).** LDW is a camera-based system that recognizes lane markings and is activated when a driver is about to leave a lane without using the turning signal. The types of warnings can be either acoustic, for example stereo beeps generated from a speaker positioned in the direction of the lane departure, or haptic, for example pulse-like steering torque, which would guide the vehicle back into the lane [6].

Even if a car was equipped with all four systems described above, each system would work on its own, disregarding the situations of the other systems. There is currently no cooperation among all the systems. Different ADAS systems that are implemented in a single car are developed and evaluated independently. This means there is no clear consideration and design concept about how these systems will work together and how these systems would provide the necessary and appropriate support to the driver, according to the driver's needs. This issue is probably not a serious problem now as there are only a few systems that are implemented in cars. When drivers dislike the presentation of a system, they can simply turn it off. However, in the near future, there may be many systems competing for the driver's attention when they are implemented in a real driving situation, if the systems are not complementary to each other.

The most common HMI designs for ADAS systems, including many systems available on the market, are to alert drivers of upcoming potential dangers. An active safety warning is a warning that requires immediate intervention by the driver to avoid a possible accident. Normally an effective warning system should integrate the subsystems of detection of extreme events, management of hazardous information, and public response. Neuroscience has indicated that certain brain functions and responses can be performed in parallel, while certain functions such as reasoning are performed serially. Therefore, when it comes to the design of warning systems, it is important to be able to selectively alert the driver based on the driver's state and the surrounding situations.

When we look at the current design of warning systems, some serious problems are very difficult to solve:

1. **The time issue.** When will the warning signal be presented to the driver? There is no common agreement among researchers and automotive manufactures. If the warning information is presented too early to the driver, it may cause distraction and discomfort. If it is presented too late, then drivers do not have time to react to it.
2. **Modality issues.** Three modalities are commonly used in HMI design: visual, auditory, and haptic. There is no common agreement on which modality should be used to present what information from the ADAS systems. Cognitive psychologists have indicated that a driver responds differently to certain modalities of stimulus (e.g. auditory, visual or haptic), depending on their current cognitive load and sensory stimulation.
3. **Information transparency.** The information presentation of current ADAS systems is often binary on/off signals, telling the driver only whether there is a hazard or not, but not what it is or how dangerous it is. However, the binary warning signal is very hard for the driver to comprehend and react to when the stimulus is activated.
4. **False alarm and driver acceptance.** As all the ADAS systems are computer-based and depend on sensor technology, data fusion, and image analysis, failure or error of the system cannot be avoided. A failure of the system can have a different impact on the interaction between the driver and the system that can cause misuse or disuse of the systems.

Warning the driver of potential dangers is not the same as letting the driver know the traffic situation. Warning designs do not take the real needs of the driver's driving into consideration. In this chapter, we will review the state of the art of the HMI on ADAS systems and how we intend to design them to let the driver know better the traffic situation during driving.

10.2 The Time Issue

The time issue is always critical for the driver. It is a well-known fact that 95% of traffic accidents globally are due to human factor problems, which means it is the drivers' or other road users' problem, rather than a technical failure of the vehicle. Taking a closer look into the "human factors" that contribute to traffic accidents, we can see that the driver's inattention to the sources of danger is the main factor. Inattention means that the driver does not pay attention early enough to potentially dangerous situations on the road. When they finally detect the problem, it is too late for them to react or to take corrective action for the situation.

In order to understand the situation better, we need to have a good knowledge of the process of driver perception-response time (PRT).

The perception-response time can be divided into four stages [7]: detection, identification, decision, and response. Figure 10.4 shows the process.

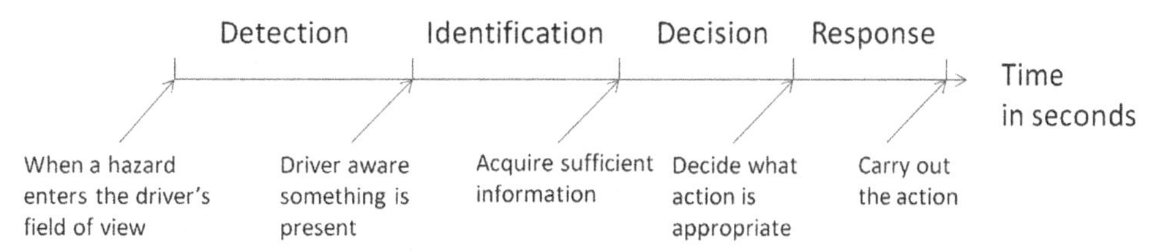

Figure 10.4: The Stages and Process of Perception-Response Time.

As the PRT is a very important parameter for designers to decide what time a driver should be warned about a hazard, there are many studies about driver PRT in the literature and a good summary can be found in Olson [7]. In this paper, he pointed out that there are many factors that can influence the driver's PRT; therefore, it is not correct to use only one value. Also, different values are reported in different studies, from 0.2 to 6.5 seconds, depending on the study setting, the situation, the driver's experience, etc. Green [8] considered five principal factors affecting PRT: expectation, urgency, age, gender, and cognitive load. However, the real situation is much more complicated and many other factors can also affect the PRT. For example, the traffic environment may present the driver with a multitude of situations requiring a response. The context in which the stimulus occurs varies greatly, which affects the driver's choice of responses, or creates difficulties in diagnosing the situation.

A warning design is intended to solve the problem of a driver's inattention to potential hazards on the road. The time period in the safety zone in Figure 10.2 should be larger than a PRT, so a warning signal should give the driver an alert early enough for appropriate action to be taken. However, a warning design can only help the driver in the first stage of PRT: drivers need to detect potential hazards on the road as early as possible, so they can respond appropriately. After understanding the situation with a driver's PRT, we can recognize a dilemma with warning design: a warning that is too early may be ignored by drivers if the cause of the warning cannot be perceived, whereas a warning that is too late will be ineffective [9]. Also, an alarm issued at a very early stage, such as in potential collision situations, can be viewed as a false alarm [10].

10.3 Modality Issues

Three modalities are commonly used in HMI design: visual, auditory, and haptic. Cognitive psychology studies have already pointed out that different modalities have their own specification and suitability for presenting specific information, and human reaction time to different modalities is different. For example, human reaction time to visual stimulation is about 200 ms, to auditory stimulation about 150 ms, while to pain (haptic) stimulation it is about 700 ms. Therefore, normally, a life-critical alarm is usually presented by sound. Sound has another advantage over visual presentation, since people can only perceive visual information that is presented in their visual field; sound can be presented anywhere within an audible area. Sound can also carry the information as direction, distance, and movements. In warning design, the modality issue is a critical factor. The right modality can help a driver to detect the hazard object faster, and even help identify the contents of hazards.

Visual and auditory [11] modalities are the most commonly used modality for warning designs. The visual modality has been the most commonly used to alert drivers. It has been argued that, for drivers, over 90% of driving-related information is perceived visually. At the same time, infotainment systems inside cars and many other after-market devices that drivers bring in to the car, such as mobile phones and GPS, also place demands on visual attention. Therefore, in HMI design, there is a tendency to develop other modalities that can be used to present different information to the driver. As visual warning demands the same cognitive resource as the driving task, it has been argued to be less effective [11] and to add extra workload to the driver. Auditory warnings are omnidirectional, which provides less conflict with the same cognitive resources as the driving task, and may be more effective.

During evolution, human beings have been continuously exposed to sounds carrying information from different directions. We have become accustomed to listening to simultaneous sounds and extracting the information from the source we desire [12]. The ability to quickly and correctly judge the position of, for example, a threat has been of vital importance to our survival. Even though our accuracy in sound localization is rather poor, about 4–10 degrees [13], the omnidirectionality and attention capabilities are important advantages. Our ability to simultaneously interpret both the direction and content of a sound has been recognized and used already in early information systems [14]. One example is the attempt to increase the navigational safety and accuracy of ships in foggy weather.

Using an auditory information presentation to offload the visual channel can have an impact on driving efficiency, especially when visual sensory information is degraded, for example when the information is not in the visual field, or visual attention is needed somewhere else. There are many studies on 3D audio applications that have been shown to effectively complement visual directional information, e.g. Refs [15–18].

Advantages like omnidirectionality, quick perception, and intuitiveness in support of our natural sound localization ability can be further explored in vehicle interaction design. For instance, auditory events can be positioned to increase the understanding of data or simply to improve the intelligibility of the auditory information. A 3D auditory display increases the understanding of what the information represents (auditory icons) and where the information is (3D position) [15,17,19–21]. With 3D sound we can simultaneously process sound direction, content, and interpret a specific sound amongst several others [14,20,22] with varying signal-to-noise ratios (SNR) [21] and for low-sample-rate systems below 4 kHz [13]. These benefits have also been shown in studies performed in noisy environments [13,20,22].

There are also some drawbacks with sound design, as it is crucial in the auditory warning process. Bad sound design can result in some sounds being very annoying [23], particularly highly urgent warnings [24].

The use of haptic modality to present information and warn the driver has been getting more attention recently and is being applied to vehicle interaction design [6,25]. The haptic modality or the sense of touch is relatively underutilized as an information channel and can be used for information reception. A tap on the shoulder is intuitive and almost part of our everyday communication and tactile displays can mimic this. The "tap on the shoulder" is simulated by causing a stimulation of the skin where the perception of skin stimulation reveals a relationship between the skin and the outside world, much like that of an audio or visual stimulation.

The skin delivers important information about the surrounding environment, important for many manipulative and exploratory tasks, by a dynamic interchange with our somatic senses [26,27]. It is a highly detailed and sensitive sensory organ with a temporal sensitivity close to that of the auditory system. This, in combination with the receptors of the skin being spread over a large area, makes it suitable to receive information relating to orientation and navigation. An example of such a design is setting the vibration on the steering wheel for a lane departure warning. As we discussed earlier, people take longer to react to haptic stimulation; therefore, it is not likely that this modality will be used for critical information presentation.

Multimodal interaction is regarded as the most intuitive way of communication between the user and the systems. There are two main approaches to combine several modalities: the parallel approach and the integrated approach [28]. The parallel approach presupposes that all modalities convey the full information alone, without support from each other, or coordination between each other [29]. The user can choose either of the possible modalities to interact with the system. The major advantage of this approach is that the user does not have to rely on a single modality. If, for example, visual warning is unsuitable at that time, then the user can choose another modality (auditory warning) and still have access to all the information. That is why most urgent warning signals normally with both flash lights and use sounds.

The integrated approach to multimodality means that the modalities in combination convey the full information. A typical example is that in a navigation system design, where the input could be a combination of speech and pointing at a touch screen (*I want to go from here* [pointing] *to here* [pointing]). This method of combining modalities is useful in multimodal systems where one modality is more suited to a specific task than the other. The drive simulation study conducted by Donmez et al [30] has proved that driver acceptance and trust varies largely depending on display modality.

10.4 False Alarm and Driver Acceptance

A big issue for vehicle manufacturers is the failure of one or multiple ADAS systems and how this failure would be perceived by drivers. Failure modes that can occur during operation of an active safety system are [31]:

- Environment related, such as deterioration of sensor signals due to weather conditions
- Equipment related, such as faults in sensors, actuators, computer hardware, and communication systems
- Vehicle related, such as faults in drive train, suspension or other vehicle subsystems
- Software faults, such as incorrect algorithms or software bugs.

However, for the driver, it is the same: alarm failure. Here we need to make a distinction between two situations. The first of these is when an alarm should be triggered and it is not (miss), the other is when an alarm is triggered when it should not be (false alarm). In most of the literature, a clear distinction between these two situations is not made. Actually these two types of failure can be interpreted very differently. For example, when a life-critical alarm is triggered while no collision takes place, this may be necessary and welcome in certain situations, as regards showing the sensitivity of the system and high security level. However, if a potential collision or crash occurs when the alarm is not triggered, people would totally lose trust in the system at once! This concern is supported in one study [32] about interaction-assisted active safety systems. In the experiment, participants were passengers in a driving simulator, and were presented with left-turn encroachment incidents. The direction of encroachment and post-encroachment time were manipulated to produce 36 near-crash incidents. After viewing each incident, the participant rated the relative acceptability of a hypothetical alert to it. The result supports the argument that alerts that are technically false alarms, in the sense that no collision takes place, are not always undesirable and instead may be necessary and welcome in certain situations.

In general, humans dislike failure alarms. The rate of false alarms may be a key factor in driver acceptance of novel active safety systems [33]. For example, a system that has high accuracy may also lead to overreliance upon it and drivers can thus misuse it. On the other hand, if a system produces many false alarms, the driver distrusts the system and will not use it (we will discuss overreliance and distrust issues in more detail in Section 10.5). This creates a dilemma for designers of active safety systems. Recognizing drivers' expectations of the acceptance rate is a challenge (see Figure 10.5).

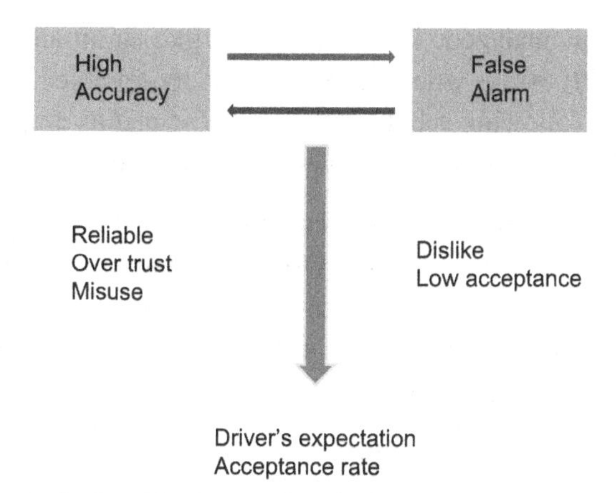

Figure 10.5: Relationship Between Driver Acceptance and False Alarm.

Historically, the goal of designing and developing an active safety system is usually to achieve the highest rate of system accuracy from an engineering perspective. Unfortunately, false alarms can be the rule when the base rate of the event to be detected is low [34]. It is difficult to estimate how often an alarm in a vehicle would be triggered in general. This rate depends entirely on whether drivers often drive in crowded big cities or in quiet countryside. This data can also be very different from country to country.

We can make an indirect estimation. For example, in the USA, there are almost 199 million licensed drivers, with 2.962 billion driving miles per year. We can estimate that a fatal motor vehicle traffic crash will occur once every 5200 driver years. A crash resulting in an injury will occur once every 107 driver years and in property damage once every 46 driver years [33]. Then, if we assume that life-critical alarms (such as forward collision warning, etc.) would be triggered probably 10 times more often than actual accidents that cause property damage crashes, that is every 4.6 driver years. This means, on average, a driver would experience the warning trigger once in every 4.6 years driven. At the same time, 90% of these alert signals are false signals! When false alarms are the rule, people either ignore them or disable the system.

There have been several different studies to evaluate drivers' acceptance and satisfaction levels toward different threshold settings of false alarms, but there is no standardized value that can be used, because the alarm effectiveness and drivers' trust varies in response to different systems and different driving contexts. It is not easy to study false alarms and driver acceptance in a simulation laboratory. The fundamental question is how often can the set up alarm be triggered in a simulation setting to make the results meaningful.

Bliss and Acton [35] conducted a study on different reliability levels of console emitted collision avoidance warning systems. The study was carried out on a simulator. The time taken by drivers to swerve, and the subjective opinion of usefulness and satisfaction of the system, were recorded and analyzed. The results show that alarm and automobile swerving reactions were significantly better when alarms were more reliable; however, drivers still failed to avoid collisions following reliable alarms. These results emphasize that alarm designers should maximize alarm reliability while minimizing alarm invasiveness.

As we discussed earlier, different ADAS systems that a single car may use have been developed and evaluated independently [36]. Driver age and driving experience may also contribute to the interpretation of the reliability of the design. The interface and interaction design between the driver and the system is another important issue that affects the driver's driving performance and acceptance of the system.

Young et al. [37] conducted a study that compared experienced and inexperienced drivers. In this study, two Intelligent Speed Adaptation (ISA) systems (informative and actively supporting) were tested on inexperienced and experienced drivers, using a driving simulator. The effectiveness and acceptability of ISA systems were investigated for experienced and inexperienced drivers. The results showed that effect of the ISA systems varies for the two different groups of drivers. Experienced drivers' subjective satisfaction ratings of the systems remained constant over the trial, and the ISA systems are more effective at reducing their speeds on certain road types. However, the inexperienced drivers' ratings changed after they became more familiar with the system.

Studies pointed out that the effectiveness and the acceptance of the ADAS systems not only depends on how accurate the systems performed their functions, but also on the effectiveness of the interaction design. Different interaction designs may provide totally different perceptions of the systems to drivers and their effectiveness will be very different. Acceptance is linked to trust such that low levels of acceptance lead to disuse. Higher levels of trust, however, do not necessarily lead to greater acceptability of technology [38]. Driver acceptance depends on a usable design of the system as well. Usability issues include ease of system use, ease of learning, perceived value, advocacy of the system or willingness to endorse, as well as driving performance [39].

10.5 Time-Critical Factors and Automation

There are strong arguments in the literature [30] that, in situations with time-critical elements (e.g. impending crash, or the "Crash" period in Figure 10.2), higher levels of automation are required [40]. The basic idea for automation design is that engineers do not believe that human beings' PRT is quick enough to deal with difficult situations, especially when they are distracted by other things, such as making phone calls, or drivers have not taken the appropriate actions. Therefore, the system needs to sense a near-fatal situation. The level of automation should be high enough to take control immediately. That is, if the driver is going to crash, the vehicle should take action.

However, when it comes to the design of the interaction between the driver and the automation system, many problems are seen. Trust is a particularly important factor that can influence the use of and the reliance on automation systems, which can result in a failure to provide the expected benefits [41]. Distrust may lead to the disuse of the automated system while mistrust can lead to a misunderstanding of the system's behavioral properties and a failure to recognize its limitations, thereby leading to inappropriate reliance on the system [42]. Mistrust can make the driver overestimate the capacity of the system and become overreliant on the system. It may lead the driver to amplify risk-taking behavior as the driver believes that the automated system will take corrective action when something happens. In such a situation, the failure of high levels of automation may lead to more severe safety problems than lower levels of automation. A study [43] on collision avoidance warning systems suggests that the warning systems at higher reliability levels led to overreliance and ultimately to maintaining shorter headways, and drivers may misuse this aid. Distrust, in turn, undermines drivers' response time to and acceptance of the system and influences overall system effectiveness [34].

High levels of automation may also lead to lower situation awareness [44]. The potential issues have several aspects. For example, Rothengatter et al. [45] suggest assistance systems could cause mental underload and loss of skills, resulting in a decrease in the driver's reliability. In another study, experts point out that in some situations it may cause another accident when the driver is alerted by the system to avoid an accident [36].

10.6 Ethnic Differences

Driving behavior is highly environmentally and culturally mediated. Lee [46] claims that culture has a significant influence on driving behavior, and can play a critical role in general driving safety. For example, there are studies on Chinese drivers' behavior showing that aggressive driving made a significant contribution to traffic accidents [47]. Statistics show that 84.6% of traffic accidents are caused by vehicle violations [48]. China, as one of the world's largest automobile markets, has attracted much attention recently. The traffic situation in China is not unique, it is similar to the situation in many developing countries. In these countries, the roads are crowded with many different road users and mixed types of transportation, and the road infrastructure and traffic are not well designed or coordinated.

Studies have indicated that many factors contribute to traffic accidents in China: infrastructural issues, such as the roads being constructed in the wrong way, poorly designed road signs or even lack of them [49]. Many studies indicate that driving behavior is considered as a major safety problem. In 2008, the report from traffic police offices in one province (there are 33 provinces in China) showed that bad driving habits, poor understanding of other road users, and poor respect for traffic regulations were factors that contributed to traffic accidents [36].

The major differences in traffic design and driver behavior between China and Western countries by implication is that the HMI design of ADAS for drivers in Western countries may not necessarily be optimal in other markets [50,51]. Lindgren et al. [52] pointed out that a system considered useful in one country can be seen as almost worthless (or even harmful) in another, and system settings feasible in one part of the world may not be suitable on the other side of the globe. Another study [53] showed different attitudes towards several types of ADAS between Swedish, American, and Chinese groups. It is believed that there is a requirement for re-design of the HMI on ADAS to adapt to the Chinese market, due to the more complex traffic situations and the driving culture in China.

Another problem, however, is that if too much account is taken of cultural differences, there is a danger that it will simply reinforce the bad driving habits of a particular culture. There is something of a "golden point" in the settings of the systems that will gain acceptance, while bringing about some important changes in driving styles to make the systems safer [36].

With the vehicle industry becoming more and more international, ADAS systems are being introduced in new and emerging markets. Often manufacturers take into account cross-cultural differences in various markets in terms of design. For example, most manufacturers have different in-car control and display designs for infotainment and navigation functions depending on whether a car is bound for the Asian, North American, or European market. However, with respect to ADAS, work to make market-specific design adjustments still seems to be in its very early stages [54]. As Western countries are the most dominant market, the development of ADAS is generally based on perceptions about the needs of drivers in those countries. An increasing amount of research has recently been published on the technical aspects of ADAS development, with many of these publications coming from China [55–57]. There is scant information, however, on attempts to understand the traffic situations, driver behavior, and driver goals among Chinese drivers, which creates a problem when designing products for a global market. The key to successful design for global use is an understanding of how needs and requirements differ around the world [58]. This may be particularly true for ADAS since not only the rules of the road, but also social environments, norms, and driver behavior may vary significantly from country to country and have a notable influence on attitudes and behaviors of drivers [59].

10.7 From Warning to Information Presentation

Currently, warning design is widely employed among ADAS systems. As we know, warning signals can only alert the driver to the urgency of a situation, not tell what the danger is and how to react. In addition, warning signals are activated in near-crash situations, resulting in drivers often not having the time to understand the warning information and react appropriately due to frustration and panic. In general, a warning signal was considered as negative feedback by drivers and therefore should be used sparingly [60]. A warning does not provide sufficient information to assess driving safety by simply telling drivers there is a danger a few seconds prior to collision. If we go back to the PRT model (see Figure 10.4), a warning signal can help the driver in the first stage of quickly detecting a potential hazard, but can provide no help during the rest of the stages in PRT.

There are so many negative issues and limited benefits from warning designs. We need to find a better solution to enhance driving safety. The solution we will argue here is to support the driver with good traffic situation awareness, so that drivers can be helped to keep their car within the comfort zone during driving, as indicated in Figure 10.2.

Endsley [61] gives the following definition of Situational Awareness (SA): "The perception of elements in the environment within a volume of time and space, the comprehension of their meaning, and the projection of their status in the near future." Thus, SA has three levels: Level 1, perception of the elements in the environment; Level 2, comprehension of the current situation; and Level 3, projection of future status. SA is like the head of a train leading the decision-making process and executing a decision with the right performance. Compared with the PRT stages in Figure 10.4, to support drivers' SA means to provide information to support the stages of identification, decision, and response.

Taking the approach of a design information system to support drivers with good traffic situation awareness, we will be able to reach the goal of "letting drivers know what is going on in traffic". The idea is to present upcoming traffic information to the driver continuously during driving, and provide the driver with an overall understanding of the current situation. By doing this, the driver will always have a good understanding of the traffic situation. When "surprises" and unexpected traffic conflicts occur, the driver will be able to quickly identify potential hazards, take the right decision, and quickly respond.

Driving can be understood as a task involving continuous adaptation to a changing traffic environment. In order to drive safely and efficiently, we need information from the surrounding traffic environment that makes it possible for us to foresee events that may negatively affect safety. We compare this external information with our internal representations of the situation and use this to predict the future state of our vehicle. According to Summala [62], a driver's normal goal is to drive without discomfort, which is staying within the driver's comfort zone (see Figure 10.2). In a situation in which the comfort zone boundary is exceeded and thus the safety margin is violated, the driver will experience discomfort and try to adapt his or her behavior in terms of corrective actions. A number of ADAS systems that vehicle manufacturers have introduced to the market [63] aim to support adaptation to changes in the driving environment and thus reduce the risk of exceeding the safety zone boundary. However, in order to create a successful design, it is necessary to find out as much as possible about the targeted users, the drivers and the situations in which a system is to be used [64]. It is thus important to understand what type of drivers may use a system, their goals, and how they will use the system. An Ecological Interface Design (EID) approach has been adopted throughout the work published by Lindgren [65]. The fundamental premise of EID is that a person's goal-directed behavior will be enhanced by revealing goal-relevant constraints of the environment in a way that is compatible with the perceptual capabilities of the person [66]. The HMI design for ADAS is not only affected by the rules of the road, but also social environment, norms, and driver behavior, and may vary significantly from country to country and have a notable influence on the attitudes and behaviors of drivers [59].

To reach the aim of "letting the driver know what is going on in traffic", a key component is to let the driver know about the vehicle's current position in relation to its destination, the upcoming road condition, the relative positions and behavior of other road users, especially the potential hazards, as well as how these critical variables are likely to change in the near future [67]. Continuous knowledge of these factors enables effective decisions to be made in real time and for the driver to be "coupled to the dynamics of their environment".

The following aspects should be considered when designing information presentation for the driver:

1. The information should not only be to simply inform drivers that there is a hazard coming, but also to provide information on the nature of the hazard, the geographic location of the hazard, and how dangerous it is to the driver. To achieve this, the system should filter out noise or irrelevant information from the traffic environment, reproduce and enhance the useful and critical information cues. These cues are, for instance, movements of pedestrians, cyclists, motorcyclists, vehicles, etc.
2. The information presentation should be designed in an intuitive and natural way, for example through auditory icons or visual symbols, to help the driver quickly identify the characteristics of other road users and not add extra mental workload to the driver. Such design enables the driver to respond quickly and accurately to critical situations when they occur.
3. The focus of information presentation should not be only on near-crash situations, but in normal driving situations, discerning whether or not the driver will become involved in a conflict in the near future. If we can assist the driver to avoid the incident and staying in the comfort zone (Figure 10.2), then the possibility of these situations evolving into accidents would be reduced. Through various ways of presenting information, the driver would be aware of the surrounding traffic, like someone walking on the street and having a good awareness of their surroundings. Even if involved in any traffic conflict,

drivers could identify the situation in advance and predict what might happen in the near future, therefore having enough time to make the correct decision and take appropriate action.

Here we can take Blind Spot Information Systems as an example to illustrate the differences between warning design and information presentation. Figure 10.3 shows a design for a signal from BLSI. The warning lights are located at respective left/right A pillars. The blinking light represents the presence of a road user in the blind spot. Current BLSI systems only present On/Off information of whether there is an object or not. They do not provide information on the characteristics of this object. The same presentation applies whether driving in the city or on a motorway. For a driver, this is insufficient information. For instance, in city driving, it is very important for the driver to know whether there is a vehicle or pedestrian near to his/her car. Drivers may have different strategies to cope with different road users. If the driver plans to change to the left lane when a vehicle is appearing in the blind zone of that lane, the driver would need to estimate what kind of vehicle it is and the speed of the other vehicle, so they could decide either to accelerate or decelerate in order to execute the desired movement. The strategy would be very different if it was a personal car, a truck, or a motorcycle in the blind spot. However, if it is in a city situation and the pedestrian shows up in the right blind spot while the driver was intending to make a right turn, the driver could stop the car and let the pedestrian pass first.

One of the challenges is how to design appropriate information presentation that supports the driver with relatively continuous traffic information without creating any undue distraction to drivers. As we have discussed in Section 10.3, the design of the information presentation will influence drivers' acceptance and satisfaction. In order to design efficient and pleasurable information presentation, there are two aspects that need to be handled carefully: the driver's need for information and the information overflow consideration. We need to identify drivers' dynamic needs under different traffic situations, and understand how to prioritize the information under certain conditions. We should avoid information overflow and causing undue distraction due to inappropriate design.

One of the problems with modern cars is that once a driver is inside the car, his/her visual and aural perceptions are incomplete. There are areas surrounding the car that the driver is not able to see. The driver's sight is blocked by many other obstacles as well, such as a hidden pedestrian behind another vehicle, a bicycle coming out from a side street, or simply the driver is occupied by other tasks, or lost in his/her own thoughts. The noise-insulated car shell blocks the driver's audio perception of outside traffic. As a result, the driver does not know what is going on outside of the car. Horswill and McKenna [68] reported that drivers who perceived quieter internal car noise chose to drive faster than those who perceived louder car noises. Also, quieter cars tend to encourage reduced headway and riskier gap acceptance. More recently, research has also shown that drivers of older vehicles show better situation awareness, presumably due to increased noise, vibration, and sounds from the environment [69,70].

Research in this area is very limited. Here we will provide an overview of the results of these few studies. Mainly, there are two modalities that are considered for presenting traffic situation information to the driver: visual and auditory.

10.7.1 Visual Information Design

The literature on visual information presentation from ADAS is mainly from Lindgren and Chen [71,72]. Their work can be divided into three parts: requirement study, interface design, and evaluation of drive simulation. The studies were carried out both in Sweden and China, which made the results much more interesting when cultural differences were taken into consideration.

Part 1: Driver Needs, Goals, and Requirements

To assess how drivers understand traffic situations in which they may need assistance and how these understandings may differ between cultures, driver interviews and expert evaluations were carried out. Participants in the interviews and evaluators were shown video recordings of potentially dangerous driving situations recorded during normal driving in China. The reason for choosing video recordings from China was because it covers many variables of incidents. Two series of semi-structured interviews were conducted with 20 (12 male and eight female) Chinese drivers and 20 drivers (10 male and 10 female) from Sweden. The five people participating in the expert evaluation (four males and one female) were all Swedish and working in the area of HMI and vehicle safety.

The results of the evaluations showed that the Chinese driving in the selected videos was considered to be much more risk-taking and aggressive than in Sweden and Scandinavia. These large differences in driver behavior are problematic when designing ADAS systems, such as LDWs and lane changing assist systems, as they may be of limited use in the Chinese market, especially with Northern European system settings. In addition, differences in infrastructure between countries and continents obviously affect people's driving and, as a result, the type of traffic situations that occur. The experts found these infrastructural differences to be a significant issue in designing ADAS systems. Along with a serious need to improve road and lane markings, they found warning signs near roadwork to be insufficient and difficult to notice. Meanwhile, the experts suggested using ADAS to provide drivers with feedforward information about upcoming roadwork areas. This could help them to plan their driving more effectively and could be of special help in situations with bad visibility and weather conditions, which the participants in the Chinese driver interviews found problematic. This type of precautionary information has been proposed for systems that detect possible dangers but where no urgent action is necessary [73–79] and could also be implemented to warn the driver about vehicles ahead traveling at slower speeds going in the same direction, such as the road obstacle situation. Today's critical collision avoidance systems are set to give a warning based on the distance and relative speed between two objects. It could therefore be useful to complement these systems by having a more precautionary system that provides drivers with feedforward information, giving them more time to think and react.

Generally, considering the other problems identified by the participants in the interview studies, it can be concluded that most of the problems identified by the Swedish participants were also identified by the Chinese. However, it was evident that the situations presented were much more commonly recognized as problems in Sweden than in China. Also, a majority of the Chinese found the situations to be very common while the Swedish found the opposite. This was evident in all driving situations and shows the strong cultural differences between China and Sweden as, even though the traffic regulations are similar, social norms and institutions differ extensively. In general, in HMI design, social and cultural factors are important in the development of products [58,80].

In the area of ADAS, however, little consideration has been given to these fundamental questions. A system designed for the Swedish market, where these types of incidents seldom occur as the basic rate is low, is unlikely to be accepted in China, where warnings would be triggered several times a day, in situations that Chinese drivers do not find very stressful or dangerous. This may cause irritation and could ultimately result in drivers either ignoring the systems or shutting them off.

Part 2: Designing an ADAS Interface

The success of interface design depends on tight coupling between information and action, and this is where the user enters the picture [81]. The research approach for gathering requirements for interface design was largely based on understanding drivers and their goals in different traffic situations presented in Part 1. In the design process of the interface, an Ecological Interface Design (EID) approach was adopted. EID is a relatively new approach to designing human–machine interfaces, and the term "ecological design" refers to an interface that is designed to reflect a work environment's different constraints in a perceptually available way for those who use it. By showing the user the constraints of the system when it is working appropriately, any break in the constraints indicates that an event is taking place [82]. However, what is not specifically targeted in EID is how an interface actually should look and how the design will be understood by the intended end-users of the system. Therefore, the traditional parts of the EID process were complemented by more user-centered methodologies to evaluate the design at different stages of the design process.

The prototype of the interface was generated by iterative means. Normal drivers and design experts were included as participants in the design process. Only the Swedish drivers and Swedish HMI experts participated in the prototype development process. The design was created by interaction design experts and was first tested by the HMI experts and then by skillful drivers. The results showed that, in general, the evaluators agreed that the interface was easy to interpret and that it was thus also fairly easy to understand what was going on in the adjacent traffic. Further, the evaluators shared the general impression that there was a good and consistent match between the system and the real world. However, the evaluators also stated that this type of interface might distract the driver as it could be detected from the periphery. They suggested that there should not be too much information presented on the display at the same time.

Part 3: Simulator Studies

Two simulator studies were conducted to evaluate the interface concept, one in China (16 participants) and one in Sweden (14 participants). In both experiments the STISIM Drive simulator software developed by System Technology Incorporated [83] (www.systemstech.com) was used to evaluate driver behavior. Previous research has shown that this medium fidelity simulator software appears to be sensitive in objectively evaluating driving performance [84] and correlates with on-road testing [85,86]. In the simulation study, the visual presentation interface was compared with corresponding warning systems. Three ADAS warning systems were included in the experiment: Forward Collision Warning (FCW), Curve Speed Warning (CSW), and Lane Departure Warning (LDW).

These two experiments identified five main findings:

1. There were no differences in longitudinal control and the advisory display did not require any extra visual attention.
2. It was evident that the Chinese participants were involved in considerably more collisions compared to the Swedes. This indicates a difference in driver behavior, with the Chinese driving more aggressively, committing more violations, and taking greater risks.
3. Looking at the number of CSWs triggered, there was no significant decrease in the number of warnings that were triggered between the baseline and critical conditions. These results agree with results presented by Leblanc et al. [76], where no significant differences were found in speed between driving with and without a combined cautionary (haptic) and critical (auditory) CSW.
4. Participants drove with a larger standard deviation of lateral position (SDLP) when driving without warning. With increased lane deviations indicating degraded control [87], it is positive that both the critical and advisory warnings help drivers maintain better lateral control. One reason that there were no major differences between the two warning conditions may be that the drivers did not consider a lateral deviation as something that was potentially dangerous until they were very close to either the road center or the curb and

therefore neglected the advisory information because they thought it was redundant. Further, the Chinese group showed significantly larger lane deviations during all three conditions. This may be because the driving scenarios were developed based on Swedish road and traffic environments. A majority of the Chinese participants commented on the low traffic density and low complexity of the scenarios. This may have been a reason why, for example, they cut curves and drove more actively and thereby had larger deviations.

5. According to the Swedish participants' subjective ratings, higher situation awareness was associated with the advisory condition compared to the critical one. Participants found the advisory warnings more helpful in detecting dangers and believed that the advisory warnings made them more aware of the surrounding traffic situation. As there was no indication of increased annoyance associated with the warnings, these results suggest that a complementary advisory display warning is a promising design alternative to the critical warnings available today.

In general, the results of this study show that continuous information given in the advisory display were appreciated by the participants and perceived as non-obtrusive. In accordance with the results of Seppelt and Lee [88], this study also indicates that this type of continuous information can enhance performance, even in a domain such as driving, where demands on visual attention are high.

While the Swedish participants found the advisory display preferable, their opinions were not shared by the Chinese participants, who preferred the critical sound warnings over the advisory display. In contrast to the Swedish participants, they did not experience the sound warnings as distracting or irritating. Instead they thought of them as being a better help in identifying potential dangers compared to the advisory display. One reason for this may be the way drivers in China use sound in the traffic environment. One example is the use of the horn when driving. Lajunen et al. [89] found that "honking" clearly reflects aggression in Scandinavia, whereas drivers in more Eastern cultures use their horns frequently to give a great variety of messages. It would therefore be normal for them to receive sounds as indicators in traffic, and with heavier traffic and smaller safety margins. Chinese drivers may experience warnings that are considered critical in Scandinavia as being of a more advisory nature. Problems concerning the cross-cultural differences and their effects on the choice of information presentation and interface modalities will be discussed in the next section.

10.7.2 Auditory Information Presentation

There are even fewer publications regarding using auditory information systems [90]. Therefore, here we will present some of our ongoing studies.

We have carried out a large study to investigate drivers' attitudes and requirements towards using 3D Auditory Traffic Information Systems (3D ATIS). In the study, a concept of 3D sound systems was introduced into the driver's compartment so to increase the driver's awareness of the traffic situation by hearing different road users' locations, distance, moving speed, and moving direction related to the vehicle. The 3D ATIS was intended to be designed in an intuitive and natural way, so that the auditory information cues enable the driver to gain an overall SA. The system did not intend to indicate the immediate threats rather than providing continuous information that would be relevant to the driver. This design can possibly guide the driver's attention in the right direction, when necessary, and without causing discomfort. The conceptual system displays the traffic situation surrounding the driver's car. For instance, if there is a pedestrian passing from left to right in front of the car, the auditory system will mimic the movement of the pedestrian, such as the direction of movement, speed, and relative position. Therefore, a driver can better foresee the movement of the pedestrian, and have a better driving plan.

In order to generate this concept, we developed a prototype system using synthesized 3D sound to simulate the traffic environment for the driver. The purpose of the study is to find which type of sounds drivers prefer to hear: abstracted auditory earcons, or auditory icons, and drivers' attitudes and requirements towards information presentation design. Four different traffic scenarios – city, highway, roundabout, and residential area – were selected. Several video clips were selected from field operative tests in China corresponding to the scenarios. The experimental settings tried to mimic the environment by using sounds to simulate the scenarios from video clips, inside the cabin with 3D ATIS. Focus group discussions with seven groups of Swedish drivers and six groups of Chinese drivers were carried out. During the study, each video clip was played to the group with different combinations of sound design, and a group discussion was performed to discuss the design of each.

In conclusion, the acceptance of 3D ATIS was generally positive among both Swedish and Chinese drivers. The results of the study support the idea that using natural sound (meaningful sound) is more appreciated by drivers than abstract sounds. It is easy to associate a natural sound with its representative meaning and it requires little learning effort. How detailed the traffic information should be depends on the experience level of the drivers and their driving routines. The conceptual design of 3D ATIS can support the driver with visual searching direction and increase the awareness of the traffic situation surrounding the driver.

However, the studies reported here can be considered as pilot studies. Many unsolved research questions remain that require some solid studies both in the laboratory and in the field before implementation of the technology can be considered in real cars. The first question is when and what the road users should be presented with via 3D sounds. The study has defined and generated several information models and prioritization rules. These rules need to be investigated further and evaluated in simulation studies.

The second question is whether there are possible distractions and annoyances induced by the 3D sound. This question is closely connected to what kind of sounds should be used and how often, how long the duration and how loud the sound should be. At the same time, individual experience and preference may be very different as well. There is no theory or standards that can be applied for the design, so some solid research work is also necessary. Long-term laboratory work may be necessary to answer these questions. More experience in real road driving with integrated 3D sound systems may give more realistic answers. One conclusion that can be taken into serious consideration from the present studies is that the use of natural sounds can give clear identification of an object. This is preferred by all subjects who were involved in the studies. The advantage of using natural sounds is that it provides the feeling of "amplification of the real auditory world" that we are familiar with. Using natural sounds can reduce the learning and recognition process of sound identification.

The third question regards the sound design. The "natural sounds" that may possibly be used inside the vehicle are still synthetic sounds and cannot represent every detail of the various types of road users. For example, a footstep will sound different between an old man walking or a young lady walking. Such detailed perceptions of sounds may differ between individuals as well. What might happen if a driver hears the footsteps of a young lady while seeing an old man in front of his vehicle? The levels of simulation fidelity need to be carefully considered.

The fourth question is how individual differences of 3D sound localization capacity affect hazard detection when it is in the blind spot? It is well known that people can localize horizontal presented sounds with 3–6 [db] accuracy [12,91] with large individual differences. When it comes to the applications in an in-car environment, the localization accuracy should not be a great obstacle for consideration, since the sound source that will be presented to the driver is dynamic, and the listener, the drivers, can also move their heads to actively allocate the sound source. This movement facilitates localization and the best localization performance can be achieved when the head is turned to face the sound source [92]. Movement also aids interaural cues, reducing the localization blur by bringing the auditory event closer to the region of sharpest hearing [93]. The auditory location accuracy should not be a critical problem, since if the driver can distinguish if the sound is from the front or from the back, as well as if it is on the right side or on the left side, then that is good enough.

10.8 Conclusion

At present, there are very limited benefits from using ADAS systems due to the fact that the information therefrom is provided to drivers mainly through warning signals. Warning design has many negative issues, such as the time issue, the modality issue, information transparency, and the false alarm issue. It is regarded as an unpleasant stimulation. If we consider the time perspective from normal driving to crash, we can divide driving into three phases: normal driving in a comfort zone, pre-crash in a safety zone, and automation when the driver loses control. Most of the time, drivers are inside their comfort zone whilst driving. The rate of alarms triggered in Western countries is generally very low. To enhance the safety of driving, it is suggested to design the ADAS HMI to support the driver with traffic situation awareness by providing a presentation of continuous traffic information during normal driving. This advisory information can help drivers to continue driving in their comfort zone. It can also help drivers to predict oncoming traffic events and prepare to take appropriate action when the unpredictable occurs. The challenge for supporting information design requires the designer to have a thorough understanding of the driving task. Ecological interface design concepts can guide problem analysis, and human-centered design provides a good method for interaction design. The ADAS information to support normal driving can be presented in visual and auditory forms. Different approaches are also discussed in the chapter.

References

[1] M. Peden, et al., World Report on Road Traffic Injury Prevention, World Health Organization, Geneva, 2004.
[2] J. Nilsson, On the Interaction Between Driver Assistance Systems and Drivers in Situations of System Failure, Chalmers University of Technology, 2011.
[3] N. Floudas, et al., Review and Taxonomy of IVIS/ADAS Applications, Information Society Technologies (IST) Programme, 2004.
[4] M. Maurer, Forward collision warning and avoidance, in: A. Eskandarian (Ed.), Handbook of Intelligent Vehicles, Springer, London, 2012, pp. 659–683.
[5] W. Prestl, et al., The BMW active cruise control ACC, in: SAE 2000 World Congress, 2000.
[6] K. Suzuki, H. Jansson, An analysis of driver's steering behaviour during auditory or haptic warnings for the designing of lane departure warning system, JSAE Review 24 (2003) 65–70.
[7] P. Olson, Driver perception-response time, in: R.E. Dewar, P.L. Olson (Eds.), Human Factors in Traffic Safety, Lawyer & Judges Publishing Company, 2002.
[8] M. Green, "How long does it take to stop?" Methodological analysis of driver perception-brake times, Transportation Human Factors 2 (3) (2000) 195–216.
[9] J.D. Lee, D.V. McGehee, T.L. Brown, M.L. Reyes, Collision warning timing, driver distraction, and driver response to imminent rear-end collisions in a high-fidelity driving simulator, Human Factors 44 (2) (2002) 314–334.
[10] T. Dingus, Human Factors Design Issues for Crash Avoidance Systems, Lawrence Erlbaum Associates, 1998.
[11] C.D. Wickens, J.G. Hollands, in: N. Roberts, B. Webber (Eds.), Engineering Psychology and Human Performance, third ed., Prentice-Hall, 1999.
[12] D. Withington, The use of directional sound to improve the safety of auditory warnings, Proceedings of the Human Factors and Ergonomics Society Annual Meeting 44 (22) (2000) 726–729.
[13] D.R. Begault, Auditory and non-auditory factors that potentially influence virtual acoustic imagery, in: AES 16th International Conference on Spatial Sound Reproduction, Citeseer, 1999, pp. 1–14.
[14] M.A. Ericson, R.L. McKinley, The intelligibility of multiple talkers spatially separated in noise, in: R. Gilkey, T. Anderson (Eds.), Binaural and Spatial Hearing in Real and Virtual Environments, 1997, pp. 701–724.
[15] E.C. Haas, Utilizing 3-D auditory display to enhance safety in systems with multiple radio communications, in: S. Kumar (Ed.), Advances in Occupational Ergonomics and Safety, IOS Press, Amsterdam, 1998, pp. 715–718.

[16] W.T. Nelson, et al., Spatial audio displays for speech communications: A comparison of free field and virtual acoustic environments, in: Proceedings of Human Factors and Ergonomics Society 43rd Annual Meeting, Human Factors and Ergonomics Society, 1999, pp. 1202–1205.
[17] J.A. Veltman, A.B. Oving, A.W. Bronkhorst, 3-D audio in the fighter cockpit improves task performance, International Journal of Aviation Psychology 14 (3) (2004) 239–256.
[18] O. Carlander, L. Eriksson, M. Kindström, Horizontal localisation accuracy with COTS and professional 3D audio display technologies, in: Proceedings of the International Ergonomics Association 2006 Conference, International Ergonomics Association, Maastricht, 2006.
[19] D.R. Begault, E.M. Wenzel, Headphone localization of speech, Journal of Human Factors and Ergonomics 35 (2) (1993) 361–376.
[20] W.T. Nelson, et al., Monitoring the simultaneous presentation of spatialized speech signals in a virtual acoustic environment, Security 166 (June) (1998) 159–166.
[21] N.L. Vause, W. Grantham, Speech intelligibility in adverse conditions in recorded virtual auditory environments, in: International Conference on Auditory Display, Glasgow, 1998.
[22] T.J. Doll, T.E. Hanna, J.S. Russotti, Masking in three-dimensional auditory displays, Human Factors 34 (3) (1992) 255–265.
[23] B. Berglund, K. Harder, A. Preis, Annoyance perception of sound and information extraction, Journal of the Acoustical Society of America 95 (3) (1994) 1501–1509.
[24] E.E. Wiese, J.D. Lee, Auditory alerts for in-vehicle information systems: The effects of temporal conflict and sound parameters on driver attitudes and performance, Ergonomics 9 (2004) 965–986.
[25] P. Griffiths, R. Brent Gillespie, Shared control between human and machine: Haptic display of automation during manual control of vehicle heading, in: Proceedings of the 12th International Symposium on Haptic Interfaces for Virtual Environment and Teleoperator Systems, 2004.
[26] N.R. Carlson, Physiology of Behaviour, seventh ed., Allyn & Bacon, Needham Heights, MA, 2001.
[27] J.B.F. van Erp, Tactile Displays for Navigation and Orientation: Perception and Behaviour, Utrecht University, 2007.
[28] F. Chen, et al., Application of speech technology in vehicles, in: F. Chen, K. Jokinen (Eds.), Speech Technology Theory and Applications, Springer, New York, 2010, pp. 195–219.
[29] B. Bringert, R. Cooper, Multimodal dialogue system grammars, in: Proceedings of DIALOR, Nancy, 2005.
[30] B. Donmez, et al., Drivers' attitudes toward imperfect distraction mitigation strategies, Transportation Research Part F: Traffic Psychology and Behaviour 9 (6) (2006) 387–398.
[31] O.J. Gietelink, et al., VEHIL: Test facility for fault management testing of advanced driver assistance systems, in: Proceedings of the 10th ITS World Congress, Madrid, 2003, p. 13.
[32] K. Smith, J.-E. Kallhammer, Driver acceptance of false alarms to simulated encroachment, Human Factors 52 (3) (2010) 466–476.
[33] J. Kallhammer, K. Smith, J. Karlsson, Shouldn't cars react as drivers expect?, in: Proceedings of the Fourth International Driving Symposium on Human Factors in Driver Assessment, Training and Vehicle Design (2007) 9–15.
[34] R. Parasuraman, P.A. Hancock, O. Olofinboba, Alarm effectiveness in driver-centred collision-warning systems, Ergonomics 40 (3) (1997) 390–399.
[35] J.P. Bliss, S. Acton, Alarm mistrust in automobiles: How collision alarm reliability affects driving, Applied Ergonomics 34 (6) (2003) 499–509.
[36] A. Lindgren, F. Chen, et al., Requirements for the design of advanced driver assistance systems – The differences between Swedish and Chinese drivers, International Journal of Design 2 (2) (2008) 41–54.
[37] K.L. Young, et al., Intelligent speed adaptation – Effects and acceptance by young inexperienced drivers, Accident Analysis and Prevention 42 (3) (2010) 935–943.
[38] M. Siegrist, The influence of trust and perceptions of risks and benefits on the acceptance of gene technology, Risk Analysis 20 (2) (2000) 195–203.
[39] M. Stearns, W. Najm, L. Boyle, A methodology to evaluate driver acceptance, in: The Transportation Research Board, 81st Annual TRB Meeting, Washington, DC, 2002.

[40] N. Moray, T. Inagaki, Attention and complacency, Theoretical Issues in Ergonomics Science 1 (4) (2000) 354–365.
[41] R. Parasuraman, V. Riley, Humans and automation: Use, misuse, disuse, abuse, Human Factors 39 (2) (1997) 230–253.
[42] J.D. Lee, K.A. See, Trust in automation: Designing for appropriate reliance, Human Factors 46 (1) (2004) 50–80.
[43] M. Maltz, D. Shinar, Imperfect in-vehicle collision avoidance warning systems can aid distracted drivers, Transportation Research Part F: Traffic Psychology and Behaviour 10 (4) (2007) 345–357.
[44] M.R. Endsley, Measurement of situation awareness in dynamic systems, Human Factors 37 (1) (1995) 65–84.
[45] T. Rothengatter, et al., The driver, in: J.A. Michon (Ed.), Generic Intelligent Driver Support, A Comprehensive Report on GIDS, Taylor & Francis, London, 1993, pp. 33–52.
[46] J.D. Lee, Driving safety, in: R.S. Nickerson (Ed.), Review of Human Factors, Human Factors and Ergonomics Society, Santa Monica, 2006, pp. 172–218.
[47] C.-qiu Xie, D. Parker, A social psychological approach to driving violations in two Chinese cities, Transportation Research Part F Traffic Psychology and Behaviour 5 (4) (2002) 293–308.
[48] Q. Zhan, Attention to road safety - first Chinese Symposium on transport and road safety, World Auto 5 (2007) 26–27.
[49] A. Lindgren, F. Chen, et al., Cross cultural issues and driver requirement for advanced driver assistance system, in: 2nd International Conference on Applied Human Factors and Ergonomics, Las Vegas, 2008.
[50] F. Chen, et al., How shall we design the future vehicle for the Chinese market?, in: Proceedings of the 18th Congress of the International Ergonomics Association, 2011.
[51] L. Duan, F. Chen, The future of Advanced Driving Assistance System development in China, in: 2011 IEEE International Conference on Vehicular Electronics and Safety (ICVES), Beijing, 2011, pp. 238–243.
[52] A. Lindgren, F. Chen, et al., Do you need assistance?, Naturalistic Case Studies Investigating the Need of Advanced Driver Assistance Systems During Normal Driving, 2007.
[53] A. Lindgren, R. Broström, et al., Driver attitudes towards Advanced Driver Assistance System – cultural differences and similarities, in: D. de Waard, et al. (Eds.), Human Factors Issues in Complex System Performance, Shaker Publishing, Maastricht, 2007, pp. 205–215.
[54] D.M. Krum, et al., All roads lead to CHI: Interaction in the automobile, in: CHI '08 Extended Abstracts on Human Factors in Computing Systems, ACM, New York, 2008, 2387–2390.
[55] J. Gong, et al., High speed lane recognition under complex road conditions, in: IEEE Intelligent Vehicles Symposium, IEEE, Piscataway, 2008, pp. 566–570.
[56] H. Zhao, et al., Driving safety and traffic data collection – A laser scanner based approach, in: IEEE Intelligent Vehicles Symposium, IEEE, Piscataway, 2008, pp. 329–336.
[57] X. Wu, et al., Color vision-based multi-level analysis and fusion for road area detection, in: 2008 IEEE Intelligent Vehicles Symposium, 2008, pp. 602–607.
[58] B. Shneiderman, C. Plaisant, Designing the User Interface, Addison-Wesley, 2005.
[59] D.M. Zaidel, A modeling perspective on the culture of driving, Accident Analysis and Prevention 24 (6) (1992) 585–597.
[60] M. Roetting, et al., When technology tells you how you drive – Truck drivers' attitudes towards feedback by technology, Transportation Research Part F: Traffic Psychology and Behaviour 6 (4) (2003) 275–287.
[61] M.R. Endsley, Toward a theory of situation awareness in dynamic systems, Human Factors 37 (1) (1995) 32–64.
[62] H. Summala, Towards understanding motivational and emotional factors in driver behaviour: Comfort through satisficing, in: P.C. Cacciabue (Ed.), Modelling Driver Behaviour in Automotive Environments: Critical Issues in Driver Interactions with Intelligent Transport Systems, Springer, London, 2007, pp. 189–207.
[63] M. Ljung Aust, J. Engström, A conceptual framework for requirement specification and evaluation of active safety functions, Theoretical Issues in Ergonomics Science 12 (1) (2011) 44–65.
[64] P. Banerjee, About Face 2.0: The Essentials of Interaction Design, Wiley, 2004.

[65] A. Lindgren, Driving Safe in the Future – HMI for Integrated Advanced Driver Assistance Systems, Chalmers University of Technology, 2009.

[66] J. Rasmussen, K.J. Vicente, Coping with human errors through system design: Implications for ecological interface design, International Journal of Man–Machine Studies 31 (4) (1989) 517–534.

[67] L.J. Gugerty, Situation awareness during driving: Explicit and implicit knowledge in dynamic spatial memory, Journal of Experimental Psychology: Applied 3 (1) (1997) 42–66.

[68] M.S. Horswill, F.P. McKenna, The development, validation, and application of a video-based technique for measuring an everyday risk-taking behavior: Drivers' speed choice, Journal of Applied Psychology 84 (6) (1999) 977–985.

[69] G.H. Walker, N.A. Stanton, M.S. Young, The ironies of vehicle feedback in car design, Ergonomics 49 (2) (2006) 161–179.

[70] G.H. Walker, N.A. Stanton, M.S. Young, What's happened to car design? An exploratory study into the effect of 15 years of progress on driver situation awareness, Engineering 45 (2007) 266–282.

[71] F. Chen, Reducing secondary-task workload while driving through interactive interfaces, in: 17th World Congress on Ergonomics, Beijing, 2009.

[72] A. Lindgren, et al., Driver behaviour when using an integrated advisory warning display for advanced driver assistance systems, IET Intelligent Transport Systems 3 (4) (2009) 390.

[73] E.D. Bekiaris, S.I. Nikolaou, Towards the development of design guidelines handbook for driver hypovigilance detection and warning – The awake approach, in: 12th IEE International Conference on Road Transport Information and Control, RTIC, 2004, pp. 314–320.

[74] T. Brunetti-Sayer, et al., Assessment of a river interface for lateral drift and curve speed warning systems: Mixed results for auditory and haptic warnings, in: L.N. Boyle, et al. (Eds.), 3rd International Driving Symposium on Human Factors in Driver Assessment, Training, and Vehicle Design, University of Iowa Public Policy Center, 2005, pp. 218–224.

[75] P. Cacciabue, M. Martinetto, A user-centred approach for designing driving support systems: The case of collision avoidance, Cognition Technology Work 8 (3) (2006) 201–214.

[76] D.J. LeBlanc, et al., Field test results of a road departure crash warning system: Driver utilization and safety implications, in: Proceedings of the 4th International Driving Symposium on Human Factors in Driver Assessment. Training and Vehicle Design, University of Iowa Public Policy Center, 2007.

[77] J.D. Lee, D.V. McGehee, T.L. Brown, M.L. Reyes, Collision warning timing, driver distraction, and driver response to imminent rear-end collisions in a high-fidelity driving simulator, Human Factors and Ergonomics Society 44 (2) (2002) 314–334.

[78] N.D. Lerner, Brake perception-reaction times of older and younger drivers, Human Factors and Ergonomics Society Annual Meeting Proceedings 37 (1993) 206–210.

[79] J. Shutko, An Investigation of Collision Avoidance Warnings on Brake Response Times of Commercial Motor Vehicle Drivers, Citeseer (1999).

[80] H. Sharp, Y. Rogers, J. Preece, Interaction Design: Beyond Human–Computer Interaction, John Wiley, New York, 2002.

[81] J.M. Flach, C.O. Dominingues, Use-centered design: Integrating the user, instrument, and goal, Ergonomics in Design 3 (3) (1995) 19–24.

[82] C.M. Burns, J.R. Hajdukiewicz, Ecological Interface Design, CRC Press, 2004.

[83] R.W. Allen, et al., A low cost, part task driving simulator based on microcomputer technology, in: 69th Annual Meeting of the Transportation Research Board, Washington, DC, 1990.

[84] B. Freund, M. Risser, C. Cain, Simulated driving performance associated with mild cognitive impairment in older adults, Journal of American Geriatric Society 49 (4) (2001) 151–152.

[85] B. Freund, et al., Evaluating driving performance of cognitively impaired and healthy older adults: A pilot study comparing on-road testing and driving simulation, Journal of the American Geriatrics Society 50 (7) (2002) 1309–1310.

[86] H.C. Lee, D. Cameron, A.H. Lee, Assessing the driving performance of older adult drivers: On-road versus simulated driving, Accident Analysis and Prevention 35 (5) (2003) 797–803.

[87] E. Johansson, et al., Review of Existing Techniques and Metrics for IVIS and ADAS Assessment, AIDE, 2004.

[88] B.D. Seppelt, J.D. Lee, Making adaptive cruise control (ACC) limits visible, International Journal of Human–Computer Studies 65 (3) (2007) 192–205.

[89] T. Lajunen, D. Parker, H. Summala, The Manchester Driver Behaviour Questionnaire: A cross-cultural study, Accident Analysis and Prevention 36 (2004) 231–238.

[90] F. Chen, et al., Listening to the traffic – Introduce 3D-sounds into truck cockpit for traffic awareness, in: Applied Human Factors and Ergonomics 2008, 2nd International Conference, Las Vegas, 2008.

[91] A.W. Bronkhorst, Localization of real and virtual sound sources, Journal of the Acoustical Society of America 98 (5) (1995) 2542–2553.

[92] J.C. Middlebrooks, D.M. Green, Sound localization by human listeners, Annual Review of Psychology 42 (1) (1991) 135–159.

[93] J. Blauert, in: J.S. Allen (Ed.), Spatial Hearing: The Psychophysics of Human Sound Localization, MIT Press, London, 2001.

Index

Note: Page numbers with "f" denote figures; "t" tables.

Printed in the United States
By Bookmasters